Emerging Concepts in Neuro-Oncology

Colin Watts
Editor

Emerging Concepts in Neuro-Oncology

Editor
Colin Watts
Department of Clinical Neurosciences
University of Cambridge
Cambridge
United Kingdom

ISBN 978-0-85729-457-9 ISBN 978-0-85729-458-6 (eBook)
DOI 10.1007/978-0-85729-458-6
Springer London Heidelberg New York Dordrecht

Library of Congress Control Number: 2012952521

Printed on acid-free paper

Springer is part of Springer Science+Business Media (www.springer.com)

Preface

The clinical and scientific field of neuro-oncology is one of the most exciting and rapidly changing areas of oncology. The heterogeneity of glial cancers is being addressed at the level of molecular genetics and gene expression profiling. This is paving the way for functionalizing the genome and individualizing therapy for patients. The cell biology of glial cancers is in hot pursuit as our understanding of glial ontogeny and function enters a new era and the lessons begin to percolate translational science.

Transgenic technologies coupled with evolving biological concepts facilitate the development of evermore sophisticated models of disease. These developments will enable better preclinical data generation and better therapies tailored to individual patients.

Yet the statistics remain grim. Central nervous system (CNS) malignancies account for 2 % of cancers but 7 % of cancer deaths. Emerging biomarkers are difficult to introduce into routine clinical practice for political, economic, and technical reasons. Early detection remains a challenge and patient recruitment is fraught with difficulties.

For those scientists, clinicians, and allied specialists, this is not new. What is new is an emerging sense of identity and enthusiasm across a broad spectrum of clinical and scientific endeavor. Against this background it is essential to facilitate communication and understanding of new ideas and concepts. This book is written with this in mind: to promote collaboration across traditional boundaries and promote translational research for patient benefit.

The first two chapters review our current understanding of how we organize and classify the glial cancers. The genetic and epigenetic characteristics that shape the clinical phenotypes seen by clinicians are rapidly evolving, and a snapshot of where we are now highlights new questions for further research. The following two chapters seek to address the vexed question of where glial cancers come from and how they evolve. Given that the brain is, to a first approximation, amitotic, we could ask: "Why are glial cancers so common?"

A key element in the manifest failure of pharmacotherapy is the relatively poor models of disease currently available for drug development. In vitro and in vivo models are discussed in Chaps. 5 and 6 outlining current state-of-the-art thinking and what key issues need to be addressed going forward.

The second part of the book begins to address the issues around patients and how we can treat them. This begins with an overview of novel approaches to one of the mainstays of treatment: ionizing radiation. Novel ionizing species are discussed and some clinical data presented. This is followed by a comprehensive overview of recent developments in imaging both structure and function of glial cancers.

The next two chapters address pragmatic issues of patient management: surgery and radiation oncology. New developments are highlighted emphasizing the broad spectrum of evolution of neuro-oncology. These chapters are followed by a review of how we can manage the elderly patient. There is a current lack of consensus among clinicians about how best to manage this difficult group. The problem is compounded by a paucity of good-quality robust scientific data on which to base clinical decision making.

The final part of the book examines two key questions going forward: How can we detect brain cancers sooner; and how can we improve clinical trial recruitment? Both will be central to the development of neuro-oncology in the future.

I hope that the clinical and scientific data reviewed in this book stimulate new ideas and collaborations. That is the best tribute we can pay to all those suffering from brain cancer.

June 2012
Cambridge, UK

Colin Watts

Contents

Contributors

Colin Watts, MBBS, Ph.D. (Cantab), FRCS (SN) Department of Clinical Neurosciences, University of Cambridge, Cambridge, UK

Sebastian Brandner, M.D., FRCPath Division of Neuropathology, UCL Institute of Neurology, London, UK

Anthony J. Chalmers, B.A. (Hons), BM BCh, MRCP, FRCR, Ph.D. Institute of Cancer Sciences, University of Glasgow, Glasgow, UK

Kathryn Graham, B.Sc. (Hons), MB ChB, MRCP, Ph.D., FRCR Department of Clinical Oncology, Beatson West of Scotland Cancer Centre, Glasgow, UK

Monika E. Hegi, Ph.D. Laboratory of Brain Tumor Biology and Genetics, Service of Neurosurgery, Department of Clinical Neurosciences, Lausanne University Hospital, Lausanne, Switzerland

Sarah J. Jefferies, MBBS, B.Sc., MRCP, FRCR, Ph.D. Oncology Department, Addenbrooke's Foundation Trust, Cambridge, UK

Raj Jena, M.D., MRCP, FRCR Department of Oncology, Cambridge University Hospitals NHS Foundation Trust, Cambridge, UK

Akiko Nishiyama, M.D., Ph.D. Department of Physiology and Neurobiology, University of Connecticut, Storrs, CT, USA

Sara G.M. Piccirillo, Ph.D. Department of Clinical Neurosciences, Cambridge Centre for Brain Repair, University of Cambridge, Cambridge, UK

Stephen J. Price, B.Sc., MBBS, Ph.D., FRCS (Neuro. Surg.) Neurosurgery Division and Wolfson Brain Imaging Centre, Department of Clinical Neurosciences, University of Cambridge, Cambridge, UK

David J. Ryan, MB, BCh, BAO, BMedSc, MRCS (Ireland) Mouse Cancer Genetics, Wellcome Trust Sanger Institute, Cambridge, UK

Davide Sciuscio, Ph.D. Service of Neurosurgery, Department of Clinical Neurosciences, Lausanne University Hospital, Lausanne, Switzerland

Susan C. Short, MBBS, MRCP, FRCR, Ph.D. Department of Oncology, Leeds Institute of Molecular Medicine, Leeds, UK

Walter Stummer Department of Neurosurgery, University of Münster, Münster, Germany

Adam D. Waldman, Ph.D., FRCP, FRCR Department of Imaging, Imperial College, London, UK

David A. Walker, BMedSci, BM, BS, FRCP, FRCPHCH Children's Brain Tumour Research Centre, University of Nottingham, Nottingham, UK

Pieter Wesseling, M.D., Ph.D. Department of Pathology, Radboud University Nijmegen Medical Centre, Nijmegen, The Netherlands

Canisius Wilhelmina Hospital Nijmegen, Nijmegen, The Netherlands

VU University Medical Center, Amsterdam, The Netherlands

Sophie H. Wilne, BA Hons (Cantab), MBBS, M.D. Department of Paediatric Oncology, Nottingham Children's Hospital, Nottingham University Hospitals NHS Trust, University of Nottingham, Nottingham, UK

Part I

Scientific Foundations

Classification of Gliomas

1

Pieter Wesseling

Abstract

For almost a century, histopathological evaluation of gliomas has provided the gold standard for classification of these neoplasms. Indeed, the (neuro) pathologist is able to render an unequivocal diagnosis of glioma in most specimens and to indicate low- or high-grade malignant character of the lesion. It is increasingly clear, however, that the traditional histopathological diagnosis lacks the robustness and specificity that is needed for more tailored treatment of glioma patients. Even for the experienced neuropathologist, at least three factors may hamper reaching an unequivocal histopathological diagnosis on glioma tissue: (a) tissue quantity and quality, (b) lack of unequivocal histopathological criteria, and (c) incomplete representation of biology by morphology. Smart integration of information on the underlying molecular aberrations in the diagnosis of gliomas will undoubtedly result in a more sophisticated classification of these tumors. Modern neuropathology is thus rapidly moving toward a combined morphological and molecular approach, the challenge being to implement this approach in an affordable way that optimally serves the individual patients suffering from these neoplasms.

Keywords

Glioma • Astrocytoma • Oligodendroglioma • Ependymoma • Histopathology • Classification • Molecular diagnosis

Introduction

Among all neoplasms in the human body, the diversity and complexity of tumors of the central nervous system (CNS) are considered by some to be unrivaled [1]. A simplified representation of the relative frequency in the general population of primary tumors of the CNS (i.e., tumors originating in the central nervous system tissue itself

P. Wesseling, M.D., Ph.D.
Department of Pathology,
Radboud University Nijmegen Medical Centre,
Nijmegen, The Netherlands

Department of Pathology, Canisius Wilhelmina Hospital
Nijmegen, Nijmegen, The Netherlands

Department of Pathology, VU University Medical
Center, Amsterdam, The Netherlands
e-mail: p.wesseling@pathol.umcn.nl

C. Watts (ed.), *Emerging Concepts in Neuro-Oncology*,
DOI 10.1007/978-0-85729-458-6_1, © Springer-Verlag London 2013

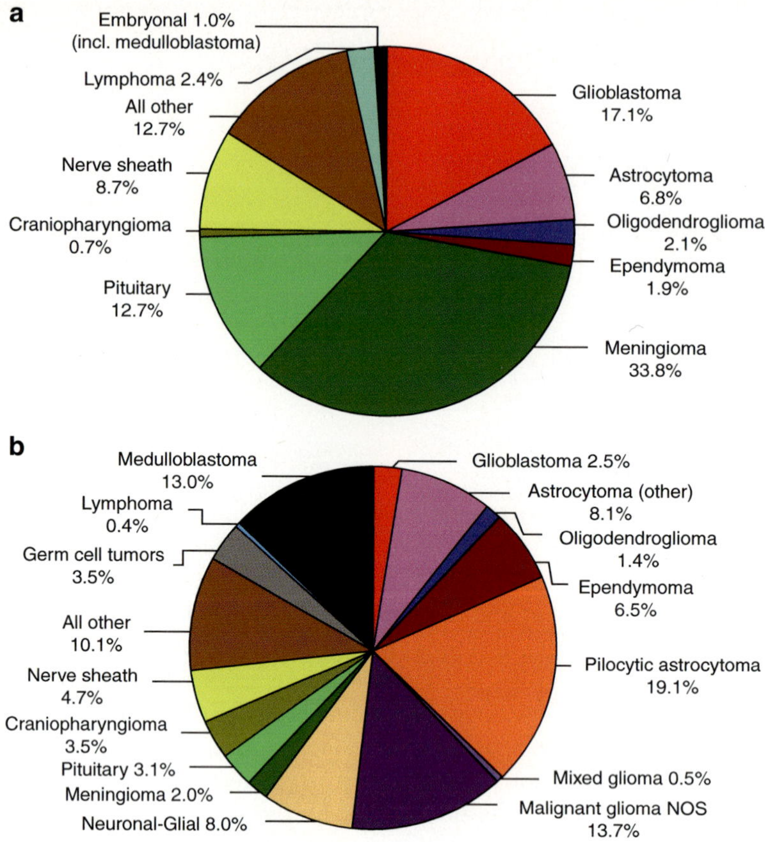

Fig. 1.1 (**a**) Simplified representation of the relative frequency of primary brain tumors of the CNS. Over half of these tumors are benign, major representatives being meningioma, schwannoma, and pituitary adenoma. Glial tumors form a major fraction of the rest. (**b**) In the pediatric age group, the relative frequency is substantially different, with much less meningiomas, but many more high-grade malignant "embryonal tumors" (the most frequent lesion in this category being medulloblastoma), glioneuronal tumors, and "variants" of glial tumors such as pilocytic astrocytoma, as well as a larger number of (malignant) glial tumors that fall outside the mainstream glioma categories and/or are more difficult to classify. Information in these charts for overall numbers is based on 158,088 patient diagnoses in 2004–2006, for tumors in the pediatric age group on 7,767 patient diagnoses in children 0–14 years of age in the same time period (*Source*: CBTRUS Statistical Report [2])

or from its coverings) is depicted in Fig. 1.1a [2]. Over half of these tumors are benign, major representatives being meningioma, schwannoma, and pituitary adenoma. The group of glial tumors (gliomas) forms a major fraction of the rest. In pediatric patients, however, meningiomas are rare, while pilocytic astrocytomas, high-grade malignant "embryonal tumors" (the most frequent example being medulloblastoma), glioneuronal tumors, and glial tumors outside the "mainstream glioma categories" are much more frequent (Fig. 1.1b). Compared to, e.g., cancers of lung, breast, colon, and prostate, the incidence rate of gliomas in the general population is low (about 7 new patients per 100.000 individuals per year) [3]. Of note, patients with CNS tumors are reported to suffer from the highest number of average "years of life lost" [4]. This can be explained by the fact that so far most patients with gliomas are incurable and that (malignant) tumors of the CNS occur relatively frequently in the pediatric age group.

Traditionally, classification of CNS tumors is based on microscopic evaluation of tumor tissue. When tissue of a patient with a CNS tumor is obtained for pathological diagnosis, the (neuro)

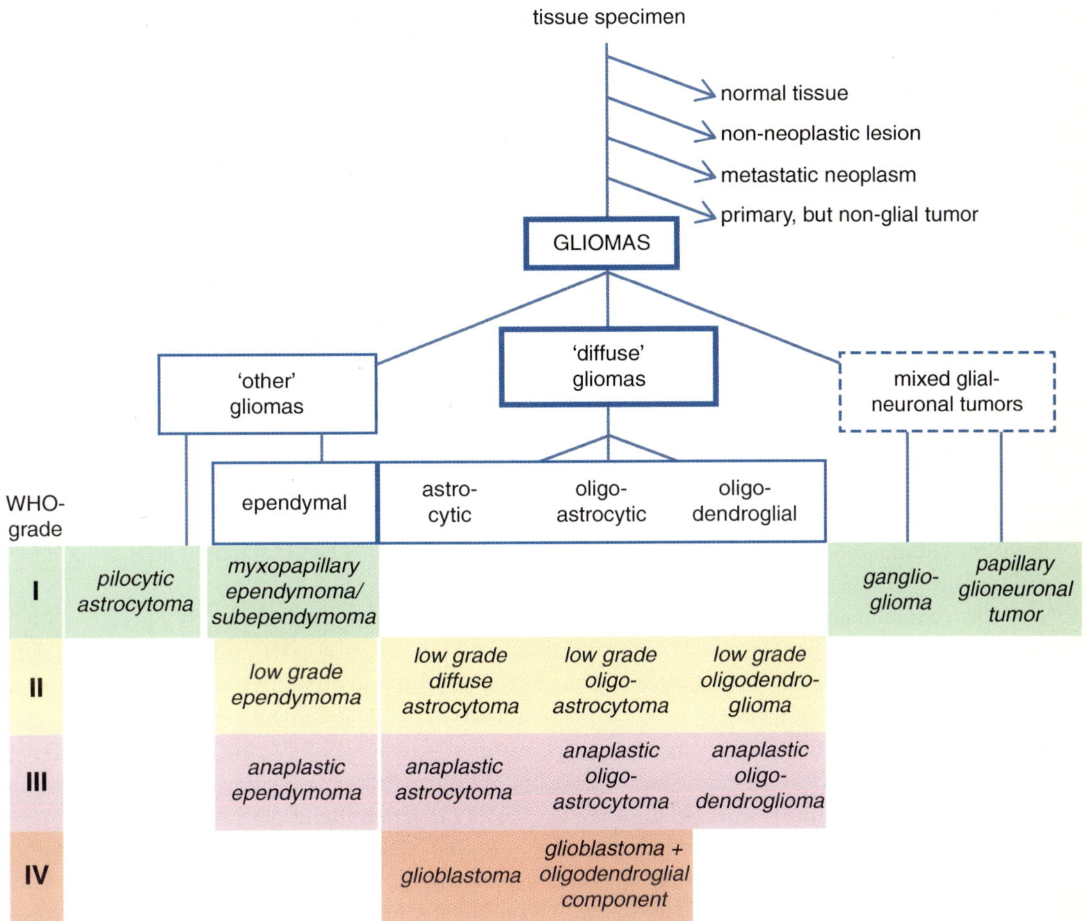

Fig. 1.2 Simplified representation of decision tree followed for histopathological diagnosis of gliomas; after excluding a number of differential diagnostic options and reaching a diagnosis of glioma, the neoplasm should be further (sub)typed and within some categories (esp. the diffuse gliomas and ependymal tumors) graded for a specific histopathological diagnosis (*in italics*). The *dashed box* surrounding the term mixed glial-neuronal tumors underscores that these neoplasms are not just glial in nature. Of note, while for some of the non-diffuse gliomas such as pilocytic astrocytomas and gangliogliomas, the histopathological diagnosis generally implies indolent biological behavior, occasionally microscopic features of more aggressive growth are found and the lesion may be diagnosed as anaplastic/WHO grade III. Also, it is important to realize that this scheme is incomplete because the WHO classification recognizes multiple other entities in the group of "other" gliomas (e.g., pilomyxoid astrocytoma, WHO grade II; angiocentric glioma, WHO grade I; chordoid glioma of the third ventricle, WHO grade II; desmoplastic infantile astrocytoma, WHO grade I; pituicytoma, WHO grade I) as well as in the group of mixed glial-neuronal tumors (e.g., desmoplastic infantile ganglioglioma, WHO grade I; rosette-forming glioneuronal tumor of the fourth ventricle, WHO grade I), but discussion of these very infrequent glial tumors falls outside the scope of the present chapter

pathologist in fact follows a decision tree: normal or abnormal tissue? Neoplastic or nonneoplastic lesion? Primary or metastatic tumor? Glial or non-glial tumor? Diffuse glioma, "variant" glioma, or glial tumor combined with other component? In case of a diffuse glioma, the tumor is generally subtyped as astrocytic, oligodendroglial, or mixed/oligo-astrocytic. Finally, within histological categories (esp. subtypes of diffuse glioma, ependymal tumors), the malignancy grade of the glial tumor has to be assessed (Fig. 1.2) [5].

Next to age and clinical condition of the patient and location of the tumor, the histopathological diagnosis carries important prognostic information and forms the basis for further patient management. For instance, even after optimal therapy, most patients with a histopathological diagnosis of glioblastoma (the most malignant and unfortunately also the most frequent glioma) die within 1–2 years after diagnosis. In contrast, many patients with a low-grade glioma survive for over 10 years [6]. For the practicing pathologist, reaching a diagnosis of glioma is generally not the most challenging part. Unequivocal (sub) typing and grading of gliomas, however, can be very difficult. Moreover, it is increasingly clear that the robustness and the level of sophistication of the classification of gliomas need to be improved for optimal implementation of more tailored therapeutic approaches.

After providing some information on the history of classification of gliomas, this chapter will describe the current pathological practice for typing and grading of these neoplasms. Thereby, the most recent 2007 World Health Organization (WHO) classification of tumors of the CNS serves as the basis [7]. Furthermore, the focus of this chapter is on the most frequent glial neoplasms, i.e., the spectrum of "diffuse gliomas," ependymal tumors, and pilocytic astrocytomas. For information on less frequent glial tumors such as mixed glioneuronal tumors, pleiomorphic xanthoastrocytoma/PXA, and subependymal giant cell astrocytoma/SEGA, the reader is referred to other textbooks [7–9]. Also, rather than providing a "cookbook" for the practicing (neuro)pathologist, this chapter is meant to explicate the essentials of the current practice of histopathological classification of glial tumors, including discussion of its strengths and weaknesses. The last part of the chapter provides some suggestions on how classification of gliomas may significantly be improved in the near future.

Some History

In the 1920s, Percival Bailey (neuropathologist) and Harvey Cushing (neurosurgeon) provided the groundwork for classification of tumors of the CNS as we know it today [10]. Gliomas were classified based on microscopic resemblance with and/or presumed derivation of tumor cells from nonneoplastic cells in the mature or developing CNS. For example, the term glioblastoma suggests derivation of a glioblast, although at that time (and even nowadays), it was not clear what exactly the nature of such a precursor cell was. Furthermore, most gliomas were subtyped as astrocytic, oligodendroglial, or ependymal because of resemblance of the tumor cells with three categories of better defined, nonneoplastic glial cells. Additionally, some glial tumors showed a mixed (e.g., oligo-astrocytic) phenotype, others a mixture of glial and neuronal differentiation. Interestingly, during the last decade, the hypothesis that gliomas are derived from glial precursor cells or stem cells rather than from mature glial cells has (re)gained enormous interest, also because this oncogenetic route may better explain the sometimes explicitly "promiscuous" expression of astrocytic, oligodendroglial, and even neuronal features in one and the same glial tumor [11, 12]. Meanwhile, typing of gliomas by resemblance of tumor cells with nonneoplastic glial cells still forms the basis of the most recent WHO classification of glial tumors [7].

It was clear that after classifying CNS tumors as proposed by Bailey and Cushing, patients with a particular histological (sub)type of glioma often still show a highly variable clinical course. Acknowledging that some microscopic features within histological (sub)types of glioma are correlated with poor prognosis, in the 1940s, James Kernohan proposed a grading scheme based on systematic comparison of the presence or absence of features such as "anaplasia" and mitotic activity [13]. In the following decades, different grading schemes were introduced for the different subtypes of gliomas, each with its own shortcomings [14]. Since about the year 2000, the WHO classification is the worldwide accepted system for pathological diagnosis of tumors of the CNS. The WHO classification of glial tumors implies (1) tumor typing, assigning the tumor to a particular histological group, and (2) tumor grading, assessing the malignancy grade of a lesion within that group.

Apart from grading systems that are designed to assess malignancy grade within certain histologically defined tumor types, in the 1970s, Karl

Joachim Zülch introduced a grading system grouping tumors of similar aggressiveness/prognosis irrespective of their histological type [15]. Roughly, this system recognized four grades of malignancy: grade I, expected survival more than 5 years ("benign"); grade II, survival between 3 and 5 years ("semi-benign"); WHO grade III, survival 1–3 years ("semi-malignant"); and WHO grade IV, survival generally <1 year ("malignant"). More recently, some histopathological information was included in the definitions of the different WHO grades: grade I, tumors with low proliferative potential and the possibility of cure following resection alone; grade II, tumors that tend to progress to higher grades of malignancy; grade III, tumors with histological evidence of malignancy, including nuclear atypia and mitotic activity; and grade IV, cytologically malignant, mitotically active, necrosis-prone neoplasms typically associated with rapid pre- and postoperative disease evolution and fatal outcome [16]. Nowadays, the grades assigned to glial neoplasms by histopathological analysis thus overlap with the WHO grades attributed to these tumors. In daily clinical practice, verbal and numeric designations of the malignancy grade are often used interchangeably (e.g., glioblastoma = astrocytoma WHO grade IV; low-grade diffuse astrocytoma = diffuse astrocytoma WHO grade II).

Histopathological Classification of Gliomas

Tumor Typing

Macroscopic evaluation of biopsies or resection specimens is generally of little help in reaching a diagnosis of glioma. In larger specimens, a gradual transition of normal appearing gray or white matter into a lesion with grayish discoloration and blurring of the preexistent anatomical structures is compatible with the presence of a diffuse glioma. In this context, necrosis indicates high-grade malignancy. In daily practice, however, typing and grading of gliomas is based on microscopic evaluation of especially hematoxylin and eosin (H&E)-stained histological sections. In this context, resemblance of the tumor cells with non-neoplastic cells in the CNS is used to type a glioma as astrocytic, oligodendroglial, ependymal, mixed (esp. oligo-astrocytic) glioma, or mixed glioneuronal tumor (Fig. 1.2) [5, 7–9].

Normal astrocytes are typically stellate cells with an oval-to-elongate, somewhat vesicular nucleus, little eosinophilic cytoplasm, and delicate eosinophilic cell processes that are often hard to identify in the neuropil (i.e., the dense network of processes of glial and neuronal cells in the brain parenchyma) without immunohistochemical staining for glial fibrillary acidic protein (GFAP). Reactive astrocytes generally show increase in the size of the perikaryon with some enlargement of the nucleus and more stout cellular processes. Not infrequently, reactive astrocytes show gemistocytic change with formation of a plump, rounded or angular, eosinophilic cell body and an eccentric nucleus.

Frequently, astrocytic tumors show a mixture of such cell types (Fig. 1.3a). Several phenotypical variants of diffuse astrocytoma are recognized, e.g., fibrillary astrocytoma (composed of tumor cells with clear fibrillary cell processes), protoplasmic astrocytoma (tumor cells with small cell bodies and few, flaccid processes), and gemistocytic astrocytoma (characterized by a plump eosinophilic cell body of the tumor cells). While the connotation "fibrillary" or "protoplasmic" in this context does not carry a clear prognostic significance, low-grade gemistocytic astrocytomas tend to show more aggressive behavior than their non-gemistocytic counterpart.

Within the most malignant astrocytic tumor (i.e., glioblastoma) category, also several phenotypical variants are recognized, the most frequent of these being giant cell glioblastoma (showing extensive presence of giant, often multinucleated tumor cells), small cell glioblastoma (with predominance of small, relatively monomorphous tumor cells with little cytoplasm), and gliosarcoma (with an extensive, highly pleiomorphic, spindle cell/sarcomatoid component). The prognostic significance of these different glioblastoma subtypes is limited [5, 7–9].

Both normal and neoplastic oligodendroglial cells typically show a round nucleus with

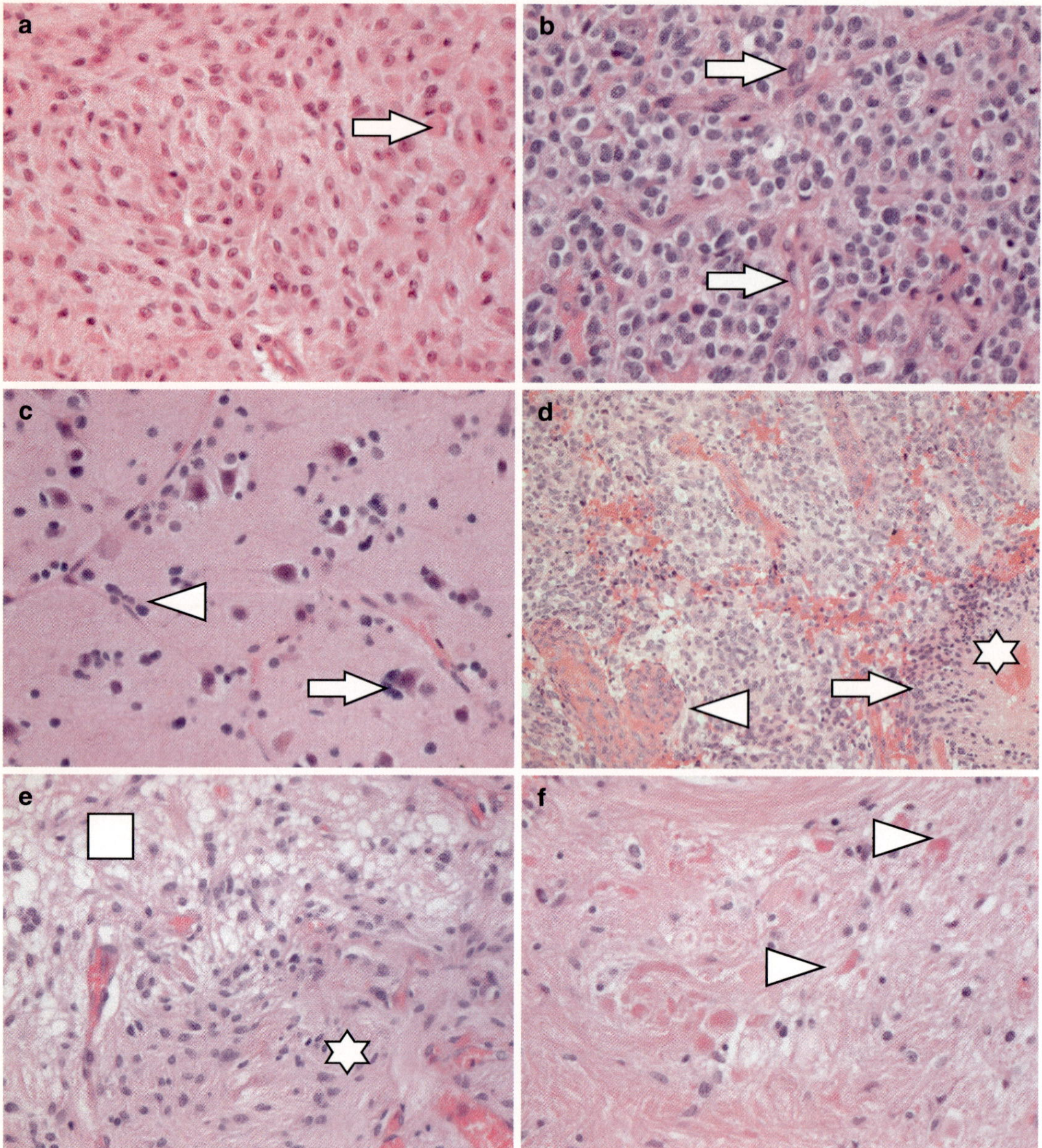

Fig. 1.3 Some microscopic examples of gliomas: (**a**) astrocytic tumor cells typically show variable amount of eosinophilic cytoplasm, often with clear eosinophilic cell processes and with variable nuclear atypia; the *arrow* indicates a gemistocytic tumor cells with a plump, rounded eosinophilic cell body and an eccentric nucleus; (**b**) in contrast, prototype oligodendroglial tumor cells are characterized by a round nucleus with a perinuclear clear halo ("fried egg appearance"); the *arrows* indicate delicate, branching tumor capillaries with some "chicken-wire-like" architectural characteristics as is not infrequently seen in esp. oligodendroglial tumors; (**c**) diffuse gliomas typically show extensive dispersion of tumor cells in the preexistent brain tissue; here, an example of very subtle perivascular (*arrow*) and perineuronal (*arrowhead*) presence of tumor cells in the outskirts of the glioma; (**d**) example of glioblastoma with necrosis (*asterisk*) surrounded by pseudopalisading tumor cells (*arrow*) and florid microvascular proliferation (*arrowhead*); (**e**) pilocytic astrocytomas are generally much more circumscribed and microscopically often show a "biphasic growth pattern" with alternation of compact (*asterisk*) and more loosely structured areas (*square*); (**f**) especially in the compact areas of pilocytic astrocytomas, Rosenthal fibers (i.e., deeply eosinophilic, often elongated structures; *arrows*) are frequently present. **a**–**f**: hematoxylin and eosin staining, original magnification in **a**–**c** and **e**, **f**×200, in **d**×100

a relatively dense chromatin pattern and a perinuclear clear halo ("fried egg appearance") (Fig. 1.3b). Oligodendroglial tumors are often highly cellular lesions with compact fields of tumor cells. Additional features indicative of oligodendroglial rather than astrocytic differentiation are the presence of a branching network of delicate capillaries ("chicken-wire pattern") and extensive calcification in the tumor. Pure oligodendrogliomas may still contain tumor cells with an astrocyte flavor like minigemistocytes (characterized by a small, round, paranuclear eosinophilic cell body) and gliofibrillary cells (showing a small eosinophilic cytoplasmic body with eosinophilic cell processes). As indicated by their name, the key criterion for the diagnosis of mixed oligoastrocytomas is the presence of a substantial component of both neoplastic astrocytic and oligodendroglial-like tumor cells. The tumor cells of both lineages may be diffusely mixed or separated [5, 7–9].

In clinical practice, the distinction of the so-called diffuse gliomas from other gliomas is of major importance. In adult patients, the vast majority of astrocytic, oligodendroglial, and mixed oligo-astrocytic tumors belong to the category of diffuse gliomas. Irrespective of their malignancy grade, the hallmark of diffuse gliomas is extensive, diffuse infiltration of individual or small groups of tumor cells in the neuropil. An important clue for diffuse infiltrative growth is the diffuse increase in glial cells in the CNS tissue with relative preservation of the original architecture. This growth pattern often results in extensive dispersion of tumor cells along white matter tracts ("intrafascicular growth") and perivascular, perineuronal, and/or subpial accumulation of tumor cells (Fig. 1.3c). These latter phenomena were coined as "secondary structures" by Hans-Joachim Scherer, a pioneer in the study of glioma growth patterns [17], and form important evidence for the diffuse infiltrative character of the glial neoplasm. In this context, crossing of tumor cells to the contralateral hemisphere via the white matter tracts of the corpus callosum may eventually result in a lesion that is radiologically recognized as "butterfly glioma." Gliomatosis cerebri is the term that is used for the most extreme form of diffuse glioma growth, with according to the WHO 2007 classification infiltration of the lesion in at least three lobes of a cerebral hemisphere, but in some patients, widespread extension to infratentorial structures (brainstem, cerebellum) and even spinal cord as well [7]. Obviously, the diffuse infiltrative growth pattern of diffuse gliomas forms a major obstacle for curative therapy [18].

In contrast to diffuse gliomas, ependymomas and other "non-diffuse" or "variant" gliomas are indeed generally more circumscribed. Of the variant astrocytic tumors, the most frequent entity is pilocytic astrocytoma, a neoplasm that is generally encountered in the posterior fossa, optic pathways, or hypothalamic region of pediatric patients and has a relatively benign clinical course. Histologically, pilocytic astrocytomas typically show alternation of compact and more loose or even (micro)cystic areas ("biphasic growth pattern"), the astrocytic tumor cells demonstrating long, slender, hairlike ("piloid") cellular processes, and with variable presence of Rosenthal fibers and eosinophilic granular bodies (i.e., eosinophilic, hyaline structures produced by astrocytic tumor cells) [7–9].

Ependymomas consist of glial cells that may have astrocytic or, less frequently, oligodendroglial features but in addition show particular arrangements of the tumor cells: radial orientation around a central lumen (true ependymal rosettes) and/or around a vessel with a zone free of nuclei immediately around the vessel wall (perivascular pseudorosettes). In a substantial number of ependymomas, true rosettes are not easy to find. Also, unequivocal identification of perivascular orientation of glial tumor cells as a perivascular pseudorosette can be difficult. Some special histological variants of ependymal tumors are recognized. Of these, the myxopapillary ependymoma and subependymoma are the most frequent and associated with a relatively benign course [7–9].

As can be expected from the wide spectrum of neuronal phenotypes that are encountered in the normal CNS, neuronal differentiation of tumor cells can also be present in many forms. Examples are the classic neuronal differentiation with a large cell body, a large vesicular nucleus and a prominent nucleolus; neurocytic morphology resembling oligodendrocytes; immature, neuroblastic cells with little cytoplasm, a small round to oval or more irregular nucleus, and a dense chromatin pattern [7–9].

Starting in the 1980s, immunohistochemistry has substantially facilitated recognition of glial versus neuronal differentiation as well as subtyping of some glial tumors [7–9, 19]. Especially the tumor cells in astrocytic and ependymal neoplasms are generally positive for GFAP and neoplastic neuronal cells for synaptophysin and/or Neu-N. A valuable marker for recognition of ependymal differentiation is epithelial membrane antigen (EMA): Ependymomas typically not only show EMA decoration of the luminal surface of the true rosettes but also dispersed, "dot-like" positivity within tumor cells, these dots ultrastructurally representing intracytoplasmatic lumina that can also be demonstrated by electron microscopy. Several markers (e.g., Leu-7, Olig-2) were introduced as helpful for immunohistochemical recognition of oligodendroglial differentiation but did not make it to the routine diagnostic panel because of less specific results than originally hoped for. Very recently, internexin-alpha (INA) was reported as a marker showing a significant correlation with oligodendroglial phenotype [20, 21].

Tumor Grading

With a few exceptions, pilocytic astrocytomas and special variants of ependymal tumors (subependymoma, myxopapillary ependymoma) are designated as WHO grade I lesions. For diffuse gliomas and other ependymal tumors, after histopathological typing of the neoplasm, a malignancy grade still has to be assigned. While features like cellularity and nuclear atypia are also considered, grading of diffuse gliomas is essentially based on assessment of mitotic activity, necrosis, and florid microvascular proliferation (Fig. 1.4) [14]. Diffuse astrocytic and mixed oligo-astrocytic tumors are graded as WHO grade II (low grade), WHO grade III (anaplastic), or WHO grade IV (glioblastoma +/– oligodendroglial component). Pure oligodendroglial and ependymal tumors are graded as WHO grade II or III [7].

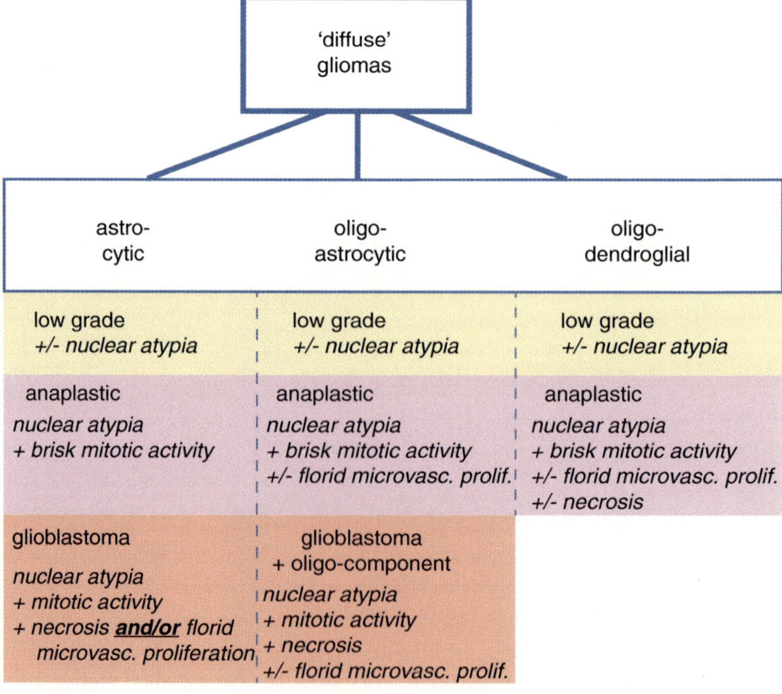

Fig. 1.4 Microscopic criteria used for grading of diffuse gliomas. Using this system, a malignancy grade can be assigned to most diffuse glioma samples with clear prognostic relevance. However, the definitions of these criteria are not precise enough to allow for unequivocal, reproducible grading of all neoplasms. For instance, the exact number of mitoses required for a diagnosis of anaplastic versus low-grade diffuse glioma and the minimum requirements for recognition of florid microvascular proliferation are not clear. Even unequivocal assessment of necrosis may be problematic in biopsy samples that are small or poorly preserved

When in surgical material of a diffuse astrocytic neoplasm, marked mitotic activity, necrosis, and florid MVP are lacking, the neoplasm is diagnosed as low-grade diffuse astrocytoma. The presence of marked mitotic activity, however, leads to the diagnosis of anaplastic astrocytoma, and the (additional) presence of necrosis and/or florid microvascular proliferation leads to a diagnosis of glioblastoma (Fig. 1.4) [7]. The same set of microscopic criteria is used for grading of oligodendroglial and mixed oligo-astrocytic tumors, but in a slightly different way, as in these tumors, necrosis and florid microvascular proliferation do not have the same unfavorable connotation as in diffuse astrocytic neoplasms. Oligodendrogliomas with florid microvascular proliferation and/or necrosis are still considered as WHO grade III lesions. In mixed oligoastrocytic tumors, the presence of microvascular proliferation is still compatible with WHO grade III, but necrosis is now considered (by some but not all neuropathologists) as reason to diagnose the tumor as glioblastoma with oligodendroglial component (GBM-O) [7, 23].

Of note, low-grade diffuse gliomas show a strong tendency for progression to a high-grade malignant lesion in the course of years. Glioblastomas that arise via malignant progression of a less malignant precursor lesion are coined as "secondary glioblastomas." In many (esp. older) patients, however, such a precursor lesion cannot be demonstrated, and the tumor is considered as "primary" (or "de novo") glioblastoma [22].

In diffuse gliomas, the labeling index of tumor cell nuclei as determined by the (immunohistochemical) Ki-67/MIB1 staining increases with malignancy grade, roughly being up to 5 % in low-grade lesions, between 5 % and 10 % in anaplastic gliomas, and (at least in some areas) much more than 10 % in glioblastomas. However, the MIB1 labeling index is not routinely incorporated into a grading system because of substantial overlap of this marker for the different malignancy grades and differences in staining results between different laboratories [14].

For assessment of the malignancy grade of ependymal tumors, a similar approach is followed as for pure oligodendroglial neoplasms. However, the meaning of necrosis is less clear, and the association between tumor grade and prognosis is less stringent for ependymal neoplasms than for, e.g., astrocytic tumors [7]. Typing a tumor as pilocytic astrocytoma generally implies that the lesion should be considered as WHO grade I. Of note, pilocytic astrocytomas frequently show florid microvascular proliferation and occasionally even necrosis. In contrast to diffuse gliomas, however, the presence of these features in pilocytic astrocytomas in itself does not signify high-grade malignancy, illustrating the importance of adequate tumor typing before assessment of the malignancy grade. A small subset of patients with pilocytic astrocytoma, however, does suffer from an unexpected aggressive behavior. The presence of especially brisk mitotic activity is an important clue for such aggressive behavior and may lead to a diagnosis of anaplastic pilocytic astrocytoma [24].

Practical Problems

Tissue Quantity and Quality

Gliomas frequently show marked phenotypical heterogeneity with spatial differences in cellular phenotype and malignancy grade. Such heterogeneity is encountered in an extreme form in glioblastomas, hence the epithet "multiforme" in the traditional name glioblastoma multiforme. In such heterogeneous neoplasms, the size and exact origin of the tissue specimens may thus have a major impact on the diagnosis that is rendered by the pathologist [25, 26].

The chance of sampling effect is inversely correlated with the size of the tissue specimens obtained for histopathological diagnosis. Also, dependent on the size and exact nature of the tissue samples submitted, the neoplastic and/or diffuse infiltrative nature of the glial lesion is more or less easily appreciated. Consequently, especially in small biopsies (e.g., in case of a neuronavigation-guided biopsy procedure of deep-seated intracerebral lesions or biopsies of the brain stem or spinal cord), the differential diagnosis of diffuse (low-grade) astrocytoma, pilocytic astrocytoma, and reactive astrocytosis may be very challenging, and the malignancy grade of diffuse gliomas may well be underestimated. For this reason and according to the WHO 2007 classification, evaluation of mitotic activity should be performed in the context of sample size: While a single mitosis in a large resection specimen is not sufficient for the diagnosis of anaplastic glioma, the identification of a single mitosis in a small biopsy fragment may well indicate high proliferative activity and thus a WHO grade III lesion or worse [7]. A spectrum of other preoperative, surgical, and/or pathological factors may negatively influence tissue quantity and quality [27]. Obviously, suboptimal preservation of the morphology of tumor tissue in surgical material may result in a less specific or even false diagnosis ("garbage in → garbage out").

Lack of Unequivocal Criteria

Another problem is the lack of histological and cytological criteria that are sufficiently precise to allow for unequivocal, reproducible (sub)typing and grading of glial neoplasms. Some microscopic features are suggestive but not pathognomonic for a particular diagnosis. For example, Rosenthal fibers are a hallmark of pilocytic astrocytomas but are also found in other "variant" gliomas (ganglioglioma, pleiomorphic xanthoastrocytoma) and can even be present in reactive gliosis elicited by (therapy of) otherwise prototype diffuse gliomas. Also, an oligodendrocyte-like phenotype of tumor cells is not only encountered in oligodendroglial or oligo-astrocytic tumors, but morphologically similar cells constitute neurocytomas and can be present in (clear cell) ependymoma and pilocytic astrocytoma. Even the results of immunohistochemical stainings may be confusing. For instance, a subset of bona fide oligodendroglial tumors is positive for synaptophysin, blurring the differential diagnosis with (extraventricular) neurocytoma [28]. Of note, the presence of perinuclear halos in oligodendrogliomas is in fact a fixation artifact and can thus be absent in specimens that are more promptly fixed or used for frozen section diagnosis.

In diffuse gliomas, the precise extent of an oligodendroglial versus astrocytic component is often hard to define, e.g., because a substantial number of tumor cells are not easily recognized as either astrocytic or oligodendroglial. Exact criteria for the minimum proportion of the different components needed for a diagnosis of mixed glioma are lacking. Also, diffuse gliomas in which the tumor cells lack the prototype oligodendroglial and astrocytic phenotype are often "by default" considered as astrocytic neoplasms rather than diagnosed as "glioma not otherwise specified." Of note, especially in the peripheral, diffuse infiltrative part of oligodendroglial tumors and in anaplastic oligodendrogliomas, the neoplastic cells may acquire a nondescript or even more astrocytic phenotype, but in this context, the presence of such areas is often considered as still being acceptable for the diagnosis of a pure oligodendroglial tumor.

Cell density and nuclear atypia are not necessarily correlated with malignancy grade of gliomas. Low-grade oligodendrogliomas are often highly cellular (Fig. 1.3b), while (parts of) glioblastomas may show moderate cellularity. Marked nuclear atypia may be encountered in otherwise low-grade gliomas while not being a prominent feature in glioblastomas. More importantly, it is unclear what the exact number of mitoses is that is required for a diagnosis of anaplastic versus low-grade diffuse glioma and how exactly mitotic activity should be weighted in the context of sample size [7].

Florid microvascular proliferation is a term used for the presence of multilayered microvessels with hypertrophy and hyperplasia of endothelial cells and pericytes in the vessel walls, but the minimum requirements for recognition of this phenomenon are not clear [23, 29]. Even unequivocal assessment of necrosis may be troublesome in biopsy samples that are small or poorly preserved. This situation leads to substantial interobserver variation in the classification of diffuse gliomas, also among experienced neuropathologists, and may well have undesirable clinical consequences [30–34].

Last but not least, in the course of time, the ideas about the consequences of the presence of particular histological features for grading of glial tumors have changed. For instance, the presence of florid microvascular proliferation in a mitotically active diffuse astrocytic tumor has not always been enough reason to consider the neoplasm as glioblastoma. Also, mitotic activity in an otherwise low-grade diffuse astrocytic tumor was for some time considered as reason to grade the lesion as anaplastic, while more recent WHO classifications require "brisk" or "marked" mitotic activity for the diagnosis of anaplastic astrocytoma (Fig. 1.5). This means that histopathological diagnoses made in the past not necessarily have the same connotation as those that are made according to the present, i.e., WHO 2007 classification.

Incomplete Representation of Biology by Morphology

Even when an unequivocal histopathological diagnosis can be delivered, the biological behavior of tumors with the same histological type and malignancy grade often varies considerably from patient to patient. In this context, it is important to realize that, while evaluation of the histopathology of a tumor provides an enormous amount of information (in fact, representing a snapshot of the interaction of thousands of genes), it would be unrealistic to expect that microscopic criteria will ever fully cover the molecular heterogeneity in tumor classes that are relatively uniform in their morphological composition. Meanwhile, with more tailored treatments targeting specific molecular aberrations in these neoplasms approaching [35], it is clear that traditional histopathological classification of gliomas does not meet the need that is increasingly felt in the clinic for a more sophisticated and robust diagnosis of gliomas in individual patients.

Way(S) To Go

Integration of Molecular Information

Starting in the first half of the 1990s, knowledge on molecular aberrations driving the development and progression of gliomas has enormously increased. There is now ample evidence that certain molecular changes can be used as biomarkers that provide clinically useful information. In this respect, different types of markers are recognized: diagnostic markers providing information that is helpful for classification of the tumor, e.g., in situations where the neoplasm has ambiguous histological features; prognostic markers carrying information about the inherent biological aggressiveness and thus the prognosis for the patient; and predictive markers providing

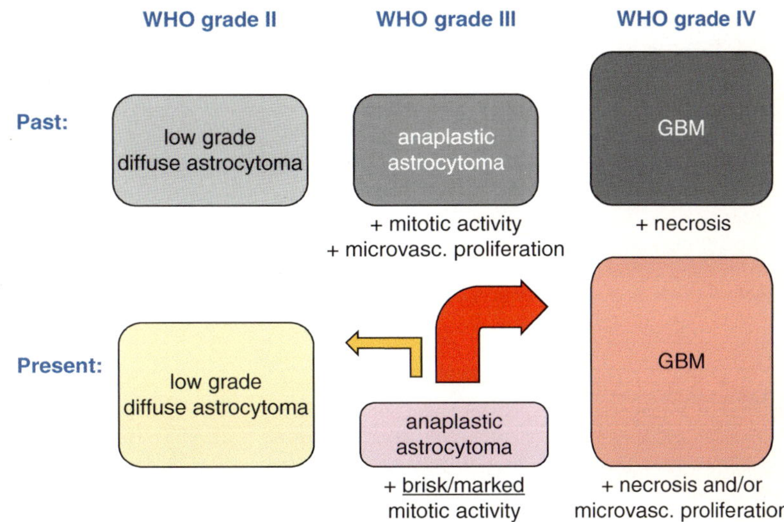

Fig. 1.5 Scheme depicting how a change in diagnostic criteria caused "deflation" of the category of anaplastic astrocytoma. Histopathological diagnoses made in the past may thus not have the same connotation as those that are made according to the present, i.e., WHO 2007 classification of tumors of the CNS

information on the response that can be expected to a particular therapeutic approach [35]. Such markers are thus attractive tools for improving unequivocal diagnosis, assessment of prognosis, and/or treatment stratification for the individual glioma patient.

Examples of molecular markers that have already elicited substantial clinical interest in glioma patients are complete co-deletion of chromosomal arms 1p and 19q, methylation of the promoter of the methyl-guanine methyl transferase gene (*MGMT*), specific mutations of the isocitrate dehydrogenase genes *IDH1* and *IDH2*, amplification of the epidermal growth factor receptor (EGFR) gene and expression of *EGFR* variant III (*EGFRvIII*), and *BRAF* aberrations [36–38].

Co-deletion of 1p and 19q was the first detected molecular marker in gliomas with diagnostic potential [39]. Loss of 1p/19q shows a strong association with classical oligodendroglial features on histology [40]. Dependent on the exact criteria that are used for discrimination of subtypes of diffuse gliomas, this aberration can be detected in up to 80 % of low-grade oligodendrogliomas and approximately 60 % of anaplastic oligodendrogliomas, whereas 30–50 % of low-grade oligoastrocytomas, 15–20 % of anaplastic oligoastrocytomas, and less than 10 % of diffuse astrocytic gliomas (including glioblastomas) carry this aberration (Fig. 1.6). Loss of 1p/19q has also been reported as a predictive marker for favorable response to alkylating chemotherapy. However, evidence is accumulating that this marker may have prognostic rather than predictive meaning [37]. Furthermore, the prognostic significance only seems to be present in tumors showing complete loss of the chromosome arms 1p and 19q, and the significance of complete 1p/19q co-deletion may be less pronounced in the presence of other prognostically unfavorable genetic alterations [41, 42].

The vast majority of diffuse low-grade and anaplastic gliomas as well as secondary glioblastomas carry *IDH1* or (much less frequently) *IDH2* mutations. These mutations represent a very early

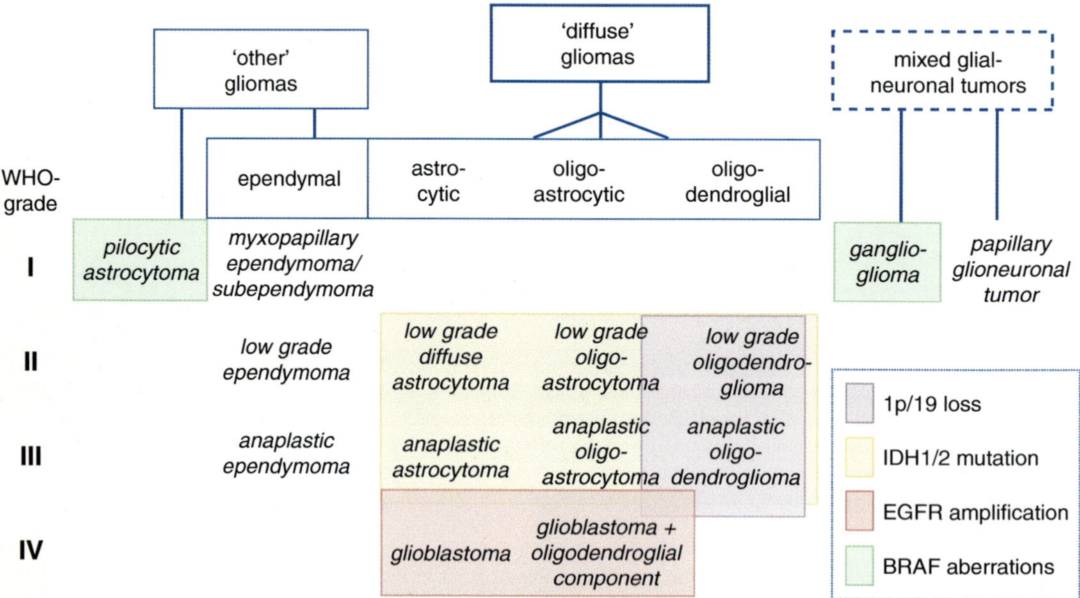

Fig. 1.6 Scheme indicating how different molecular markers may be of diagnostic use for assessment of (sub)type or malignancy grade of glial neoplasms; complete co-deletion of chromosome arms 1p and 19q is strongly associated with oligodendroglial nature of the tumor, the presence of *IDH1* or *IDH2* mutations with low-grade and anaplastic diffuse gliomas and secondary glioblastomas; *EGFR* amplification and expression of *EGFRvIII* indicate high-grade malignancy in diffuse gliomas, and *BRAF* aberrations (*KIAA154-BRAF* fusion gene, *BRAF V600E* mutation) are markers of "variant" gliomas (pilocytic astrocytomas, pleiomorphic xanthoastrocytomas) and gangliogliomas

oncogenetic event and indicate a more favorable prognosis compared to tumors in the same glioma category lacking these mutations [43–45]. *IDH1* and *IDH2* mutation analysis can be helpful in the differential diagnosis of, e.g., diffuse glioma versus pilocytic astrocytoma or ependymoma, of oligodendroglioma versus other tumors with an oligodendroglioma-like component, and of diffuse glioma versus reactive astrocytosis (Fig. 1.6) [46–48]. The fact that the protein product of the *IDH1* R132H mutation (representing about 90 % of the *IDH* mutations in diffuse gliomas) can now also be demonstrated by immunohistochemistry is a good example of how molecular information can sometimes be translated back to a simple, immunohistochemical test. Of note, *IDH* mutations are rare in pediatric patients with (low-grade) gliomas, underscoring that gliomas in the pediatric age group are generally distinct at the molecular level [49].

MGMT promoter methylation has been reported to be significantly correlated with response of glioblastomas to alkylating chemotherapy [50]. More recent studies, however, indicate that the predictive value of *MGMT* promoter methylation in diffuse gliomas is broader than for alkylating chemotherapy alone and is associated with other prognostically favorable molecular features such as 1p/19q co-deletion and *IDH1* mutations [36, 37, 51, 52].

The *EGFR* gene, located at chromosome 7p12, is the most frequently amplified and overexpressed gene in primary glioblastomas. *EGFR* rearrangements are also frequently found in these tumors, by far the most common *EGFR* variant being variant III (*EGFRvIII)* [36, 53]. Identification of *EGFR* amplification and of *EGFRvIII* is suggestive of high-grade malignancy and therefore may provide diagnostic as well as prognostic information (Fig. 1.6) [42]. Also, the *EGFRvIII* mutant may serve as an attractive target for immunotherapy [54]. Unfortunately, the efficacy of therapies targeting the EGFR pathways in gliomas using small molecules is thus far disappointing.

Aberrant activation of the *BRAF* proto-oncogene at 7q34 by gene duplication and fusion or by point mutation has recently been identified as a common genetic aberration in pilocytic astrocytomas, pleiomorphic xanthoastrocytomas, and gangliogliomas [36, 37, 55]. Testing for *BRAF* gene alterations might thus be helpful in the sometimes difficult differential diagnosis between low-grade diffuse astrocytomas and "variant" gliomas (Fig. 1.6) [56]. Additionally, as tumors with duplication or activating mutations of *BRAF* show aberrant signaling via the *BRAF* pathway, pharmacological inhibition of this pathway may prove to be a valuable therapeutic option for these neoplasms.

Except for the identification of individual (epi)genetic alterations that carry diagnostic, prognostic, and/or predictive information, the signatures produced by high-throughput profiling techniques may convey clinically relevant information. For instance, one study reported that molecular classification of gliomas on the basis of genomic profiles obtained by array CGH closely parallels histological classification and was able to distinguish, with few exceptions, between different astrocytoma grades as well as between primary and secondary glioblastomas [57]. Also, gene expression-based classification of morphologically ambiguous high-grade gliomas was reported to correlate better with prognosis than histological classification [58, 59]. Other expression profiling studies reported three or four subclasses of high-grade astrocytomas with prognostic relevance and differences in response to aggressive therapy and in underlying oncogenetic mechanisms (aberrations involving *EGFR*, *NF1*, and *PDGFRA/IDH1* being associated with, respectively, the "classical," "mesenchymal," and "proneural" subclass of gliomas) [60–62].

In 2008, the Cancer Genome Atlas (TCGA) consortium published results of an integrative analysis of DNA copy number, gene expression, and DNA methylation status in over 200 human glioblastomas [63]. This study showed that in the majority of glioblastomas, two or three of the following pathways are involved: the p53 pathway, the RB pathway, and the receptor tyrosine kinase/RAS/PI3K pathway. Promoter DNA methylation profiling revealed that a subset of patients had concerted hypermethylation at a large number of loci, indicating the existence of a glioma-CpG island methylator phenotype (G-CIMP) [64]. Also, a tight association between the *IDH1* mutation status and gene expression profiles was found, suggesting two major pathomechanisms in diffuse astrocytic gliomas: one characterized by *IDH1* mutation and a proneural expression profile (found mostly in diffuse low-grade and anaplastic astrocytomas and in secondary glioblastomas) and the other by lack of *IDH1* mutation and a mesenchymal/proliferative expression profile.

Multidisciplinary Approach

While in some cases, the pathologist may be able to make an unequivocal diagnosis (e.g., of glioblastoma) on a brain tumor biopsy without further knowledge of the context, in most situations, clinical information (esp. patient age, duration of symptoms, previous treatment) and radiological findings (including location and growth pattern of the tumor, contrast enhancement) provide important clues for narrowing down the differential diagnosis in patients with a tumor of the CNS [9].

Conventional magnetic resonance imaging (MRI) is now the gold standard for radiological assessment of (glial) tumors of the CNS [65]. Low-grade diffuse gliomas show hyperintensity on T2-weighted MRI scans but generally do not enhance in T1-weighted images when using the contrast agent gadolinium-DTPA. The absence of contrast enhancement in these tumors can be explained by incorporation ("coöption") of preexistent microvessels with only limited changes to the blood–brain barrier (BBB) and lack of neovascularization. Similarly, contrast enhancement on MRI scans of high-grade gliomas indicates disruption of the BBB of preexistent or newly formed microvessels. Many glioblastomas present radiologically with a non-enhancing, necrotic core surrounded by a contrast-enhancing ring of viable, highly cellular, and angiogenic tumor tissue (ring enhancement). Importantly, contrast enhancement in "variant" gliomas such as pilocytic astrocytomas is fully compatible with a WHO grade I character of the lesion. Obviously, the radiological findings can be very helpful in cases where the differential diagnosis of, e.g., diffuse versus "variant" glioma or of low- versus high-grade malignancy in a diffuse glioma is difficult for the pathologist.

Both radiological and pathological assessment of response to different treatment strategies for diffuse gliomas can be challenging. For example, distinguishing radiation necrosis from "spontaneous" tumor necrosis on MRI and in a biopsy of recurrent glioma can be virtually impossible. Additional MR methods such as diffusion-weighted imaging (DWI), perfusion-weighted imaging (PWI), and magnetic resonance spectroscopy (MRS), as well as positron emission tomography (PET) and single-photon emission computed tomography (SPECT) imaging, may be helpful for narrowing down differential diagnostic options and for improved evaluation of response to different treatment protocols [65, 66]. Brain tumor diagnosis and therapy is thus a multidisciplinary task that requires close collaboration of colleagues from (among others) neurology, neuroradiology, neurosurgery, (neuro)pathology, radiation oncology, and pediatric and/or medical oncology.

Conclusions and Future Perspectives

At present, the histopathological diagnosis based on the criteria defined by the WHO classification is still the gold standard for classification of glial tumors. However, it is increasingly clear that this diagnosis is not robust and specific enough to meet the increasingly refined clinical demands and to guide more targeted therapeutic strategies in the modern neuro-oncology practice. Even if the criteria for morphological typing and grading can be substantially improved, it will be difficult to capture the biological variation in strict morphological criteria. Moreover, tissue sampling can be incomplete and may lead to, e.g., underestimation of the true degree of malignancy in regionally heterogeneous tumors.

Several molecular markers that may supplement or even overrule the information provided by microscopic investigation are now ready to be translated into clinical practice, and it is to be expected that the number of informative molecular markers in glioma diagnostics will increase much further. Up till now, however, very few molecular markers are actually being used routinely in daily clinical practice, also because there is a danger that after enthusiastic introduction of particular markers, "reality checks" reveal later on that the information provided by certain markers is not as straightforward as was originally hoped for [35]. Furthermore, assessment of glioma signatures by detailed genomic and/or expression array analysis is not yet suitable for broad introduction in the routine diagnostic panel because of the limited availability and high costs of this approach. Moreover, it remains to be proven that these signatures yield clinically relevant data for individual patients beyond the information provided by simpler tests analyzing, e.g., *IDH1* mutation, *MGMT* promoter methylation, and 1p/19q deletion. Interestingly, a recent study reported that immunohistochemical expression analysis of a nine-gene signature, which is applicable to routinely processed tissue samples, may be sufficient to predict glioblastoma outcome [67].

In conclusion, histopathological evaluation of tumor tissue still forms the basis for classification of gliomas. However, integration of information on the underlying molecular aberrations will undoubtedly result in a more refined and robust classification of these tumors, and modern neuropathology is rapidly moving toward such a combined morphological and molecular approach. The challenges in this context are to sort out which markers really provide (diagnostic, prognostic, and/or predictive) information that is useful for clinical decision-making and how to implement analysis of these markers in a reliable and affordable way in order to optimally serve the individual patients suffering from these neoplasms in a routine neuro-oncology setting.

References

1. Fuller GN, Scheithauer BW. The 2007 revised World Health Organization (WHO) classification of tumours of the central nervous system: newly codified entities. Brain Pathol. 2007;17:304–7.
2. CBTRUS. Central Brain Tumor Registry of the United States, 2010. www.cbtrus.org.
3. Kohler BA, Ward E, McCarthy BJ, Schymura MJ, Ries LA, Eheman C, Jemal A, Anderson RN, Ajani UA, Edwards BK. Annual report to the nation on the status of cancer, 1975–2007, featuring tumors of the brain and other nervous system. J Natl Cancer Inst. 2011;103:714–36.
4. Burnet NG, Jefferies SJ, Benson RJ, Hunt DP, Treasure FP. Years of life lost (YLL) from cancer is an important measure of population burden – and should be considered when allocating research funds. Br J Cancer. 2005;92:241–5.
5. Wesseling P, Kros JM, Jeuken JWM. The pathological diagnosis of diffuse gliomas: towards a smart synthesis of microscopic and molecular information in a multidisciplinary context. Diagn Histopathol. 2011;17:486–94.
6. Ohgaki H, Kleihues P. Population-based studies on incidence, survival rates, and genetic alterations in astrocytic and oligodendroglial gliomas. J Neuropathol Exp Neurol. 2005;64:479–89.
7. Louis DN, Ohgaki H, Wiestler OD, Cavenee WK. WHO classification of tumours of the central nervous system. 3rd ed. Lyon: IARC Press; 2007.
8. Burger PC, Scheithauer BW. Tumors of the central nervous system, vol. 4. 7th ed. Washington: AFIP Atlas of tumor pathology; 2007.
9. Perry A, Brat DJ. Practical surgical neuropathology. Philadelphia: Churchill Livingstone; 2010.
10. Bailey PC, Cushing H. A classification of the tumors of the glioma group on a histogenetic basis with a correlated study of prognosis. Philadelphia: J.B. Lippincott Co; 1926.
11. Stiles CD, Rowitch DH. Glioma stem cells: a midterm exam. Neuron. 2008;58:832–46.
12. Venere M, Fine HA, Dirks PB, Rich JN. Cancer stem cells in gliomas: identifying and understanding the apex cell in cancer's hierarchy. Glia. 2011;59:1148–54.
13. Kernohan JW, Mabon RF, Svien HJ, Adson AW. A simplified classification of the gliomas. Proc Staff Meet Mayo Clin. 1949;24:71–5.
14. Kros JM. Grading of gliomas: the road from eminence to evidence. J Neuropathol Exp Neurol. 2011;70:101–9.
15. Scheithauer BW. Development of the WHO classification of tumors of the central nervous system: a historical perspective. Brain Pathol. 2009;19:551–64.
16. Louis DN, Ohgaki H, Wiestler OD, Cavenee WK, Burger PC, Jouvet A, Scheithauer BW, Kleihues P. The 2007 WHO classification of tumours of the central nervous system. Acta Neuropathol. 2007;114:97–109.
17. Scherer HJ. Structural development in gliomas. Am J Cancer. 1938;34:333–51.
18. Claes A, Idema AJ, Wesseling P. Diffuse glioma growth: a guerilla war. Acta Neuropathol. 2007;114:443–58.
19. Dunbar E, Yachnis AT. Glioma diagnosis: immunohistochemistry and beyond. Adv Anat Pathol. 2010;17:187–201.

20. Ducray F, Mokhtari K, Crinière E, Idbaih A, Marie Y, Dehais C, Paris S, Carpentier C, Dieme MJ, Adam C, Hoang-Xuan K, Duyckaerts C, Delattre JY, Sanson M. Diagnostic and prognostic value of alpha internexin expression in a series of 409 gliomas. Eur J Cancer. 2011;47:802–8.
21. Durand K, Guillaudeau A, Pommepuy I, Mesturoux L, Chaunavel A, Gadeaud E, Porcheron M, Moreau JJ, Labrousse F. Alpha-internexin expression in gliomas: relationship with histological type and 1p, 19q, 10p and 10q status. J Clin Pathol. 2011;64:793–801.
22. Ohgaki H, Kleihues P. Genetic pathways to primary and secondary glioblastoma. Am J Pathol. 2007;170: 1445–53.
23. Scheithauer BW, Fuller GN, VandenBerg SR. The 2007 WHO classification of tumors of the nervous system: controversies in surgical neuropathology. Brain Pathol. 2008;18:307–16.
24. Rodriguez FJ, Scheithauer BW, Burger PC, Jenkins S, Giannini C. Anaplasia in pilocytic astrocytoma predicts aggressive behavior. Am J Surg Pathol. 2010;34: 147–60.
25. Burger PC, Heinz ER, Shibata T, Kleihues P. Topographic anatomy and CT correlations in the untreated glioblastoma multiforme. J Neurosurg. 1988;68:698–704.
26. Burger PC, Kleihues P. Cytologic composition of the untreated glioblastoma with implications for evaluation of needle biopsies. Cancer. 1989;63:2014–23.
27. Groenen PJ, Blokx WA, Diepenbroek C, Burgers L, Visinoni F, Wesseling P, van Krieken JH. Preparing pathology for personalized medicine: possibilities for improvement of the pre-analytical phase. Histopathology. 2011;59:1–7.
28. Perry A, Scheithauer BW, Macaulay RJ, Raffel C, Roth KA, Kros JM. Oligodendrogliomas with neurocytic differentiation. A report of 4 cases with diagnostic and histogenetic implications. J Neuropathol Exp Neurol. 2002;61:947–55.
29. Wesseling P, Schlingemann RO, Rietveld FJ, Link M, Burger PC, Ruiter DJ. Early and extensive contribution of pericytes/vascular smooth muscle cells to microvascular proliferation in glioblastoma multiforme: an immuno-light and immuno-electron microscopic study. J Neuropathol Exp Neurol. 1995;54:304–10.
30. Coons SW, Johnson PC, Scheithauer BW, Yates AJ, Pearl DK. Improving diagnostic accuracy and interobserver concordance in the classification and grading of primary gliomas. Cancer. 1997;79:1381–93.
31. Giannini C, Scheithauer BW, Weaver AL, Burger PC, Kros JM, Mork S, Graeber MB, Bauserman S, Buckner JC, Burton J, Riepe R, Tazelaar HD, Nascimento AG, Crotty T, Keeney GL, Pernicone P, Altermatt H. Oligodendrogliomas: reproducibility and prognostic value of histologic diagnosis and grading. J Neuropathol Exp Neurol. 2001;60:248–62.
32. Kros JM, Gorlia T, Kouwenhoven MC, Zheng PP, Collins VP, Figarella-Branger D, Giangaspero F, Giannini C, Mokhtari K, Mørk SJ, Paetau A, Reifenberger G, van den Bent MJ. Panel review of anaplastic oligodendroglioma from European Organization For Research and Treatment of Cancer Trial 26951: assessment of consensus in diagnosis, influence of 1p/19q loss, and correlations with outcome. J Neuropathol Exp Neurol. 2007;66:545–51.
33. van den Bent MJ. Interobserver variation of the histopathological diagnosis in clinical trials on glioma: a clinician's perspective. Acta Neuropathol. 2010;120: 297–304.
34. Tabatabai G, Stupp R, van den Bent MJ, Hegi ME, Tonn JC, Wick W, Weller M. Molecular diagnostics of gliomas: the clinical perspective. Acta Neuropathol. 2010;120:585–92.
35. Mellinghoff IK, Lassman AB, Wen PY. Signal transduction inhibitors and antiangiogenic therapies for malignant glioma. Glia. 2011;59:1205–12.
36. Yip S, Iafrate AJ, Louis DN. Molecular diagnostic testing in malignant gliomas: a practical update on predictive markers. J Neuropathol Exp Neurol. 2008;67:1–15.
37. Riemenschneider MJ, Jeuken JW, Wesseling P, Reifenberger G. Molecular diagnostics of gliomas: state of the art. Acta Neuropathol. 2010;120:567–84.
38. von Deimling A, Korshunov A, Hartmann C. The next generation of glioma biomarkers: MGMT methylation, BRAF fusions and IDH1 mutations. Brain Pathol. 2011;21:74–87.
39. Reifenberger J, Reifenberger G, Liu L, James CD, Wechsler W, Collins VP. Molecular genetic analysis of oligodendroglial tumors shows preferential allelic deletions on 19q and 1p. Am J Pathol. 1994;145: 1175–90.
40. Aldape K, Burger PC, Perry A. Clinicopathologic aspects of 1p/19q loss and the diagnosis of oligodendroglioma. Arch Pathol Lab Med. 2007;131: 242–251.
41. Idbaih A, Marie Y, Pierron G, Brennetot C, Hoang-Xuan K, Kujas M, Mokhtari K, Sanson M, Lejeune J, Aurias A, Delattre O, Delattre JY. Two types of chromosome 1p losses with opposite significance in gliomas. Ann Neurol. 2005;58:483–7.
42. Jeuken JW, Sijben A, Bleeker FE, Boots-Sprenger SH, Rijntjes J, Gijtenbeek JM, Mueller W, Wesseling P. The nature and timing of specific copy number changes in the course of molecular progression in diffuse gliomas: further elucidation of their genetic "life story". Brain Pathol. 2011;21:308–20.
43. Parsons DW, Jones S, Zhang X, Lin JC, Leary RJ, Angenendt P, Mankoo P, Carter H, Siu IM, Gallia GL, Olivi A, McLendon R, Rasheed BA, Keir S, Nikolskaya T, Nikolsky Y, Busam DA, Tekleab H, Diaz Jr LA, Hartigan J, Smith DR, Strausberg RL, Marie SK, Shinjo SM, Yan H, Riggins GJ, Bigner DD, Karchin R, Papadopoulos N, Parmigiani G, Vogelstein B, Velculescu E, Kinzler KW. An integrated genomic analysis of human glioblastoma multiforme. Science. 2008;321:1807–12.

44. Yan H, Parsons DW, Jin G, McLendon R, Rasheed BA, Yuan W, Kos I, Batinic-Haberle I, Jones S, Riggins GJ, Friedman H, Friedman A, Reardon D, Herndon J, Kinzler KW, Velculescu VE, Vogelstein B, Bigner DD. IDH1 and IDH2 mutations in gliomas. N Engl J Med. 2009;360:765–73.
45. Hartmann C, Hentschel B, Wick W, Capper D, Felsberg J, Simon M, Westphal M, Schackert G, Meyermann R, Pietsch T, Reifenberger G, Weller M, Loeffler M, von Deimling A. Patients with IDH1 wild type anaplastic astrocytomas exhibit worse prognosis than IDH1-mutated glioblastomas, and IDH1 mutation status accounts for the unfavorable prognostic effect of higher age: implications for classification of gliomas. Acta Neuropathol. 2010;120:707–18.
46. Capper D, Weissert S, Balss J, Habel A, Meyer J, Jäger D, Ackermann U, Tessmer C, Korshunov A, Zentgraf H, Hartmann C, von Deimling A. Characterization of R132H mutation-specific IDH1 antibody binding in brain tumors. Brain Pathol. 2010;20:245–54.
47. Capper D, Reuss D, Schittenhelm J, Hartmann C, Bremer J, Sahm F, Harter PN, Jeibmann A, von Deimling A. Mutation-specific IDH1 antibody differentiates oligodendrogliomas and oligoastrocytomas from other brain tumors with oligodendroglioma-like morphology. Acta Neuropathol. 2011;121: 241–52.
48. Camelo-Piragua S, Jansen M, Ganguly A, Kim JC, Cosper AK, Dias-Santagata D, Nutt CL, Iafrate AJ, Louis DN. A sensitive and specific diagnostic panel to distinguish diffuse astrocytoma from astrocytosis: chromosome 7 gain with mutant isocitrate dehydrogenase 1 and p53. J Neuropathol Exp Neurol. 2011;70: 110–5.
49. Jones DT, Mulholland SA, Pearson DM, Malley DS, Openshaw SW, Lambert SR, Liu L, Bäcklund LM, Ichimura K, Collins VP. Adult grade II diffuse astrocytomas are genetically distinct from and more aggressive than their paediatric counterparts. Acta Neuropathol. 2011;121:753–61.
50. Hegi ME, Diserens AC, Gorlia T, Hamou MF, de Tribolet N, Weller M, Kros JM, Hainfellner JA, Mason W, Mariani L, Bromberg JE, Hau P, Mirimanoff RO, Cairncross JG, Janzer RC, Stupp R. MGMT gene silencing and benefit from temozolomide in glioblastoma. N Engl J Med. 2005;352:997–1003.
51. Weller M, Stupp R, Reifenberger G, Brandes AA, van den Bent MJ, Wick W, Hegi ME. MGMT promoter methylation in malignant gliomas: ready for personalized medicine? Nat Rev Neurol. 2010;6:39–51.
52. Olson RA, Brastianos PK, Palma DA. Prognostic and predictive value of epigenetic silencing of MGMT in patients with high grade gliomas: a systematic review and meta-analysis. J Neurooncol. 2011;105:325–335.
53. Jeuken J, Sijben A, Alenda C, Rijntjes J, Dekkers M, Boots-Sprenger S, McLendon R, Wesseling P. Robust detection of EGFR copy number changes and EGFR variant III: technical aspects and relevance for glioma diagnostics. Brain Pathol. 2009;19:661–71.
54. Choi BD, Archer GE, Mitchell DA, Heimberger AB, McLendon RE, Bigner DD, Sampson JH. EGFRvIII-targeted vaccination therapy of malignant glioma. Brain Pathol. 2009;19:713–23.
55. Cin H, Meyer C, Herr R, Janzarik WG, Lambert S, Jones DT, Jacob K, Benner A, Witt H, Remke M, Bender S, Falkenstein F, Van Anh TN, Olbrich H, von Deimling A, Pekrun A, Kulozik AE, Gnekow A, Scheurlen W, Witt O, Omran H, Jabado N, Collins VP, Brummer T, Marschalek R, Lichter P, Korshunov A, Pfister SM. Oncogenic FAM131B-BRAF fusion resulting from 7q34 deletion comprises an alternative mechanism of MAPK pathway activation in pilocytic astrocytoma. Acta Neuropathol. 2011;121: 763–74.
56. Korshunov A, Meyer J, Capper D, Christians A, Remke M, Witt H, Pfister S, von Deimling A, Hartmann C. Combined molecular analysis of BRAF and IDH1 distinguishes pilocytic astrocytoma from diffuse astrocytoma. Acta Neuropathol. 2009;118: 401–5.
57. Roerig P, Nessling M, Radlwimmer B, Joos S, Wrobel G, Schwaenen C, Reifenberger G, Lichter P. Molecular classification of human gliomas using matrix-based comparative genomic hybridization. Int J Cancer. 2005;117:95–103.
58. Nutt CL, Mani DR, Betensky RA, Tamayo P, Cairncross JG, Ladd C, Pohl U, Hartmann C, McLaughlin ME, Batchelor TT, Black PM, von Deimling A, Pomeroy SL, Golub TR, Louis DN. Gene expression-based classification of malignant gliomas correlates better with survival than histological classification. Cancer Res. 2003;63:1602–7.
59. Gravendeel LA, Kouwenhoven MC, Gevaert O, de Rooi JJ, Stubbs AP, Duijm JE, Daemen A, Bleeker FE, Bralten LB, Kloosterhof NK, De Moor B, Eilers PH, van der Spek PJ, Kros JM, Sillevis Smitt PA, van den Bent MJ, French PJ. Intrinsic gene expression profiles of gliomas are a better predictor of survival than histology. Cancer Res. 2009;69:9065–72.
60. Phillips HS, Kharbanda S, Chen R, Forrest WF, Soriano RH, Wu TD, Misra A, Nigro JM, Colman H, Soroceanu L, Williams PM, Modrusan Z, Feuerstein BG, Aldape K. Molecular subclasses of high-grade glioma predict prognosis, delineate a pattern of disease progression, and resemble stages in neurogenesis. Cancer Cell. 2006;9:157–73.
61. Verhaak RG, Hoadley KA, Purdom E, Wang V, Qi Y, Wilkerson MD, Miller CR, Ding L, Golub T, Mesirov JP, Alexe G, Lawrence M, O'Kelly M, Tamayo P, Weir BA, Gabriel S, Winckler W, Gupta S, Jakkula L, Feiler HS, Hodgson JG, James CD, Sarkaria JN, Brennan C, Kahn A, Spellman PT, Wilson RK, Speed TP, Gray JW, Meyerson M, Getz G, Perou CM, Hayes DN. Integrated genomic analysis identifies clinically relevant subtypes of glioblastoma characterized by

abnormalities in PDGFRA, IDH1, EGFR, and NF1. Cancer Cell. 2010;17:98–110.

62. Huse JT, Phillips HS, Brennan CW. Molecular subclassification of diffuse gliomas: seeing order in the chaos. Glia. 2011;59:1190–9.
63. TCGA, The Cancer Genome Atlas Research Network. Comprehensive genomic characterization defines human glioblastoma genes and core pathways. Nature. 2008;455:1061–8.
64. Noushmehr H, Weisenberger DJ, Diefes K, Phillips HS, Pujara K, Berman BP, Pan F, Pelloski CE, Sulman EP, Bhat KP, Verhaak RG, Hoadley KA, Hayes DN, Perou CM, Schmidt HK, Ding L, Wilson RK, Van Den Berg D, Shen H, Bengtsson H, Neuvial P, Cope LM, Buckley J, Herman JG, Baylin SB, Laird PW, Aldape K. Identification of a CpG Island methylator phenotype that defines a distinct subgroup of glioma. Cancer Cell. 2010;17:510–22.
65. Faehndrich J, Weidauer S, Pilatus U, Oszvald A, Zanella FE, Hattingen E. Neuroradiological viewpoint on the diagnostics of space-occupying brain lesions. Clin Neuroradiol. 2011;21:123–39.
66. Chen W, Silverman DH. Advances in evaluation of primary brain tumors. Semin Nucl Med. 2008;38: 240–50.
67. Colman H, Zhang L, Sulman EP, McDonald JM, Shooshtari NL, Rivera A, Popoff S, Nutt CL, Louis DN, Cairncross JG, Gilbert MR, Phillips HS, Mehta MP, Chakravarti A, Pelloski CE, Bhat K, Feuerstein BG, Jenkins RB. Aldape K A multigene predictor of outcome in glioblastoma. Neuro Oncol. 2010;12:49–57.

Epigenetics and Brain Cancer

2

Davide Sciuscio and Monika E. Hegi

Abstract

Epigenetic alterations have been recognized as important mechanisms in neoplastic transformation, malignant progression of cancer, and response to therapy. Epigenetic modifications include DNA methylation and post-translational modifications of histone proteins that influence the chromatin structure. Moreover, with the identification of the RNA interference machinery, a new layer of gene regulation has been added to the definition. The coordinated interaction of these processes regulates gene expression activity. The disruption of these mechanisms of control is involved in a wide variety of pathologies, including but not restricted to cancer. Although epigenetic changes are somatically inheritable, they are reversible and hence may represent actionable targets for novel therapies.

Here we will discuss the current understanding of alterations in the epigenetic landscape that occur in the evolution of brain tumors and their potential impact on patient therapy.

Keywords

DNA methylation • Histone modification • MicroRNAs • Epigenetic gene silencing • Epigenetic therapy

The Epigenetic Control

The term "epigenetics" (literally "upon" genetics) was coined by Conrad Waddington in the early 1940s. It was initially used to explain why genetic variations sometimes do not lead to phenotypic variations and how genes might interact with their environment to yield a phenotype [1]. Currently epigenetics is defined as the study of mitotically and/or meiotically heritable changes in gene expression that involve molecular and structural changes of DNA but do not alter the

D. Sciuscio, Ph.D.
Service of Neurosurgery, Department of Clinical Neurosciences, Lausanne University Hospital, Lausanne, Switzerland

M.E. Hegi, Ph.D. (✉)
Laboratory of Brain Tumor Biology and Genetics, Service of Neurosurgery, Department of Clinical Neurosciences, Lausanne University Hospital, Rue du Bugnon 46, Lausanne, 1011, Switzerland
e-mail: monika.hegi@chuv.ch

C. Watts (ed.), *Emerging Concepts in Neuro-Oncology*,
DOI 10.1007/978-0-85729-458-6_2, © Springer-Verlag London 2013

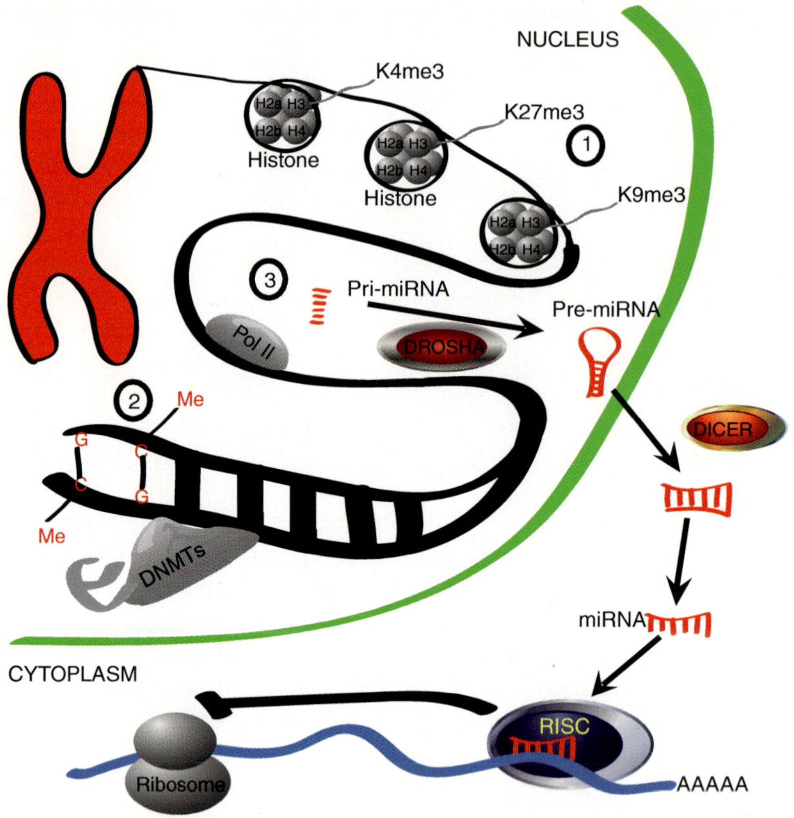

Fig. 2.1 Epigenetic regulation. Epigenetic regulation acts at different levels and with different molecular mechanisms like posttranslational modifications (PTMs) of the histone tails (*1*) DNA methylation of CpG dinucleotides (*2*) and microRNA expression (*3*) PTMs like tri-methylation (me3) of lysine number 4 (K4) of histone 3 (H3) (H3K4me3), H3K27me3, H3K9me3, etc. allow a more relaxed or compact chromatin status resulting in expression or repression of the genes under the modified histones. DNA methylation is catalized by DNMTs (DNA methyltransferases) and leads to gene silencing directly by recruitment of methyl-CpG-binding domain (MBD) proteins that in turn recruit histone-modifying and chromatin-remodeling complexes to the methylated sites or indirectly by precluding the recruitment of DNA-binding proteins. MicroRNAs are transcribed from intragenic or intergenic regions by RNA polymerase II (PolII), and after several maturation steps mediated by the ribonucleases DROSHA and DICER, they are integrated in the RNA-induced silencing complex (RISC) that blocks translation of specific mRNAs

DNA sequence (Fig. 2.1) [2]. Epigenetic regulation ensures that the right genes are expressed at the right time to allow for cell-type-specific programs in development and differentiation, and adaptation to environmental cues that are not encoded in the DNA.

Epigenetic aberrations in cancer involve global DNA demethylation (hypomethylation) affecting intergenic regions, DNA repetitive sequences and gene bodies, and de novo methylation of CpG islands (hypermethylation) in promoter regions of tumor suppressor genes (Fig. 2.2). It has been largely established that epigenetic silencing of key genes mediated by promoter methylation plays an important role in cancer [3]. In addition, dynamic regulation of the chromatin state is mediated by mechanisms such as covalent modifications of chromatin including histone acetylation, methylation, phosphorylation, and ATP-dependent chromatin remodeling. The latter is mediated by enzyme complexes using ATP-hydrolysis to slide away histones along the DNA, which may expose transcription factor binding sites and thus facilitate their association with

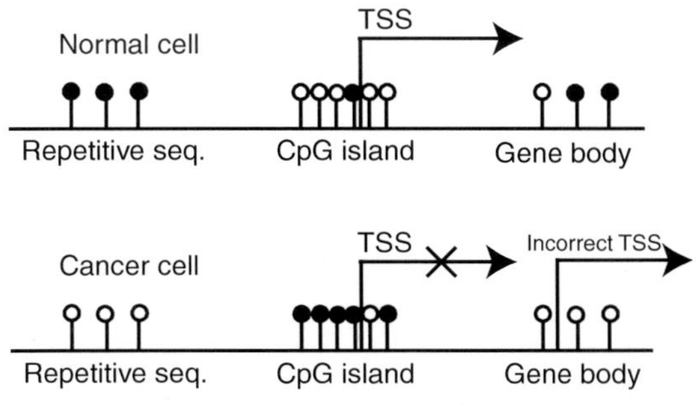

Fig. 2.2 Aberrant DNA methylation in cancer. In normal cells DNA methylation at CpG sites is mainly present in gene bodies and repetitive sequences and contributes to chromosomal stability. In cancer cells we observe global DNA demethylation prominent at repetitive sequences and specific hypermethylation of CpG islands of promoters that frequently affects suppressor genes. Aberrant hypomethylation of gene bodies may unblock alternative transcription start sites

regulatory sequences. The best-known chromatin remodelers of this type belong to the family of SWI/SNF complexes [4]. The three processes of DNA methylation, histone modification, and nucleosomal remodeling are intimately linked, and their alterations result in reprogramming of cancer-relevant genes (reviewed in [5]) (Fig. 2.1). Recent data in cancer biology emphasize the importance of epigenetic processes and illustrate that genetic and epigenetic phenomena cooperate at all stages of cancer development.

DNA Methylation

DNA methylation is the most intensively studied regulatory mechanism involved in the epigenetic control. It occurs predominantly on cytosine residues in CpG dinucleotides. So far, three enzymes that catalyze DNA methylation have been described: the DNA methyltransferases DNMT1, DNMT3a, and DNMT3b. All of them use the substrate *S*-adenosyl-L-methionine as source of methyl groups. DNMT1 preferentially methylates hemi-methylated DNA and is responsible for maintenance of the methylation patterns during DNA replication. DNMT3a and DNMT3b act on unmethylated DNA substrates and are responsible for de novo methylation [6, 7]. CpG dinucleotides are not evenly distributed across the human genome but are concentrated in short CpG-rich DNA stretches called "CpG islands." They are preferentially located at the 5′ end of genes and are present in about 60 % of human gene promoters [8] or reside in regions of large repetitive genomic sequences [9, 10]. DNA methylation of repetitive sequences has been proposed as a mechanism to prevent chromosomal instability by suppressing events such as homologous recombination [11], while gene body methylation is thought to prevent uncontrolled transcription initiation (reviewed by Portela et al. [5]). DNA hypermethylation of CpG islands located in the promoter regions has been associated with loss of expression (Fig. 2.2). Epigenetic gene silencing following CpG island methylation is mediated through recruitment of methyl-CpG-binding domain (MBD) proteins that in turn recruit histone-modifying and chromatin-remodeling complexes to the methylated sites [12, 13] or indirectly by precluding the recruitment of DNA-binding proteins from their target sites [14]. Normally, most CpG islands remain unmodified during development and in differentiated tissues [15]. However, there are some exceptions like the CpG island methylation occurring during X-chromosome inactivation and those for imprinted genes [9]. Recent findings also suggest that extensive DNA methylation changes caused by differentiation take place at CpG island "shores," regions of comparatively low CpG density close to CpG islands [16–18].

Although CpG methylation is the most studied epigenetic modification, it is not the only one that can occur at the DNA level. Recently other regulatory chemical modifications have been described like the methylation at non-CpG sites like CHG and CHH (where H is A, C, or T) or the 5-hydroxymethylcytosine (5-hmC). Methylated CHG and CHH have been found in stem cells and seem to be enriched in gene bodies directly correlated with gene expression, while they are depleted in protein binding sites and enhancers [19]. The levels of non-CpG methylation decrease during differentiation and are restored in induced pluripotent stem cells (iPS), suggesting a key role in the maintenance of pluripotency [19, 20]. The function of 5-hmC is not yet understood and poorly studied at the moment especially because this modification cannot be easily distinguished technically from the classic 5-methylcytosine [21].

Detection Methods for Methylated DNA

In the last decade, the study of DNA methylation has become essential for the understanding of regulatory processes in biology, and more recently aberrantly methylated genes have been identified as biomarkers in cancer with clinical applications. This has led to the development of many methods for its detection using various technical strategies that are associated with different resolution. Choice of technology depends on the purpose, ranging from diagnostic tests for individual genes for patient selection to genome-wide methylation profiling allowing for an unbiased comprehensive view of DNA methylation.

One of the most common methods to differentiate between methylated and unmethylated CpG sites uses a bisulfite treatment. This step converts unmethylated cytosine—but not 5-methylcytosine—in the DNA to uracil [22] that after amplification by polymerase chain reaction (PCR) is replaced by thymidine. Subsequently, the altered sequence can be identified by any technology allowing sequence-specific readouts that differentiates between cytosine and thymidine (Fig. 2.3). A popular method is methylation-specific PCR (MSP) that uses distinct sets of primers, and each set is designed to bind either only to completely methylated or unmethylated sequences, respectively [23]. Each primer typically interrogates a series of three to five CpGs. Quantitative versions of MSP, QMSP, allow definition of cutoff, standardization, and high-throughput analysis [24, 25]. Other quantitative/semiquantitative methods comprise methylation-specific pyrosequencing and methylation-specific clone sequencing [26]. For genome-wide analysis of bisulfite-treated DNA, high-density bead chip arrays are available (e.g., Infinium 450 K Methylation-Bead Chip, Illumina) for high-throughput analysis, while deep sequencing technology (MethylC-seq) allows for unbiased evaluation of the methylome [27].

Methods not depending on bisulfite conversion for differentiating methylated from unmethylated CpGs take advantage of methylation-sensitive restriction endonucleases that recognize and cleave sequence-specific either methylated or unmethylated CpGs only, followed by amplification for detection and quantification of characteristic restriction fragments. Other methods enrich methylated DNA fragments using antibodies against methylated CpGs (MeDIP) or affinity columns loaded with recombinant peptides derived from DNA methylation-binding proteins, such as the methyl-CpG-binding protein 2 (MeCP2). These enriched methylated DNA fragments are then used as input for detection methods such as deep sequencing or DNA microarrays that allow quantification of captured methylated DNA fragments [28–30].

The detailed comparison of the different technologies is beyond the scope of this chapter. Comparison of different technologies used to determine the methylation status of marker genes such as *MGMT* has been reviewed in Weller et al. [31], and the assessment of different technologies for unbiased genome-wide DNA methylation analysis has been published recently [28–30].

Posttranslational Modification of Chromatin

The eukaryotic genome is packaged into chromatin, a highly ordered structure that contains DNA, RNA, histones, and other chromosomal proteins. Chromatin was originally classified into two domains, euchromatin and heterochromatin, based on the density of staining of the nucleic acid in micrographs [32, 33]. The definition of these domains has since been expanded. Euchromatin is gene-rich, transcriptionally active, hyperacetylated, and hypomethylated chromatin. Conversely, heterochromatin is gene-poor, transcriptionally inactive, hypoacetylated, and hypermethylated chromatin [32–34]. The basic unit of chromatin is the nucleosome, which is composed of two copies of the histones H2A, H2B, H3, and H4 wrapped with 146 base pairs of DNA [33, 35]. The ability of chromatin to condense can be regulated in

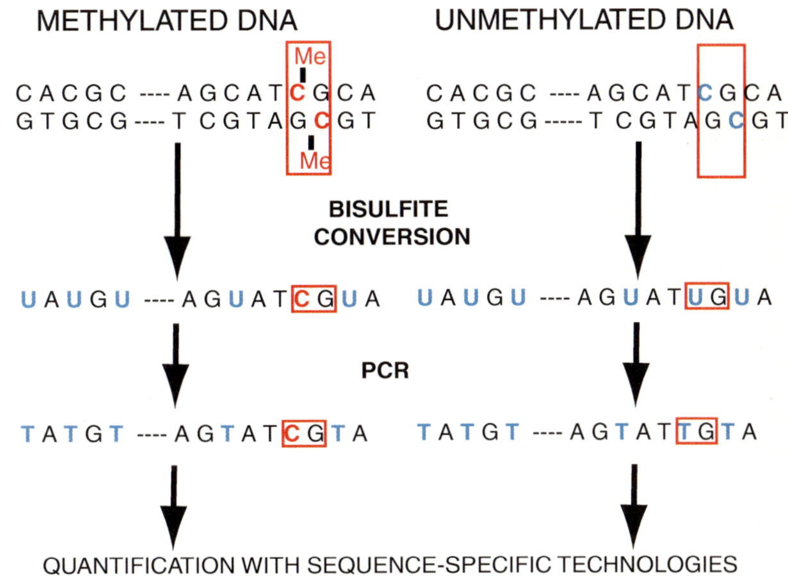

Fig. 2.3 Discrimination between methylated and unmethylated CpGs using bisulfite conversion. Methylated cytosines (5-methylcytosines) are resistant to bisulfite conversion, whereas unmethylated cytosines are converted into uracil. Sequence-specific technologies allow then sequence-specific discrimination and quantification of methylated and unmethylated CpGs, respectively

part by posttranslational modification (PTM) of the N-terminal tails of the histones which include acetylation, methylation, phosphorylation, sumoylation, poly(ADP)-ribosylation, and ubiquitination. These modifications regulate key cellular processes such as transcription, replication, and repair. So far over 60 different modifications on histones have been described [34] defining the so-called "histone code" that refers to the patterns of modifications where different combinations of histone modifications designate or regulate specific cellular processes and events [36–38]. Active genes have been associated with particular modifications also called active histone marks, e.g., tri-methylation of lysine 4 (H3K4me3) and acetylation of lysine 9 (H3K9ac). In contrast histone marks for inactive genes may comprise H3K9me2, H3K9me3, H3K27me2, and H3K27me3. However, many active and inactive genes have overlapping patterns of histone modifications. In fact bivalent histone marks are a hallmark of embryonic stem cells that is thought to keep the genes in a "transcription-ready" state and may predispose important regulatory genes to inactivation by aberrant DNA hypermethylation that results in heritable gene silencing during malignant transformation and tumor progression [39]. For almost each modification, enzymes exist which either lay down the appropriate mark or remove it. Histone acetyltransferases (HATs) and histone methyltransferases (HMTs) add acetyl and methyl groups, respectively, whereas histone deacetylases (HDACs) and histone demethylases (HDMs) remove them [40, 41]. These histone-modifying enzymes interact with each other as well as other DNA regulatory mechanisms to tightly link chromatin state and transcription. Although there is an intimate relationship between DNA methylation and PTM of histones, the former is considered to be relatively stable, while PTMs of histones are more dynamic, balanced by the activities of the histone-modifying enzymes removing or adding respective modifications. In cancer cells this equilibrium is disturbed by deregulated expression of HMTs and HDMs and overexpression of HDACs. Deregulated expression of histone-modifying enzymes makes them potential targets for therapy to normalizing their equilibrium.

Like DNA methylation the study of the post-transcriptional modification of chromatin led to the development of several methods of analysis. Most of them are based on immunoprecipitation of the chromatin cross-linked to DNA using specific antibodies against the different PTM of the chromatin. Coprecipitated DNA is subsequently analyzed and quantified by PCR (ChIP-PCR), on DNA chips (ChIP on CHIP), or by genome-wide deep sequencing (ChIP-seq) to identify and quantify the chromatin status at loci of interest.

MicroRNAs

MicroRNAs are endogenously expressed short noncoding RNAs, 18–25 nucleotides in length, that repress protein translation through binding to target mRNAs [42]. More than 1,000 human microRNAs have been discovered to date, and recent studies have estimated that they are responsible for the regulation of up to one-third of all human genes [43]. MicroRNAs are mostly transcribed from intragenic or intergenic regions by RNA polymerase II into primary transcripts called pri-microRNAs [44, 45]. The primary transcripts undergo further processing usually by a ribonuclease named DROSHA resulting in a hairpin intermediate of about 70–100 nucleotides, called pre-microRNA [46, 47]. The pre-microRNA is then transported out of the nucleus to the cytoplasm by exportin 5 [48]. In the cytoplasm, the pre-microRNA is processed by another ribonuclease, DICER, into a mature double-stranded microRNA [49, 50]. After strand separation, the guide strand or mature microRNA is incorporated into an RNA-induced silencing complex (RISC), whereas the passenger strand is degraded [50–53] (Fig. 2.1). RISC comprises also argonaute proteins that have a crucial role in microRNA biogenesis, maturation, and miRNA effector functions [51–53]. The mature guide strand is important for target recognition and for the incorporation of specific target mRNAs into RISC [50–53]. The specificity of microRNA targeting is defined by Watson–Crick complementarities between positions 2 and 8 from the 5′ miRNA (also known as the seed), with the 3′ untranslated region (UTR) of their target mRNAs [53]. When microRNA and its target mRNA sequence show perfect complementarities, the RISC induces mRNA degradation. Should an imperfect microRNA–mRNA target pairing occur, translation into a protein is blocked [53] (Fig. 2.1). Analyzing the complementarities between microRNA and mRNA has revealed that each microRNA can potentially target multiple mRNAs [50, 54–56], while a single mRNA can be targeted by several different microRNAs [54, 55]. Many of these predictions have been validated experimentally, suggesting that microRNAs might cooperate with each other to regulate gene expression [56]. Similar to promoter methylation of genes, expression of these regulatory RNAs may also be silenced by aberrant CpG methylation.

Besides the canonical mechanisms of microRNA gene regulation, other "noncanonical" microRNA-mediated mechanisms of mRNA expression modulation are emerging [50, 57–61]. Some microRNAs have been shown to bind to the open reading frame or to the 5′ UTR of the target genes, and, in some cases, they have been shown to activate rather than to inhibit gene expression [57, 58]. Moreover, some studies have recently reported that microRNAs can also regulate gene expression at the transcriptional level by binding directly to the DNA [50, 59–61].

Epigenetic Deregulation in Cancer

The cancer epigenome is characterized by extensive aberrations at any level of epigenetic control. The integrity of epigenetic regulation including maintenance of appropriate patterns of histone modifications, DNA methylation, and microRNA expression is not only crucial for normal development and differentiation but is also intimately associated with tumor initiation and progression [62]. It is also becoming clear that epigenetic deregulation may precede classical transforming events like mutations in cancer-relevant genes and genomic instability [63]. Disruption of the epigenetic machineries, either by mutation, deletion, or altered expression of any of their components, contributes to epigenetic deregulation. Aberrant promoter methylation of genes may complement mutation or deletion of the second allele, as postulated by the two-hit model for inactivation of tumor suppressor genes, or even provide both hits by methylation of both alleles [18, 64, 65]. Identification of new "epimutations" is rapidly increasing with the availability of more performing technologies. Similar to genetic alterations, tumor-type-specific patterns of epigenetic alterations are observed [66]. The current challenge is to differentiate drivers from passenger alterations and identify those that are actionable for future treatment approaches or select the ones already druggable by available therapies [67] and develop respective biomarkers for patient selection.

Epigentic Deregulation in Glioma

Silencing by Promoter Methylation of O6-Methylguanine-DNA Methyltransferase Gene (MGMT)

In glioma probably the best-known epigenetic alteration is promoter hypermethylation of the repair gene that encodes the O6-methylguanine-DNA methyltransferase (MGMT) that has become the first epigenetic biomarker in this disease [31, 68]. MGMT rapidly reverses alkylation (including methylation) at the O6 position of guanine by transferring the alkyl group to the active site of the enzyme, in a suicide reaction [69], hence annihilating the therapeutic effect of alkylating agents such as temozolomide. Consequently, epigenetic inactivation of the *MGMT* gene by promoter methylation renders tumor cells more sensitive to alkylating agents. The clinical relevance of epigenetic silencing of the *MGMT* promoter for benefit from alkylating agent therapy was shown in a randomized trial for newly diagnosed glioblastoma (GBM) [70–72]. Patients whose tumors contained a methylated *MGMT* promoter had a clear survival benefit from the addition of the alkylating agent temozolomide (TMZ) to standard radiotherapy (RT) with a median overall survival (OS) of 23.4 months as compared to 12.6 months in patients with an unmethylated *MGMT*, while in the radiotherapy arm, OS was 15.3 months in the *MGMT* methylated and 11.8 in the unmethylated patients, respectively [70]. A predictive effect of *MGMT* methylation for benefit from TMZ is suggested, however, the result for OS is confounded by the fact that TMZ was given to 60 % of the patients in the RT arm at relapse. The predictive effect is supported by the data from progression-free survival (PFS). Patients with a methylated *MGMT* had a median PFS of 10.3 months, as compared with 5.9 months for patients who received radiotherapy alone. In contrast, patients with an unmethylated *MGMT* did not show such a benefit from the addition of TMZ with a PFS of 5.3 months, as compared with 4.4 months for patients who were treated with radiotherapy alone [68].

Subsequent to this trial, the *MGMT* methylation status has been evaluated in many studies, revealing that frequencies are specific to the glioma subtype and malignancy grade, ranging from 40 % in GBM to over 80 % in anaplastic oligoastrocytoma (WHO grade III), while *MGMT* promoter methylation in pilocytic astrocytoma and most non-glial brain tumors is infrequent (Fig. 2.4) [31, 73–83].

Surprisingly, the methylation status of the *MGMT* promoter has only a prognostic as opposed to a predictive effect in anaplastic gliomas. In two studies, it was shown that the prognostic significance of *MGMT* promoter methylation was similar in the RT as compared to the chemoradiotherapy arm [84, 85]. The underlying reason for this puzzling result became more clear when it was shown that in contrast to GBM, in anaplastic glioma (WHO grade III), *MGMT* methylation is associated with good prognostic factors such as 1p/19q co-deletions and mutations of the isocitrate dehydrogenase 1 gene (IDH1) [84]. Furthermore, it is of note that 80 % of GBMs have loss of one copy of chromosome 10 which combined with *MGMT* promoter methylation that is located on 10q26 leads to complete loss of MGMT function. This is not the case in anaplastic glioma. They do not frequently exhibit loss of chromosome 10; hence, *MGMT* promoter methylation may affect both or only one allele, resulting either in complete loss or just reduced "gene dosage," respectively. Taken together, the impact of *MGMT* methylation on response to alkylating agent therapy needs to be established for each tumor type.

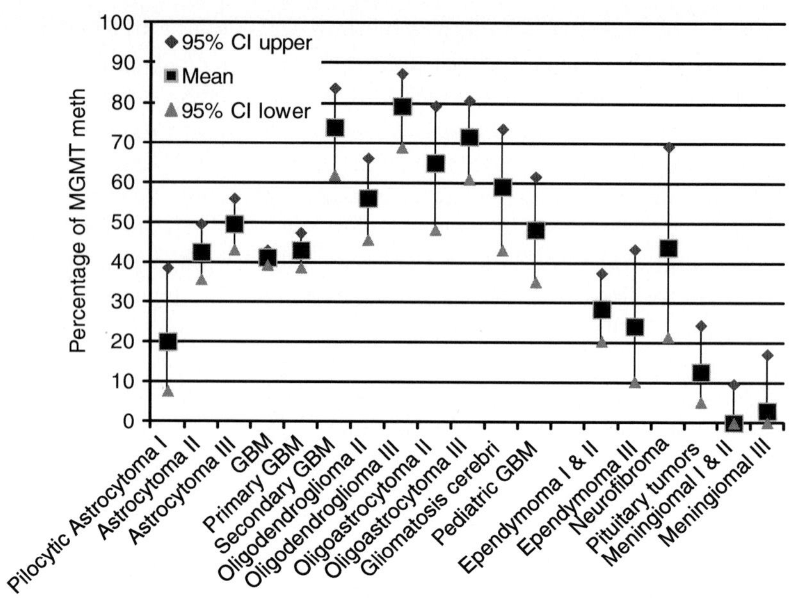

Fig. 2.4 Frequency of methylated *MGMT*. Meta-analysis of *MGMT* methylation frequencies compiled from the literature (updated from Weller et al. [31] and [73–83]). Mean values and confidence intervals (*CI*) of the *MGMT* methylation frequencies are given for different brain tumors. Large confidence intervals usually mean few and small series published

Epigenetic Deregulation of Cancer-Relevant Pathways in Gliomas

Besides inactivation of DNA repair as exemplified by *MGMT* silencing, DNA methylation analyses have revealed that silencing of negative regulators of mitogenic pathways or activators of apoptosis is common in cancer showing tumor type-specific patterns. In GBM the WNT pathway may be activated through promoter methylation of negative regulators such as the WNT inhibitory factor 1, the family of secreted frizzled-related proteins (sFRPs), dickkopf (DKK), and naked (NKDs) [86, 87]. Another example is the ras pathway that in a subset of GBM is deregulated by silencing of the negative regulators Ras association (RalGDS/AF-6) domain family members RASSF1A and RASSF10 [88, 89]. RASSF1 is methylated in many tumor types and is thought to contribute to ras signaling [90]. Examples of genes and respective affected pathways are given in Table 2.1 [86–89, 91–98].

Glioma CPG Island Methylator Phenotype (G-CIMP)

Improvement of technology in the last few years allows comprehensive analysis of genome-wide DNA methylation on high-throughput platforms. Large-scale analysis in GBM on aberrant DNA methylation at CpG sites has unraveled a plethora of genes that are affected. A project of the "The Cancer Genome Atlas" (TCGA) has classified GBM into three distinct DNA methylation GBM subgroups [93]. A striking pattern with highly concordant DNA methylation was identified in 8 % of GBM, indicative of a glioma CpG island methylator phenotype (G-CIMP) [93] (Fig. 2.5) [93, 99, 100]. Patients with G-CIMP tumors are younger at the time of diagnosis and experience significantly improved outcome. G-CIMP tumors constitute a subgroup of the proneural subtype as defined by the Verhaak gene expression-based classification of GBM [101]. Furthermore, G-CIMP is highly correlated with *IDH1* gene mutations. Hence, G-CIMP is also associated with secondary GBM, arising from lower-grade glioma [102]. In grade II and grade III glioma, G-CIMP was also commonly identified with a strong association with *IDH1/2* mutations, suggesting an early event in the evolution of these tumors [91, 93, 103] (Table 2.1). In anaplastic glioma, G-CIMP has also been reported as good prognostic factor [103]. These observations further support the hypothesis that primary glioblastoma with low frequencies of *IDH1* mutations and G-CIMP

Table 2.1 Methylated genes and affected pathways in gliomas (without MGMT)

Tumor type	Pathway affected	Gene	% methylation	Ref.
Glioblastoma				
	RAS pathway	RASSF1A	47	[88]
		RASSF10	65	[89]
	WNT pathway	WIF1	26	[86]
		SFRP1	80	[87]
		SFRP2	77	[87]
		SFRP4	50	[87]
		SFRP5	20	[87]
		NKD1	0	[87]
		NKD2	81	[87]
		DKK1	0	[87]
		DKK3	0	[87]
		FZD9	72	[91]
	Metalloproteinase	MMP14	75	[91]
		MMP2	97	[91]
	PI3K pathway	PTEN	9	[92]
	G-CIMP		8	[93]
	Others			
		HOXA9	50	[91]
		HOXA11	75	[91]
		TMS1/ASC	21	[94]
		HIC1	100	[95]
Secondary glioblastoma				
	RAS pathway	RASSF1A		[88]
		RASSF10	69	[89]
	WNT pathway	SFRP1	5	[87]
		SFRP2	12	[87]
		SFRP4	0	[87]
		SFRP5	40	[87]
		NKD1	0	[87]
		NKD2	13	[87]
		DKK1	83	[87]
		DKK3	100	[87]
	PI3K pathway	PTEN	82	[92]
	G-CIMP		75	[93]

(continued)

Table 2.1 (continued)

Tumor type	Pathway affected	Gene	% methylation	Ref.
Anaplastic astrocytoma				
	RAS pathway	RASSF1A	64	[88]
		RASSF10	80	[89]
	WNT pathway	SFRP1	10	[87]
		SFRP2	0	[87]
		SFRP4	25	[87]
		SFRP5	20	[87]
		NKD1	0	[87]
		NKD2	0	[87]
		DKK1	17	[87]
	PI3K pathway	PTEN	68	[92]
	G-CIMP		52	[93]
	Others	HIC1	100	[95]
Diffuse astrocytoma				
	RAS pathway	RASSF1A	75	[88]
		RASSF10	60	[89]
	WNT pathway	SFRP1	5	[87]
		SFRP2	10	[87]
		SFRP4	25	[87]
		SFRP5	20	[87]
		NKD1	0	[87]
		NKD2	6	[87]
		DKK	0	[87]
	Cell proliferation	TES		[96]
	PI3K pathway	PTEN	43	[92]
	G-CIMP		80	[93]
		TET2	14	[97]
	Others	CDH1	65	[98]
		HIC1	100	[95]
Pilocytic astrocytoma				
	RAS pathway	RASSF1A	20	[88]
		RASSF10	0	[89]

have a different pathogenetic/epigenetic origin than secondary glioblastoma and should be classified separately.

The correlation of the neomorphic IDH1/2 mutants with a DNA methylator phenotype was also observed in acute myeloid leukemia (AML). This provided an important mechanistic link, together with the fact that *IDH1*/2 mutations in leukemia were exclusive with tet oncogene family member 2 (*TET2*) mutations. The oncometabolite D-2-hydroxy glutarate (D-2HG) produced by neomorphic IDH mutants accumulates to high concentrations in the tumor tissues and has been shown to be a competitive inhibitor of α-KG-dependent dioxygenases, hence reducing the activities of the families of histone demethylases and TET 5-methylcytosine hydroxylases, including TET2. This leads to a genome-wide increase of DNA methylation (5-methylcytosine) and reduction of 5-hydroxymethylcytosine [104, 105]. Consequently suggesting a functional link between IDH1/2 mutations and the development of a methylator phenotype—metabolism meets epigenetics! Evidence for this functional link was provided recently by introducing an *IDH1* mutant into primary human astrocytes that leads to extensive hypermethylation reminiscent of patterns identified in G-CIMP-positive low-grade

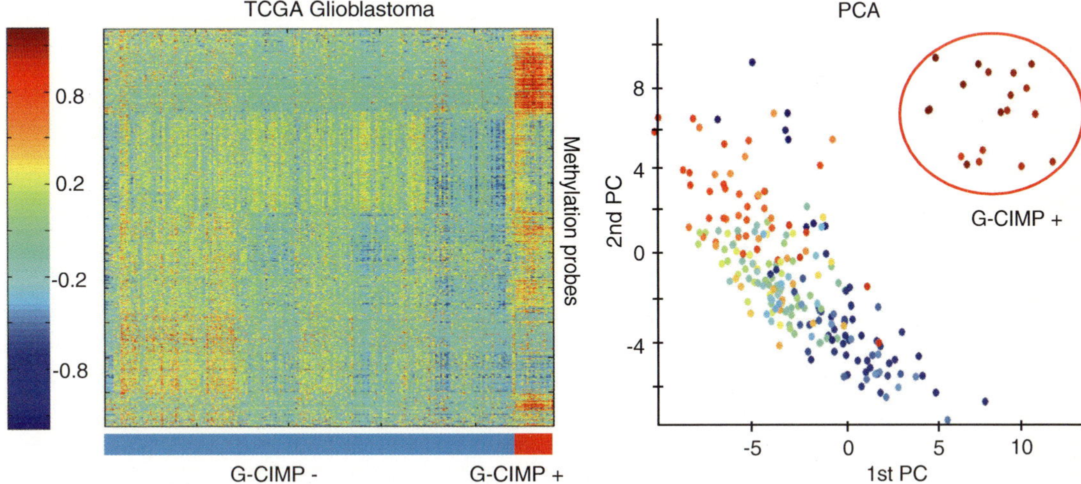

Fig. 2.5 Glioma CpG island methylator phenotype (*G-CIMP*). The heatmap shows a panel of 241 GBM clustered according to the 2,285 most variable Infinium DNA methylation probes. The structure of the heatmap identifies several GBM methylation subtypes, of which one group exhibits a highly coordinated gene methylation pattern that has been denominated as G-CIMP. The striking difference of these G-CIMP-associated GBM from the G-CIMP-negative GBM is well reflected in the principal component analysis (*PCA*) on the *right side*. The DNA methylation data has been downloaded from The Cancer Genome Atlas database [93], subjected to a variation filter (SD>0.19) and clustered using coupled two-way clustering (*CTWC*) [99, 100]. *Red*, high methylation; *blue*, low methylation

glioma, hence establishing *IDH* mutations as the molecular bases of CIMP in glioma [106].

Posttranslational Modifications of Histones

Epigenetic alterations in glial tumors frequently involve proteins controlling the PTM of histones. In particular the enhancer of zeste human homolog 2 gene (EZH2), which is the catalytic component of the polycomb repressive complexes 2 (PRC2) and PRC3, has been demonstrated to play an important role in gliomas. It is involved in setting the H3K27me3 marks and also links different layers of epigenetic control. Indeed, EZH2 may indirectly control DNA methylation through providing a platform for recruiting DNA methyltransferases [107]. EZH2 is overexpressed in most astrocytic and oligodendroglial tumors and even more highly expressed in GBM [108]. Nevertheless, proteins belonging to PRCs are not the only histone modifiers altered in gliomas. Expression of some histone deacetylases (HDAC) has been reported to be altered in GBM especially class II and IV HDACs [109]. Moreover, large-scale sequencing analysis in GBMs uncovered mutations in many genes encoding proteins involved in epigenetic regulation, including histone deacetylases HDAC2 and HDAC9, histone demethylases JMJD1A and JMJD1B, histone methyltransferases SET7, SETD7, MLL, and MLL4, and methyl-CpG-binding domain protein 1 (MBD1), although they have not been confirmed yet as drivers of glioma genesis [110].

Aberrant Expression of MicroRNAs

Many studies have shown that glial tumors are also characterized by strong alterations in microRNA content. One of the best characterized alterations in GBM is represented by the miR-21. Identified targets of miR-21 are *TP53*, *TGFB*, the mitochondrial apoptotic pathway, and probably the tumor suppressor gene *PTEN* [111]. Its expression levels have been correlated with overall and disease-free survival and suggested to be a biomarker for chemoresistance in other types of cancer including leukemia and pancreatic and lung cancer [112, 113]. A growing number of other microRNAs have been recently linked with gliomagenesis; striking examples are miR-10b [114] and miR-196 glioma involved in glioma progression [115]. In particular miR-10b is often upregulated in both low-grade and high-grade glioma and seems to downregulate *BCL2L11*/BIM, *TFAP2C*/AP-2γ, *CDKN1A*/p21, and *CDKN2A*/p16 that normally protect cells from uncontrolled growth. Furthermore, the use of high-throughput technologies has allowed identification of expression signatures of microRNAs that characterize GBM subtypes or exert a prognostic value for survival in GBM [116, 117]. Again using TCGA data, microRNA expression profiles yielded biologically meaningful subclassification of GBM. Five subclasses were proposed that relate to developmental patterns, of which three overlap substantially with three of the four subclasses defined by the Verhaak gene expression classification [101]. The "oligoneural" microRNA profile was associated with "proneural," the "radial glial" with "classical," and the "astrocytic" with "mesenchymal" gene expression-defined classification [116].

Epigenetic Deregulation in Other Brain Tumors

Ependymal Tumors

We know little about the mechanisms involved in initiation, maintenance, or progression of ependymal tumors. This is in part due to the heterogeneity and the low incidence of these tumors. Most of the epigenetic studies on these tumors have used a candidate gene approach with genes mostly selected looking at their methylation status in other brain tumor types [118]. Despite the limited information available, a number of aberrantly methylated genes have been identified. The most commonly methylated gene in ependymomas seems to be *RASSF1A* with a reported incidence of 86 % [118, 119]. Another gene commonly methylated in ependymomas is *HIC1* with an incidence of 83 % that has been associated with a non-spinal localization [120]. Finally, Rousseau and colleagues have shown promoter hypermethylation of *CDKN2A*, *CDKN2B* [118, 121]. A non-exhaustive list of methylated genes and their pathways identified in different brain tumors is available in Table 2.2 [88, 118–136]

Pineal Tumors

The first gene identified in sporadic pituitary tumors affected by promoter methylation has been *CDKN2A* [122, 137]. Subsequent studies have described methylation-mediated gene silencing in multiple other genes including *RB1*, fibroblast growth factor receptor 2 (*FGFR2*), death-associated protein kinase (*DAPK*), and galectin 3 [122]. High-throughput technologies enormously boosted the discoveries in the field that led to the observation that *MEG3a* and *GADD45γ* are frequently inactivated in pituitary tumors by promoter hypermethylation [122, 126, 138] (Table 2.2). Finally, alterations involving PTM of histones have also been described for pituitary tumors. The MLL-p27(Kip1) pathway, for instance, is often downregulated in pituitary adenomas [139].

Medulloblastoma

Among the embryonal brain tumors, medulloblastoma (MB) is probably the most studied at both genetic and epigenetic level. Epigenetic inactivation of specific genes by DNA methylation has been found for *HIC1*, *RASSF1*A, and *CASP8* [88, 127–129]. More recently also *SFRP1*, *SFRP2*, and *SFRP3* have been found to be methylated in primary MB [133] (Table 2.2). BMI1, a component of the polycomb repressive complexe (PRC) 1 is also significantly upregulated in medulloblastoma. Recently BMI1 expression has been associated with poor survival [140, 141]. Furthermore, recent evidence suggests that microRNAs play an important role in medulloblastoma. The miR-124 has been one of the first microRNAs indentified as important in medulloblastoma. It is able to modulate the cell cycle, and its expression is significantly decreased in medulloblastoma [142]. Moreover, microRNA expression profiles from medulloblastoma overexpressing

Table 2.2 Methylated genes and pathways in selected non-glial brain tumors

Tumor type	Pathway affected	Gene	% methylation	Refs.
Ependymoma				
	RAS pathway	RASSF1A	86	[118, 119]
	Cell cycle regulation	CDKN2A (p16/ARF)	21	[118, 121]
		CDKN2B	32	[118, 121]
	Others	HIC1	83	[120]
Pituitary tumors				
	Cell cycle regulators	RB1	26	[122, 123]
		FGFR2	45	[122, 124]
		INK4a (p16)	34	[125]
		GADD45γ	58	[126]
	Apoptosis	CASP-8	54	[125]
		THBS1	43	[125]
Medulloblastoma				
	RAS pathway	RASSF1A	100	[88, 127–131]
	Apoptosis	CASP8	62	[132]
	WNT pathway	SFRP1	24	[133]
		SFRP2	4	[133]
		SFRP3	16	[133]
	Others	HIC1	85	[134]
Meningioma WHO II				
	RAS pathway	RASSF1A	64	[135]
	Metalloproteinase	TIMP3	67	[136]
	Others	TP73	82	[135]
Meningioma WHO III				
	RAS pathway	RASSF1A	43	[135]
	Metalloproteinase	TIMP3	22	[136]
	Others	TP73	71	[135]

either Her2 or c-Myc allowed the identification of specific microRNA signatures in each group of medulloblastoma. Expression of miR-10b, miR-135a, miR-135b, miR-125b, miR-153, and miR-199b was altered in Her2-overexpressing tumors, whereas c-Myc-overexpressing medulloblastomas had expression changes in miR-181b, miR-128a, and miR-128b [143]. Finally, the miR-17–92 cluster has been found to functionally collaborate with the sonic hedgehog pathway in medulloblastoma development [144].

Meningeal Tumors

Like for ependymal tumors, little is known about epigenetic alterations in meningeal tumors. What seems to be clear is that *MGMT* is not methylated in this tumor type [82]. Nevertheless, some epigenetic alterations have been observed. *RASSF1A, TIMP3,* and *TP73*, for instance, are frequently methylated in meningiomas (Table 2.2). Moreover, downregulation of miR-200a in meningioma seems to promote growth by reducing E-cadherin and activating the WNT/beta-catenin signaling pathway [145].

Epigenetic Treatments

Epigenetic therapies have already been FDA approved for leukemia and comprise DNA demethylating agents and HDAC inhibitors, and combinations thereof have been tested in clinical trials (see review by Kelly et al. [146]). In glioblastoma HDAC inhibitors have entered clinical trials (see http://clinicaltrials.gov/), while demethylating agents have not been considered. This is likely due to the fact that the methylated *MGMT* promoter sensitizes the tumors to alkylating agents and that the alkylating agent TMZ is part of the current standard of care for GBM [70] (see respective paragraph above). Furthermore, TMZ-containing treatment schedules are tested or are already used for most other glioma subtypes. Demethylating agents such as 5-Aza-cytidine or 5-Aza-2′-deoxcytidine lock DNMT enzymes on to the DNA, thereby inhibiting further DNA methylation. Consequently, demethylating agents require cell division for activity, hence targeting rapidly dividing cells. Due to their unspecific mechanism, demethyling agents may lead to reexpression not only of tumor suppressor genes but also of oncogenes. Furthermore, the treatment may induce expression of alternative transcripts due to demethylation of gene bodies and further accentuate hypomethylation of repetitive sequences leading to increased genomic instability.

The HDAC inhibitor vorinostat (SAHA) has shown modest benefit as single agent in a phase II trial for recurrent GBM [147]. Analysis of respective tumor tissues for histone acetylation and RNA expression profiles indicated that the tested dose schedule affected targeted pathways. At present, vorinostat is tested in phase I/II trials for recurrent GBM in combination with various drugs or in newly diagnosed GBM in combination with standard chemoradiotherapy. Combination therapies with vorinostat are also tested in embryonal tumors of the CNS. In contrast, the HDAC inhibitor romidepsin was reported as ineffective in a phase I/II study for patients with recurrent GBM as single agent at the standard dose and schedule [148]. Interestingly, the treatment of GBM patients with valproic acid as antiepileptic drug has shown a survival advantage in combined chemoradiotherapy [149]. Valproic acid is considered to have weak HDAC inhibitor properties and is currently tested in a phase 2 trial for newly diagnosed GBM in combination with standard chemoradiotherapy (NCT00302159). Other HDAC inhibitors (entinostat, panobinostat phenylbutyrate) are in clinical evaluation for recurrent high-grade glioma or refractory pediatric brain tumors and neuroblastoma. Drugs attempting to interfere with histone methylation that are expected to deplete PRC2 components are in preclinical testing: They comprise drugs like SL11144 that inhibits lysine (K)-specific demethylase 1A (KDM1A) and DZNep, an inhibitor of S-adenosylhomocysteine hydrolase [146]. Targeting of DNA–histone H1 complexes with a 131-iodine conjugated monoclonal antibody (Cotara) delivered by convection-enhanced delivery is under investigation in a phase II study for recurrent GBM [150].

Outlook

New concepts suggest that resistance to therapy may be partly mediated by epigenetic changes, based on the observation that acquired drug resistance was associated with alterations in the chromatin structure [151]. Indeed, treatment with HDAC inhibitors resensitized the drug-resistant cells, hence, providing evidence that development of drug resistance may be reversible in nature. This would explain the clinical observation of re-treatment response of tumors after "drug holidays." Consequently, new drug schemes are suggested adding concomitant HDAC inhibitors to therapies to prevent or at least delay acquisition of epigenetically mediated treatment resistance. Respective trials are ongoing.

Targeting of aberrantly overexpressed microRNAs as a therapeutic option has become technically feasible using locked nucleic acid (LNA)-modified phosphorothioate oligonucleotide technology that renders them more stable. First phase II trials using this technology are performed in hepatitis C infection [152, 153]. The question if microRNAs are actionable in brain tumors remains to be determined and tested preclinically.

The hypothesis that the neomorphic mutants of IDH1/2 by means of inhibition of DNA demethylases through production of high concentrations of the oncometabolite D-2HG are the underlying cause of G-CIMP makes them a prominent drug target. It remains to be seen if inhibition of the neomorphic function of IDH1/2 mutants is sufficient to reverse the methylator phenotype in these tumors. It is not known if the IDH mutants are required for maintenance of the tumors. Efforts aim at developing the oncometabolite D-2HG as biomarker detectable by magnetic resonance spectroscopy that would provide a noninvasive diagnostic tool to identify IDH1/2 mutant gliomas [154]. In contrast, the detection of the oncometabolite in the serum of patients afflicted with IDH1/2 mutant gliomas has been reported not to be successful [155].

Finally, mining epigenomics in cancer, as uncovered by large-scale analyses of DNA methylation profiles and chromatin structure, has just started. Insights into the molecular mechanisms and pathways affected by epigenetic cancer-related changes will provide new targets. The challenge will be to identify changes with the quality of drivers versus passengers and to find actionable targets. Most interestingly, some of these epigenetic alterations can be converted into the "Achilles heel" of the affected tumors upon treatment with certain classes of anticancer agents. These may include DNA repair pathways as we have shown previously for GBM with a methylated *MGMT* gene that particularly benefit from treatment with the alkylating agent temozolomide.

References

1. Waddington CH. The epigenotype. Endeavour. 1942;1(1):18–20.
2. Berger SL, Kouzarides T, Shiekhattar R, Shilatifard A. An operational definition of epigenetics. Genes Dev. 2009;23(7):781–3.
3. Baylin SB, Ohm JE. Epigenetic gene silencing in cancer – a mechanism for early oncogenic pathway addiction? Nat Rev Cancer. 2006;6(2):107–16.
4. Roberts CW, Orkin SH. The SWI/SNF complex – chromatin and cancer. Nat Rev Cancer. 2004;4(2):133–42.
5. Portela A, Esteller M. Epigenetic modifications and human disease. Nat Biotechnol. 2010;28(10):1057–68.
6. Hermann A, Gowher H, Jeltsch A. Biochemistry and biology of mammalian DNA methyltransferases. Cell Mol Life Sci. 2004;61(19–20):2571–87.
7. Quina AS, Buschbeck M, Di Croce L. Chromatin structure and epigenetics. Biochem Pharmacol. 2006;72(11):1563–9.
8. Wang Y, Leung FC. An evaluation of new criteria for CpG islands in the human genome as gene markers. Bioinformatics. 2004;20(7):1170–7.
9. Bird A. DNA methylation patterns and epigenetic memory. Genes Dev. 2002;16(1):6–21.
10. Antequera F, Bird A. CpG islands. EXS. 1993;64:169–85.
11. Xu GL, Bestor TH, Bourc'his D. Chromosome instability and immunodeficiency syndrome caused by mutations in a DNA methyltransferase gene. Nature. 1999;402(6758):187–91.
12. Esteller M. Epigenetic gene silencing in cancer: the DNA hypermethylome. Hum Mol Genet. 2007;16(Spec No 1):R50–9.
13. Lopez-Serra L, Esteller M. Proteins that bind methylated DNA and human cancer: reading the wrong words. Br J Cancer. 2008;98(12):1881–5.
14. Kuroda A, Rauch TA, Todorov I, et al. Insulin gene expression is regulated by DNA methylation. PLoS One. 2009;4(9):e6953.
15. Suzuki MM, Bird A. DNA methylation landscapes: provocative insights from epigenomics. Nat Rev Genet. 2008;9(6):465–76.
16. Doi A, Park I-H, Wen B, et al. Differential methylation of tissue- and cancer-specific CpG island shores distinguishes human induced pluripotent stem cells, embryonic stem cells and fibroblasts. Nat Genet. 2009;41(12):1350–3.
17. Meissner A, Mikkelsen TS, Gu H, et al. Genome-scale DNA methylation maps of pluripotent and differentiated cells. Nature. 2008;454(7205):766–70.
18. Rodriguez-Paredes M, Esteller M. Cancer epigenetics reaches mainstream oncology. Nat Med. 2011;17(3):330–9.
19. Lister R, Pelizzola M, Dowen RH, et al. Human DNA methylomes at base resolution show widespread epigenomic differences. Nature. 2009;462(7271):315–22.
20. Laurent L, Wong E, Li G, et al. Dynamic changes in the human methylome during differentiation. Genome Res. 2010;20(3):320–31.
21. Huang Y, Pastor WA, Shen Y, Tahiliani M, Liu DR, Rao A. The behaviour of 5-hydroxymethylcytosine in bisulfite sequencing. PLoS One. 2010;5(1):e8888.
22. Wang RY-H, Gehrke CW, Ehrlich M. Comparison of bisulfite modification of 5-methyldeoxycytidine and deoxycytidine residues. Nucleic Acids Res. 1980;8(20):4777–90.
23. Herman JG, Graff JR, Myöhänen S, Nelkin BD, Baylin SB. Methylation-specific PCR: a novel PCR assay for methylation status of CpG islands. Proc Natl Acad Sci USA. 1996;93:9821–6.
24. Ogino S, Kawasaki T, Brahmandam M, et al. Precision and performance characteristics of bisulfite conversion and real-time PCR (MethyLight) for quantitative DNA methylation analysis. J Mol Diagn. 2006;8(2):209–17.
25. Vlassenbroeck I, Califice S, Diserens AC, et al. Validation of real-time methylation-specific PCR to determine O6-methylguanine-DNA methyltransferase

gene promoter methylation in glioma. J Mol Diagn. 2008;10(4):332–7.
26. Tost J, Gut IG. Analysis of gene-specific DNA methylation patterns by pyrosequencing technology. Methods Mol Biol. 2007;373:89–102.
27. Eckhardt F, Lewin J, Cortese R, et al. DNA methylation profiling of human chromosomes 6, 20 and 22. Nat Genet. 2006;38(12):1378–85.
28. Beck S. Taking the measure of the methylome. Nat Biotechnol. 2010;28(10):1026–8.
29. Harris RA, Wang T, Coarfa C, et al. Comparison of sequencing-based methods to profile DNA methylation and identification of monoallelic epigenetic modifications. Nat Biotechnol. 2010;28(10): 1097–105.
30. Bock C, Tomazou EM, Brinkman AB, et al. Quantitative comparison of genome-wide DNA methylation mapping technologies. Nat Biotechnol. 2010;28(10):1106–14.
31. Weller M, Stupp R, Reifenberger G, et al. MGMT promoter methylation in malignant gliomas: ready for personalized medicine? Nat Rev Neurol. 2010;6(1):39–51.
32. Trojer P, Reinberg D. Facultative heterochromatin: is there a distinctive molecular signature? Mol Cell. 2007;28(1):1–13.
33. Black JC, Whetstine JR. Chromatin landscape: methylation beyond transcription. Epigenetics. 2011;6(1): 9–15.
34. Kouzarides T. Chromatin modifications and their function. Cell. 2007;128(4):693–705.
35. Horn PJ, Peterson CL. Molecular biology. Chromatin higher order folding – wrapping up transcription. Science. 2002;297(5588):1824–7.
36. Strahl BD, Allis CD. The language of covalent histone modifications. Nature. 2000;403(6765):41–5.
37. Jenuwein T, Allis CD. Translating the histone code. Science. 2001;293(5532):1074–80.
38. Li B, Carey M, Workman JL. The role of chromatin during transcription. Cell. 2007;128(4):707–19.
39. Ohm JE, McGarvey KM, Yu X, et al. A stem cell-like chromatin pattern may predispose tumor suppressor genes to DNA hypermethylation and heritable silencing. Nat Genet. 2007;39(2):237–42.
40. Haberland M, Montgomery RL, Olson EN. The many roles of histone deacetylases in development and physiology: implications for disease and therapy. Nat Rev Genet. 2009;10(1):32–42.
41. Shi Y. Histone lysine demethylases: emerging roles in development, physiology and disease. Nat Rev Genet. 2007;8(11):829–33.
42. Esquela-Kerscher A, Slack FJ. Oncomirs – microRNAs with a role in cancer. Nat Rev Cancer. 2006;6(4):259–69.
43. Lewis BP, Burge CB, Bartel DP. Conserved seed pairing, often flanked by adenosines, indicates that thousands of human genes are microRNA targets. Cell. 2005;120(1):15–20.
44. Rodriguez A, Griffiths-Jones S, Ashurst JL, Bradley A. Identification of mammalian microRNA host genes and transcription units. Genome Res. 2004;14(10A): 1902–10.
45. Lee Y, Kim M, Han J, et al. MicroRNA genes are transcribed by RNA polymerase II. EMBO J. 2004;23(20): 4051–60.
46. Lee Y, Ahn C, Han J, et al. The nuclear RNase III Drosha initiates microRNA processing. Nature. 2003;425(6956):415–9.
47. Landthaler M, Yalcin A, Tuschl T. The human DiGeorge syndrome critical region gene 8 and Its D. melanogaster homolog are required for miRNA biogenesis. Curr Biol. 2004;14(23):2162–7.
48. Bohnsack MT, Czaplinski K, Gorlich D. Exportin 5 is a RanGTP-dependent dsRNA-binding protein that mediates nuclear export of pre-miRNAs. RNA. 2004; 10(2):185–91.
49. Hammond SM, Bernstein E, Beach D, Hannon GJ. An RNA-directed nuclease mediates post-transcriptional gene silencing in Drosophila cells. Nature. 2000;404(6775):293–6.
50. Garzon R, Marcucci G, Croce CM. Targeting microRNAs in cancer: rationale, strategies and challenges. Nat Rev Drug Discov. 2010;9(10):775–89.
51. Chendrimada TP, Gregory RI, Kumaraswamy E, et al. TRBP recruits the Dicer complex to Ago2 for microRNA processing and gene silencing. Nature. 2005;436(7051):740–4.
52. Hutvagner G, Zamore PD. A microRNA in a multiple-turnover RNAi enzyme complex. Science. 2002;297(5589):2056–60.
53. Bartel DP. MicroRNAs: target recognition and regulatory functions. Cell. 2009;136(2):215–33.
54. Lewis BP, Shih I, Jones-Rhoades MW, Bartel DP, Burge CB. Prediction of mammalian microRNA targets. Cell. 2003;115(7):787–98.
55. Krek A, Grun D, Poy MN, et al. Combinatorial microRNA target predictions. Nat Genet. 2005;37(5): 495–500.
56. Friedman RC, Farh KK, Burge CB, Bartel DP. Most mammalian mRNAs are conserved targets of microRNAs. Genome Res. 2009;19(1):92–105.
57. Stark A, Lin MF, Kheradpour P, et al. Discovery of functional elements in 12 Drosophila genomes using evolutionary signatures. Nature. 2007;450(7167): 219–32.
58. Eiring AM, Harb JG, Neviani P, et al. miR-328 functions as an RNA decoy to modulate hnRNP E2 regulation of mRNA translation in leukemic blasts. Cell. 2010;140(5):652–65.
59. Gonzalez S, Pisano DG, Serrano M. Mechanistic principles of chromatin remodeling guided by siRNAs and miRNAs. Cell Cycle. 2008;7(16): 2601–8.
60. Khraiwesh B, Arif MA, Seumel GI, et al. Transcriptional control of gene expression by microRNAs. Cell. 2010;140(1):111–22.
61. Kim DH, Saetrom P, Snove Jr O, Rossi JJ. MicroRNA-directed transcriptional gene silencing in mammalian cells. Proc Natl Acad Sci USA. 2008;105(42): 16230–5.

62. Chi P, Allis CD, Wang GG. Covalent histone modifications—miswritten, misinterpreted and mis-erased in human cancers. Nat Rev Cancer. 2010;10(7):457–69.
63. Feinberg AP. Cancer epigenetics is no Mickey Mouse. Cancer Cell. 2005;8(4):267–8.
64. Esteller M. Epigenetics in cancer. N Engl J Med. 2008;358(11):1148–59.
65. Jones PA, Laird PW. Cancer epigenetics comes of age. Nat Genet. 1999;21(2):163–7.
66. Esteller M, Corn PG, Baylin SB, Herman JG. A gene hypermethylation profile of human cancer. Cancer Res. 2001;61(8):3225–9.
67. Sciuscio D, Diserens AC, van Dommelen K, et al. Extent and patterns of MGMT promoter methylation in glioblastoma- and respective glioblastoma-derived spheres. Clin Cancer Res. 2011;17(2):255–66.
68. Hegi ME, Diserens AC, Gorlia T, et al. MGMT gene silencing and benefit from temozolomide in glioblastoma. N Engl J Med. 2005;352(10):997–1003.
69. Hegi ME, Sciuscio D, Murat A, Levivier M, Stupp R. Epigenetic deregulation of DNA repair and its potential for therapy. Clin Cancer Res. 2009;15(16):5026–31.
70. Stupp R, Hegi ME, Mason WP, et al. Effects of radiotherapy with concomitant and adjuvant temozolomide versus radiotherapy alone on survival in glioblastoma in a randomised phase III study: 5-year analysis of the EORTC-NCIC trial. Lancet Oncol. 2009;10(5):459–66.
71. Stupp R, Goldbrunner R, Neyns B, et al. Phase I/IIa trial of cilengitide (EMD121974) and temozolomide with concomitant radiotherapy, followed by temozolomide and cilengitide maintenance therapy in patients with newly diagnosed glioblastoma. J Clin Oncol. 2007 ASCO annual meeting proceedings Part I. 2007;25(suppl):Abstract# 2000.
72. Hegi ME, Diserens AC, Godard S, et al. Clinical trial substantiates the predictive value of O-6-methylguanine-DNA methyltransferase promoter methylation in glioblastoma patients treated with temozolomide. Clin Cancer Res. 2004;10(6):1871–4.
73. Kaloshi G, Everhard S, Laigle-Donadey F, et al. Genetic markers predictive of chemosensitivity and outcome in gliomatosis cerebri. Neurology. 2008;70(8):590–5.
74. Glas M, Bahr O, Felsberg J. NOA-05 phase 2 trial of procarbazine and lomustine therapy in gliomatosis cerebri. Ann Neurol. 2011;70(3):445–53.
75. Srivastava A, Jain A, Jha P, et al. MGMT gene promoter methylation in pediatric glioblastomas. Childs Nerv Syst. 2010;26(11):1613–8.
76. Sardi I, Cetica V, Massimino M, et al. Promoter methylation and expression analysis of MGMT in advanced pediatric brain tumors. Oncol Rep. 2009;22(4):773–9.
77. Koos B, Peetz-Dienhart S, Riesmeier B, Fruhwald MC, Hasselblatt M. O(6)-methylguanine-DNA methyltransferase (MGMT) promoter methylation is significantly less frequent in ependymal tumours as compared to malignant astrocytic gliomas. Neuropathol Appl Neurobiol. 2010;36(4):356–8.
78. Hasselblatt M, Muhlisch J, Wrede B, et al. Aberrant MGMT (O6-methylguanine-DNA methyltransferase) promoter methylation in choroid plexus tumors. J Neurooncol. 2009;91(2):151–5.
79. Gonzalez-Gomez P, Bello MJ, Arjona D. Aberrant CpG island methylation in neurofibromas and neurofibrosarcomas. Oncol Rep. 2003;10(5):1519–23.
80. McCormack AI, McDonald KL, Gill AJ, et al. Low O6-methylguanine-DNA methyltransferase (MGMT) expression and response to temozolomide in aggressive pituitary tumours. Clin Endocrinol (Oxf). 2009;71(2):226–33.
81. Salehi F, Scheithauer BW, Kros JM. MGMT promoter methylation and immunoexpression in aggressive pituitary adenomas and carcinomas. J Neurooncol. 2011;104(3):647–57.
82. Brokinkel B, Fischer BR, Peetz-Dienhart S, et al. MGMT promoter methylation status in anaplastic meningiomas. J Neurooncol. 2010;100(3):489–90.
83. de Robles P, McIntyre J, Kalra S, et al. Methylation status of MGMT gene promoter in meningiomas. Cancer Genet Cytogenet. 2008;187(1):25–7.
84. Wick W, Hartmann C, Engel C, et al. NOA-04 randomized phase III trial of sequential radiochemotherapy of anaplastic glioma with procarbazine, lomustine, and vincristine or temozolomide. J Clin Oncol. 2009;27(35):5874–80.
85. van den Bent MJ, Dubbink HJ, Sanson M, et al. MGMT promoter methylation is prognostic but not predictive for outcome to adjuvant PCV chemotherapy in anaplastic oligodendroglial tumors: a report from EORTC Brain Tumor Group Study 26951. J Clin Oncol. 2009;9:5881–6.
86. Lambiv WL, Vassallo I, Delorenzi M, et al. The Wnt inhibitory factor 1 (WIF1) is targeted in glioblastoma and has a tumor suppressing function potentially by induction of senescence. Neuro Oncol. 2011;13(7):736–47.
87. Gotze S, Wolter M, Reifenberger G, Muller O, Sievers S. Frequent promoter hypermethylation of Wnt pathway inhibitor genes in malignant astrocytic gliomas. Int J Cancer. 2010;126(11):2584–93.
88. Horiguchi K, Tomizawa Y, Tosaka M. Epigenetic inactivation of RASSF1A candidate tumor suppressor gene at 3p21.3 in brain tumors. Oncogene. 2003;22(49):7862–5.
89. Hill VK, Underhill-Day N, Krex D, et al. Epigenetic inactivation of the RASSF10 candidate tumor suppressor gene is a frequent and an early event in gliomagenesis. Oncogene. 2011;30(8):978–89.
90. Toyota M, Suzuki H, Yamashita T. Cancer epigenomics: implications of DNA methylation in personalized cancer therapy. Cancer Sci. 2009;17:17.
91. Laffaire J, Everhard S, Idbaih A. Methylation profiling identifies 2 groups of gliomas according to their tumorigenesis. Neuro Oncol. 2010;13(1):84–98.

92. Wiencke JK, Zheng S, Jelluma N. Methylation of the PTEN promoter defines low-grade gliomas and secondary glioblastoma. Neuro Oncol. 2007;9(3): 271–9.
93. Noushmehr H, Weisenberger DJ, Diefes K, et al. Identification of a CpG island methylator phenotype that defines a distinct subgroup of glioma. Cancer Cell. 2010;17:419–20.
94. Martinez R, Schackert G, Esteller M. Hypermethylation of the proapoptotic gene TMS1/ASC: prognostic importance in glioblastoma multiforme. J Neurooncol. 2007;82(2):133–9.
95. Uhlmann K, Rohde K, Zeller C, et al. Distinct methylation profiles of glioma subtypes. Int J Cancer. 2003;106(1):52–9.
96. Mueller W, Nutt CL, Ehrich M, et al. Downregulation of RUNX3 and TES by hypermethylation in glioblastoma. Oncogene. 2006;26(4):583–93.
97. Kim YH, Pierscianek D, Mittelbronn M, et al. TET2 promoter methylation in low-grade diffuse gliomas lacking IDH1/2 mutations. J Clin Pathol. 2011;20: 2011.
98. D'Urso PI, D'Urso OF, Storelli C. Retrospective protein expression and epigenetic inactivation studies of CDH1 in patients affected by low-grade glioma. J Neurooncol. 2010;104(1):113–8.
99. Getz G, Levine E, Domany E. Coupled two-way clustering analysis of gene microarray data. Proc Natl Acad Sci USA. 2000;97(22):12079–84.
100. Getz G, Domany E. Coupled two-way clustering server. Bioinformatics. 2003;19(9):1153–4.
101. Verhaak RG, Hoadley KA, Purdom E, et al. Integrated genomic analysis identifies clinically relevant subtypes of glioblastoma characterized by abnormalities in PDGFRA, IDH1, EGFR, and NF1. Cancer Cell. 2010;17(1):98–110.
102. Nobusawa S, Watanabe T, Kleihues P, Ohgaki H. IDH1 mutations as molecular signature and predictive factor of secondary glioblastomas. Clin Cancer Res. 2009;15(19):6002–7.
103. van den Bent MJ, Gravendeel LA, Gorlia T, et al. A hypermethylated phenotype in anaplastic oligodendroglial brain tumors is a better predictor of survival than MGMT methylation in anaplastic oligodendroglioma: a report from EORTC study 26951. Clin Cancer Res. 2011;17:7148–55.
104. Figueroa ME, Abdel-Wahab O, Lu C, et al. Leukemic IDH1 and IDH2 mutations result in a hypermethylation phenotype, disrupt TET2 function, and impair hematopoietic differentiation. Cancer Cell. 2010;18(6):553–67.
105. Xu W, Yang H, Liu Y, et al. Oncometabolite 2-hydroxyglutarate is a competitive inhibitor of α-ketoglutarate-dependent dioxygenases. Cancer Cell. 2011;19(1):17–30.
106. Turcan S, Rohle D, Goenka A, et al. IDH1 mutation is sufficient to establish the glioma hypermethylator phenotype. Nature. 2012;483(7390):479–83.
107. Vire E, Brenner C, Deplus R, et al. The Polycomb group protein EZH2 directly controls DNA methylation. Nature. 2006;439(7078):871–4.
108. Zheng S, Houseman EA, Morrison Z, et al. DNA hypermethylation profiles associated with glioma subtypes and EZH2 and IGFBP2 mRNA expression. Neuro Oncol. 2011;13(3):280–9.
109. Lucio-Eterovic AK, Cortez MA, Valera ET, et al. Differential expression of 12 histone deacetylase (HDAC) genes in astrocytomas and normal brain tissue: class II and IV are hypoexpressed in glioblastomas. BMC Cancer. 2008;8:243.
110. Parsons DW, Jones S, Zhang X, et al. An integrated genomic analysis of human glioblastoma multiforme. Science. 2008;321(5897):1807–12.
111. Papagiannakopoulos T, Shapiro A, Kosik KS. MicroRNA-21 targets a network of key tumor-suppressive pathways in glioblastoma cells. Cancer Res. 2008;68(19):8164–72.
112. Gao W, Yu Y, Cao H, et al. Deregulated expression of miR-21, miR-143 and miR-181a in non small cell lung cancer is related to clinicopathologic characteristics or patient prognosis. Biomed Pharmacother. 2010;64(6):399–408.
113. Hwang JH, Voortman J, Giovannetti E, et al. Identification of microRNA-21 as a biomarker for chemoresistance and clinical outcome following adjuvant therapy in resectable pancreatic cancer. PLoS One. 2010;5(5):e10630.
114. Gabriely G, Yi M, Narayan RS, et al. Human glioma growth is controlled by microRNA-10b. Cancer Res. 2011;71(10):3563–72.
115. Guan Y, Mizoguchi M, Yoshimoto K, et al. MiRNA-196 is upregulated in glioblastoma but not in anaplastic astrocytoma and has prognostic significance. Clin Cancer Res. 2010;16(16):4289–97.
116. Kim TM, Huang W, Park R, Park PJ, Johnson MD. A developmental taxonomy of glioblastoma defined and maintained by MicroRNAs. Cancer Res. 2011;71(9):3387–99.
117. Srinivasan S, Patric IR, Somasundaram K. A ten-microRNA expression signature predicts survival in glioblastoma. PLoS One. 2011;6(3):e17438.
118. Mack SC, Taylor MD. The genetic and epigenetic basis of ependymoma. Childs Nerv Syst. 2009;25(10):1195–201.
119. Hamilton DW, Lusher ME, Lindsey JC, Ellison DW, Clifford SC. Epigenetic inactivation of the RASSF1A tumour suppressor gene in ependymoma. Cancer Lett. 2005;227(1):75–81.
120. Waha A, Koch A, Hartmann W, et al. Analysis of HIC-1 methylation and transcription in human ependymomas. Int J Cancer. 2004;110(4):542–9.
121. Rousseau E, Ruchoux MM, Scaravilli F, et al. CDKN2A, CDKN2B and p14ARF are frequently and differentially methylated in ependymal tumours. Neuropathol Appl Neurobiol. 2003;29(6): 574–83.

122. Dudley KJ, Revill K, Clayton RN, Farrell WE. Pituitary tumours: all silent on the epigenetics front. J Mol Endocrinol. 2009;42(6):461–8.
123. Simpson DJ, Hibberts NA, McNicol AM, Clayton RN, Farrell WE. Loss of pRb expression in pituitary adenomas is associated with methylation of the RB1 CpG island. Cancer Res. 2000;60(5):1211–6.
124. Zhu X, Lee K, Asa SL, Ezzat S. Epigenetic silencing through DNA and histone methylation of fibroblast growth factor receptor 2 in neoplastic pituitary cells. Am J Pathol. 2007;170(5):1618–28.
125. Bello MJ, De Campos JM, Isla A, Casartelli C, Rey JA. Promoter CpG methylation of multiple genes in pituitary adenomas: frequent involvement of caspase-8. Oncol Rep. 2006;15(2):443–8.
126. Bahar A, Bicknell JE, Simpson DJ, Clayton RN, Farrell WE. Loss of expression of the growth inhibitory gene GADD45[gamma], in human pituitary adenomas, is associated with CpG island methylation. Oncogene. 2003;23(4):936–44.
127. Harada K, Toyooka S, Maitra A, et al. Aberrant promoter methylation and silencing of the RASSF1A gene in pediatric tumors and cell lines. Oncogene. 2002;21(27):4345–9.
128. Rood BR, Zhang H, Weitman DM, Cogen PH. Hypermethylation of HIC-1 and 17p allelic loss in medulloblastoma. Cancer Res. 2002;62(13):3794–7.
129. Pingoud-Meier C, Lang D, Janss AJ, et al. Loss of caspase-8 protein expression correlates with unfavorable survival outcome in childhood medulloblastoma. Clin Cancer Res. 2003;9(17):6401–9.
130. Grotzer MA, Eggert A, Zuzak TJ, et al. Resistance to TRAIL-induced apoptosis in primitive neuroectodermal brain tumor cells correlates with a loss of caspase-8 expression. Oncogene. 2000;19(40):4604–10.
131. Zuzak TJ, Steinhoff DF, Sutton LN, Phillips PC, Eggert A, Grotzer MA. Loss of caspase-8 mRNA expression is common in childhood primitive neuroectodermal brain tumour/medulloblastoma. Eur J Cancer. 2002;38(1):83–91.
132. Gonzalez-Gomez P, Bello MJ, Inda MM, et al. Deletion and aberrant CpG island methylation of Caspase 8 gene in medulloblastoma. Oncol Rep. 2004;12(3):663–6.
133. Kongkham PN, Northcott PA, Croul SE, Smith CA, Taylor MD, Rutka JT. The SFRP family of WNT inhibitors function as novel tumor suppressor genes epigenetically silenced in medulloblastoma. Oncogene. 2010;29(20):3017–24.
134. Waha A, Koch A, Meyer-Puttlitz B, et al. Epigenetic silencing of the HIC-1 gene in human medulloblastomas. J Neuropathol Exp Neurol. 2003;62(11): 1192–201.
135. Nakane Y, Natsume A, Wakabayashi T, et al. Malignant transformation-related genes in meningiomas: allelic loss on 1p36 and methylation status of p73 and RASSF1A. J Neurosurg. 2007;107(2): 398–404.
136. Barski D, Wolter M, Reifenberger G, Riemenschneider MJ. Hypermethylation and transcriptional downregulation of the TIMP3 gene is associated with allelic loss on 22q12.3 and malignancy in meningiomas. Brain Pathol. 2010;20(3):623–31.
137. Woloschak M, Yu A, Post KD. Frequent inactivation of the p16 gene in human pituitary tumors by gene methylation. Mol Carcinog. 1997;19(4):221–4.
138. Zhao J, Dahle D, Zhou Y, Zhang X, Klibanski A. Hypermethylation of the promoter region is associated with the loss of MEG3 gene expression in human pituitary tumors. J Clin Endocrinol Metab. 2005;90(4):2179–86.
139. Horiguchi K, Yamada M, Satoh T, et al. Transcriptional activation of the mixed lineage leukemia-p27Kip1 pathway by a somatostatin analogue. Clin Cancer Res. 2009;15(8):2620–9.
140. Milde T, Oehme I, Korshunov A, et al. HDAC5 and HDAC9 in medulloblastoma: novel markers for risk stratification and role in tumor cell growth. Clin Cancer Res. 2010;16(12):3240–52.
141. Leung C, Lingbeek M, Shakhova O, et al. Bmi1 is essential for cerebellar development and is overexpressed in human medulloblastomas. Nature. 2004;428(6980):337–41.
142. Pierson J, Hostager B, Fan R, Vibhakar R. Regulation of cyclin dependent kinase 6 by microRNA 124 in medulloblastoma. J Neurooncol. 2008;90(1): 1–7.
143. Ferretti E, De Smaele E, Po A, et al. MicroRNA profiling in human medulloblastoma. Int J Cancer. 2009;124(3):568–77.
144. Uziel T, Karginov FV, Xie S, et al. The miR-17 92 cluster collaborates with the Sonic Hedgehog pathway in medulloblastoma. Proc Natl Acad Sci USA. 2009;106(8):2812–7.
145. Saydam O, Shen Y, Wurdinger T, et al. Downregulated microRNA-200a in meningiomas promotes tumor growth by reducing E-cadherin and activating the Wnt/beta-catenin signaling pathway. Mol Cell Biol. 2009;29(21):5923–40.
146. Kelly TK, De Carvalho DD, Jones PA. Epigenetic modifications as therapeutic targets. Nat Biotechnol. 2010;28(10):1069–78.
147. Galanis E, Jaeckle KA, Maurer MJ, et al. Phase II trial of vorinostat in recurrent glioblastoma multiforme: a north central cancer treatment group study. J Clin Oncol. 2009;27(12):2052–8.
148. Iwamoto FM, Lamborn KR, Robins HI, et al. Phase II trial of pazopanib (GW786034), an oral multi-targeted angiogenesis inhibitor, for adults with recurrent glioblastoma (North American Brain Tumor Consortium Study 06–02). Neuro Oncol. 2010;12(8): 855–61.
149. Weller M, Gorlia T, Cairncross JG. Prolonged survival with valproic acid use in the EORTC/NCIC temozolomide trial for glioblastoma. Neurology. 2011;77(12):1156–64.

150. Hdeib A, Sloan AE. Convection-enhanced delivery of 131I-chTNT-1/B mAB for treatment of high-grade adult gliomas. Expert Opin Biol Ther. 2011;11(6):799–806.
151. Sharma SV, Lee DY, Li B, et al. A chromatin-mediated reversible drug-tolerant state in cancer cell subpopulations. Cell. 2010;141(1):69–80.
152. Broderick JA, Zamore PD. MicroRNA therapeutics. Gene Ther. 2011;18(12):1104–10.
153. Lanford RE, Hildebrandt-Eriksen ES, Petri A, et al. Therapeutic silencing of microRNA-122 in primates with chronic hepatitis C virus infection. Science. 2010;327(5962):198–201.
154. Choi C, Ganji SK, DeBerardinis RJ, et al. 2-hydroxyglutarate detection by magnetic resonance spectroscopy in IDH-mutated patients with gliomas. Nat Med. 2012;18(4):624–629.
155. Capper D, Simon M, Langhans CD. 2-hydroxyglutarate concentration in serum from patients with gliomas does not correlate with IDH1/2 mutation status or tumor size. Int J Cancer. 2011;131(3):766–8.

3 Astrocyte Differentiation from Oligodendrocyte Precursors

Akiko Nishiyama

Abstract

Oligodendrocyte progenitor cells (OPCs) are found not only in the developing central nervous system (CNS) but also exist abundantly and uniformly throughout the mature CNS. They are identified in vivo by the expression of NG2 (Cspg4 gene product) or the alpha receptor for platelet-derived growth factor (PDGFRα) and are antigenically and functionally distinct from neurons, mature oligodendrocytes, astrocytes, microglia, and neural stem cells. The majority of OPCs initially arise from committed neuroepithelial cells in the ventral germinal zone and subsequently migrate out and proliferate in the parenchyma. A different subset of later-born OPCs originates from dorsal germinal zones. While initial culture studies suggested that OPCs behave as bipotential glial progenitor cells endowed with the ability to differentiate into oligodendrocytes or astrocytes, more recent genetic fate-mapping studies have revealed that OPCs in the postnatal CNS and the majority of OPCs in the prenatal CNS are committed to the oligodendrocyte lineage, with the exception of a small subpopulation of OPCs in the embryonic ventral forebrain that generates protoplasmic astrocytes. These observations provide the basis for future molecular analyses on the mechanisms that restrict the fate of OPCs.

Astrocytes and oligodendrocytes represent two major macroglial cell types in the mammalian central nervous system (CNS) that arise from the neuroepithelium and constitute the major cell types found in primary brain tumors. While the functions of astrocytes and oligodendrocytes appear to be clearly distinct, some observations made over the last 30 years have suggested a close lineal relationship between these two cell types; other findings suggest that they are distinct cell lineages. In this chapter, a review

A. Nishiyama, M.D., Ph.D.
Department of Physiology and Neurobiology,
University of Connecticut,
75 North Eagleville Rd., Storrs, CT, 06269-3156, USA
e-mail: akiko.nishiyama@uconn.edu

C. Watts (ed.), *Emerging Concepts in Neuro-Oncology*,
DOI 10.1007/978-0-85729-458-6_3,

of the recent literature on the development and fate of oligodendrocyte progenitor cells (OPCs) will be provided with a focus on their relationship with astrocytes and some historical perspective. The chapter will be divided into three sections: identification of macroglial cells, development of astrocytes and oligodendrocytes, and lineage plasticity of astrocytes and oligodendrocytes.

Keywords
Oligodendrocyte • NG2 • PDGF • Astrocyte • Cre-loxP • Transgenic

Identification of Astrocytes and Oligodendrocytes

Early History of the Identification of Astrocytes and Oligodendrocytes

Astrocytes and oligodendrocytes represent two major macroglial cell types in the mammalian central nervous system (CNS) that arise from the neuroepithelium and constitute the major cell types found in primary brain tumors [1]. The German pathologist Rudolf Virchow introduced the term "neuroglia" to refer to the nonneuronal substance, which he thought played a cement-like role in holding the neuronal elements together [2]. The cellular elements of Virchow's "neuroglia" were subsequently described as "spongiocytes" [3], and the most common type of nonneuronal cells was named "astrocytes," reflecting their star-shaped appearance in Golgi-stained tissue. Modification of the Golgi silver impregnation method by Ramon y Cajal led to the discovery of "the third element" [4], which was later found to comprise the oligodendrocyte and the microglia [5]. Del Rio-Hortega first proposed that oligodendrocytes make the central nervous system (CNS) myelin [6]. However, it was not until the 1960s that oligodendrocytes were unequivocally established as the myelinating cells in the CNS by the demonstration of connections between oligodendroglial processes and myelin sheaths using electron microscopy [7]. Astrocytes have long been viewed as supportive cells that maintain homeostasis of ions and neurotransmitters around synapses and provide trophic support for neurons [8]. Recent studies have led to new hypotheses on the role of astrocytes as active participants of the neural network [9, 10] and as multipotent neural stem cells [11]. While the classical morphological studies have delineated astrocytes and oligodendrocytes as distinct glial lineages, subsequent studies using cell culture, immunolabeling, and in vivo cell tracing have led to new questions about the lineal relationship between these two glial cell types, as summarized below.

Identification of Astrocytes and Oligodendrocyte Lineage Cells in Culture

In the 1970s, neurochemists were interested in culturing distinct cellular populations to define their biochemical properties. The intermediate filament protein glial fibrillary acidic protein (GFAP) was one of the first cell type-specific proteins to be discovered and has come to be widely used as a marker for astrocytes [12]. The development of the hybridoma technique [13] allowed generation of monoclonal antibodies directed against cell surface antigens, which could be used to isolate distinct cell populations.

The mouse A2B5 monoclonal antibody was generated by immunizing mice with chick embryonic retinal cells and was initially found to recognize a ganglioside on neuronal cells [14]. Subsequently, in cultures from neonatal rat optic nerves, which contain glial cells but not neurons,

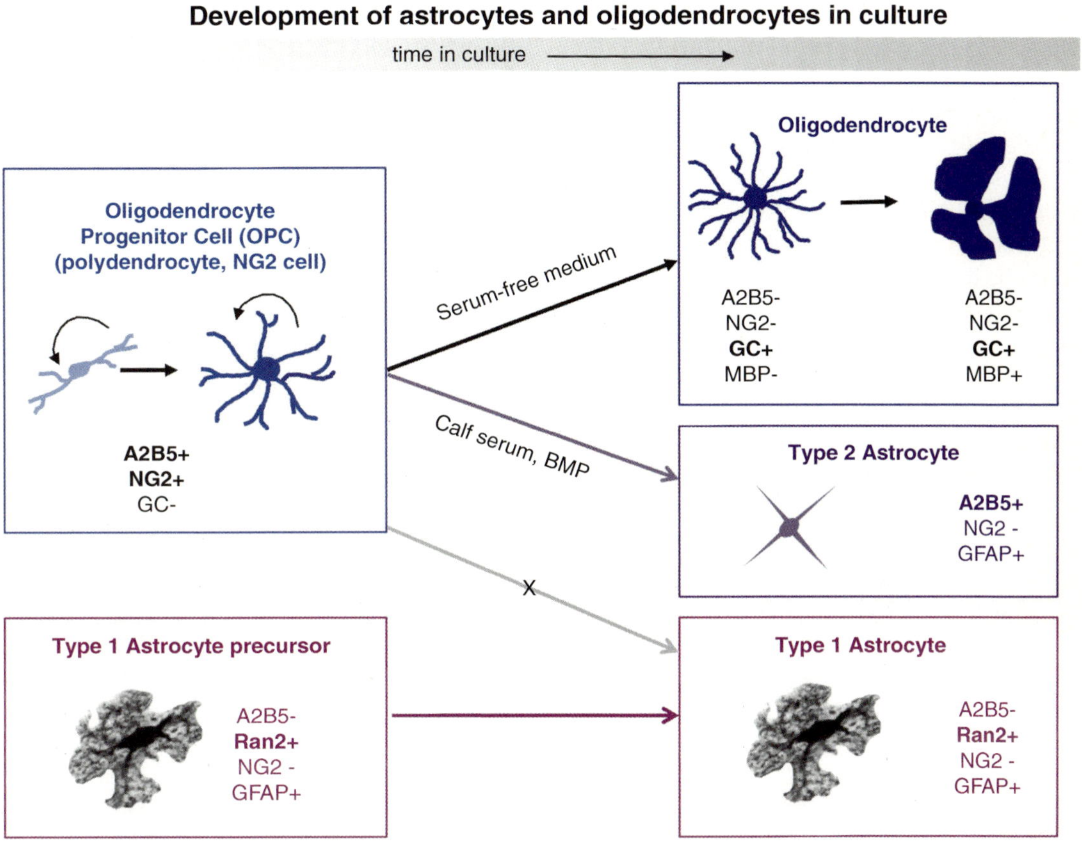

Fig. 3.1 Development of astrocytes and oligodendrocytes in vitro. A historical scheme of different macroglial lineages identified in culture, primarily from work in Martin Raff's laboratory. Oligodendrocyte progenitor cells (*OPCs*), also known as O-2A progenitor cells, express A2B5 and NG2 and differentiate into GC+ oligodendrocytes in serum-free medium. They lose the expression of the progenitor antigens A2B5 and NG2 as they differentiate. In the absence of neuronal cells, these newly differentiated oligodendrocytes further differentiate to express myelin basic protein (*MBP*) but do not differentiate to resemble fully mature myelinating oligodendrocytes. Olig2 is expressed throughout this lineage and also in some neural stem cells that have not yet committed to the oligodendrocyte lineage. In the presence of fetal calf serum, BMP, or CNTF, OPCs differentiate into type 2 astrocytes that have a stellar morphology and express both A2B5 and GFAP. Type 1 astrocytes defined by the lack of A2B5, presence of Ran2, and their flat epithelioid morphology comprise a distinct lineage and are not generated from OPCs

the A2B5 antibody was shown to recognize GFAP+ multi-process-bearing cells but not GFAP+ cells with flat, fibroblast-like morphology [15]. The stellate A2B5+ GFAP+ cells and flat A2B5− GFAP+ cells were named type 2 and type 1 astrocytes, respectively. The stellate astrocytes also expressed the newly identified NG2 (neuron-glial antigen 2) cell surface antigen [16, 17], while the Ran-2 antigen (rat neural antigen-2) was detected on type 1 but not type 2 astrocytes [18] (Fig. 3.1).

The major myelin glycolipid galactocerebroside (GC) was shown to specifically react with GFAP-negative, process-bearing cells from the optic nerve [19], which were presumed to be oligodendrocytes. When A2B5+ GFAP− cells from the optic nerves were cultured in chemically defined medium in the absence of fetal calf serum, they differentiated into GC+ oligodendrocytes within 3 days, which suggested that A2B5+ cells could give rise to both astrocytes and oligodendrocytes [20] (Fig. 3.1). It was then proposed that the A2B5+ cells represent bipotential glial progenitor cells, and thus they were named O-2A (oligodendrocyte-type 2 astrocyte) progenitor cells. Bipotential A2B5+ cells have also been isolated from adult tissues [21–23].

While both type 1 and type 2 astrocytes were found in cultures of white matter (optic nerve and corpus callosum), only type 1 astrocytes could be detected in cultures from subpial cortical gray matter tissue or the cerebellar cortex [15]. It was therefore suggested that type 1 and type 2 astrocytes correspond to protoplasmic astrocytes in the gray matter and fibrous astrocytes in the white matter, respectively [24, 25]. Miller and Raff initially used the A2B5 antibody on tissue sections and reported that fibrous astrocytes in the adult optic nerve bound the A2B5 antibody, while protoplasmic astrocytes in the cerebral cortex did not [24]. These studies marked the beginning of an era of characterizing distinct populations of glial cells in vivo using immunohistochemistry.

Identification of Astrocytes and Oligodendrocytes In Vivo

Astrocytes

While the cell surface antigens proved useful for marking and isolating different glial cell populations in vitro, there was a need to identify glial cells in tissue sections. By the end of the 1970s, GFAP expression had become the gold standard for identifying astrocytes in vivo. In the postmortem human brain, GFAP is readily detectable in both protoplasmic astrocytes in the gray matter and fibrous astrocytes in the white matter, making this a highly reliable marker for human astrocytes. By contrast, in the rodent CNS, GFAP is robustly detected only in fibrous astrocytes in the white matter, and the majority of protoplasmic astrocytes in the gray matter express little GFAP, with the exception of the hippocampus. Despite the nonuniform expression of GFAP in the rodent brain, many studies have equated astrocytes with GFAP+ cells. Other antigens such as S100β and glutamine synthetase (GS) have been used to detect gray matter astrocytes, but they are also expressed in some oligodendrocyte lineage cells and are not specific to astrocytes [26]. Recently, aldehyde dehydrogenase 1L1 (Aldh1L1) was identified by microarray analysis as an antigen that is enriched in both white matter and gray matter astrocytes [27]. Fibroblast growth factor receptor 3 (Fgfr3) is also expressed by both GFAP-negative protoplasmic astrocytes in the gray matter and by GFAP+ fibrous astrocytes in the white matter but not by oligodendrocyte lineage cells or neurons (Fig. 3.2). It is also expressed by neural stem cells in the subventricular zone (SVZ) [28, 29].

Oligodendrocytes and Oligodendrocyte Progenitor Cells

The glycolipid antigens GC and A2B5, which were successfully used to mark oligodendrocytes and their progenitor cells in culture, are often inadequately preserved for immunohistochemical detection in routinely processed tissue sections. Based on the in vitro observation that OPCs proliferate in response to platelet-derived growth factor (PDGF) through the alpha receptor for PDGF (PDGFRα) expressed on their surface [30–32], PDGFRα expression was used to examine the distribution of OPCs in vivo. In the embryonic brain and spinal cord, cells that express PDGFRα first appear in discrete regions of the germinal and subventricular zones of the ventral neural tube and subsequently expand and occupy the CNS [33, 34]. These PDGFRα + cells were later shown to express the key transcription factors required for the specification and maintenance of the oligodendrocyte lineage such as Olig1, Olig2, and Sox10 [35–38] (reviewed in [39, 40]).

Independently, Stallcup and colleagues identified an integral membrane chondroitin sulfate proteoglycan NG2 (neuron-glia antigen 2), which is expressed on O-2A progenitor cells from the optic nerve and disappears as they differentiate into oligodendrocytes or astrocytes (Fig. 3.1) [20, 41, 42]. Comparison of PDGFRα and NG2 expression by double immunofluorescence labeling revealed that there was almost a complete overlap between glial cells that expressed NG2 and those that expressed PDGFRα both in vitro [43] and in vivo [44],

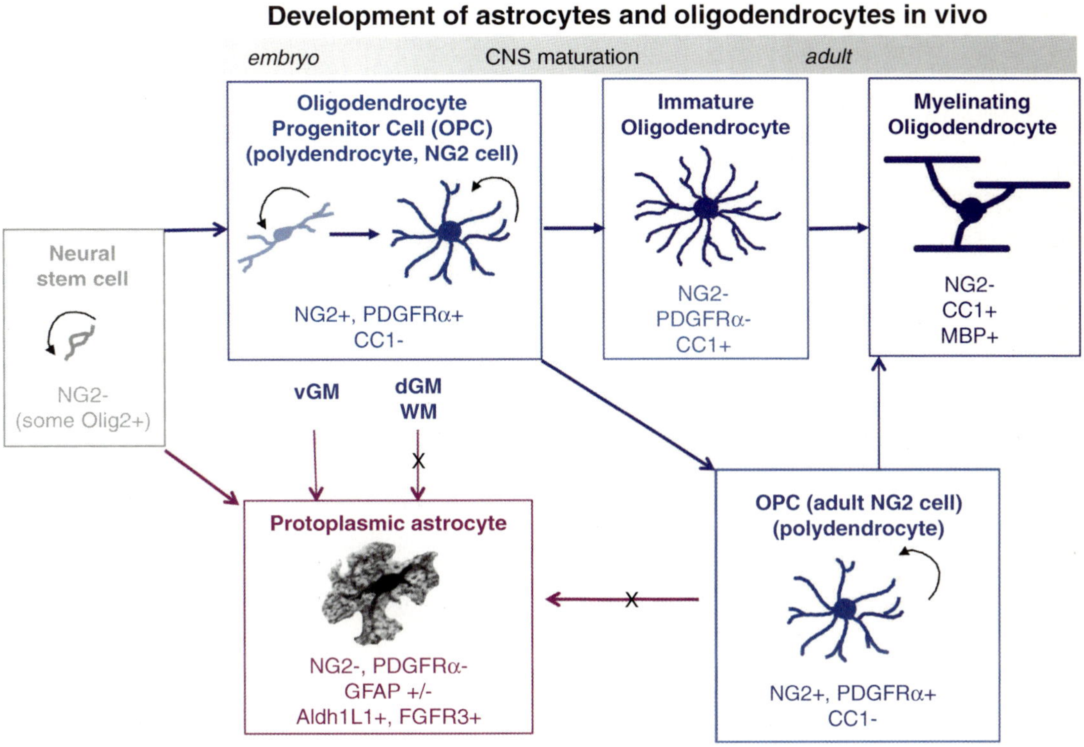

Fig. 3.2 Development of astrocytes and oligodendrocytes in vivo. The lineage of oligodendrocytes (*blue*) and astrocytes (*purple*) appear to be segregated early, and a common oligodendrocyte-astrocyte bipotential glial progenitor has not been found in vivo. Oligodendrocyte lineage cells arise from neural stem cells in the VZ/SVZ and begin to express the progenitor antigens NG2 and PDGFRα as they migrate out of the germinal zones. From late embryonic stage throughout the postnatal life, OPCs generate myelinating oligodendrocytes characterized by the sequential expression of CC1, MBP, and additional myelin proteins such as PLP and myelin oligodendrocyte glycoprotein (MOG). Some OPCs in the mature CNS remain as NG2+ OPCs throughout life (*bottom right*). All the NG2+ OPCs retain their proliferative ability (indicated by *black semicircular arrows*). NG2 and PDGFRα are lost from the cell surface as OPCs differentiate into oligodendrocytes. Olig2 is expressed throughout this lineage and also in some neural stem cells that have not yet committed to the oligodendrocyte lineage. A small fraction of OPCs in the gray matter of the embryonic ventral forebrain (vGM) generate protoplasmic astrocytes (Aldh1L1+) but those in white matter (WM), gray matter of the dorsal forebrain (dGM), or postnatal CNS do not become astrocytes, and their fate is restricted to oligodendrocytes

and NG2+ PDGFRα + cells were found in the anatomical locations where oligodendrocyte progenitor cells were expected to be found [45]. Cells that coexpress NG2 and PDGFRα appear after embryonic day 14 (E14) in the mouse forebrain, expand to occupy the entire CNS by the end of the first postnatal week after birth, and persist in the adult CNS (see below). The NG2+ PDGFRα + cells that are distributed uniformly throughout the CNS of adult mammals are distinct from astrocytes, microglia, mature oligodendrocytes, and neural stem cells [46, 47]. To distinguish them from other cell types and to reinforce the notion that they exist in the CNS not only during development but throughout the life of the animal, we have proposed to call them polydendrocytes. This name was chosen to reflect the multiple slender processes they have and their lineal relation to oligodendrocytes [48]. Neither PDGFRα nor NG2 is specific to OPCs. There is an early embryonic neuronal expression of PDGFRα, and NG2 is expressed on vascular mural cells in the CNS and elsewhere as well as on immature proliferative progenitor cells of mesenchymal lineages. Currently both NG2 and PDGFRα are used as markers for OPCs (Fig. 3.2) in both rodent and human CNS, and their enrichment in the OPCs has been confirmed by transcriptome analyses [27, 49].

The Origin of Astrocytes and Oligodendrocytes

Development of Astrocytes

In the mammalian CNS, astrocytes arise from radial glia in the ventricular zone (VZ) and from immature cells in the secondary germinal zones such as the subventricular zone (SVZ) around the lateral ventricles [50]. Radial glia (originally described as radial cells) are the direct progeny of neuroepithelial cells that retain their cytoplasmic contacts with both ventricular and pial surfaces. Morphological evidence that radial glia transform into astrocytes was reported as early as the end of the nineteenth century [3] and confirmed by more modern labeling techniques [51–53]. Studies conducted over the last decade have established that radial glia give rise not only to astrocytes but also to neurons and thus function as multipotent stem cells [54, 55]. In the cerebral cortex, transformation of radial glia into astrocytes follows the period of neurogenesis and thus is a relatively late event, occurring perinatally in rodents.

While the majority of radial glial cells disappear as the brain matures, some persist in the adult brain. The Bergmann glia in the cerebellum is an example of such a population of glia with persistent radial morphology, and they have electrophysiological properties that are different from those of protoplasmic astrocytes in the hippocampus [8]. A recent lineage tracing and fate-mapping study revealed an early birth date of the Bergmann glia in the mouse, around the same time as Purkinje cells and deep cerebellar nuclear neurons, long before other cerebellar astrocytes are generated [56].

Retroviral tracing studies have shown that progenitor cells in the perinatal SVZ generate astrocytes that migrate into the neocortex [57]. Interestingly, SVZ progenitor cells that migrate to the white matter tract of the corpus callosum mostly become oligodendrocytes but not astrocytes. Multipotent neural stem cells in the SVZ express GFAP [58], and these cells generate olfactory bulb interneurons and some oligodendrocytes [59], as well as parenchymal astrocytes. However, the relative contribution of radial glial progeny and SVZ progeny to the astrocyte population is not clear. Until recently, it was assumed that GFAP+ cells in the SVZ are neural stem cells, while GFAP+ cells in the parenchyma are differentiated astrocytes and functionally distinct from the neural stem cells. However, this tenet has been questioned by recent observations that some parenchymal astrocytes can be induced to generate neurons and thus may behave more like multipotent stem cells [11], which may explain the mixed cellular phenotypes seen in glioblastoma multiforme.

It is now well established that neural stem cells isolated from early embryonic cortex differentiate into neurons, while those isolated from late embryonic cortex differentiate into astrocytes, even in the presence of similar Wnt and JAK/STAT signaling pathways. Much effort has been put into elucidating the cell-intrinsic and cell-extrinsic mechanisms underlying the temporal change in the fate of neural stem cells during rodent gestation. Epigenetic cell-intrinsic mechanisms such as DNA methylation and the Polycomb group repressor complex are involved in regulating the activation of astrocyte-specific genes at the onset of the gliogenic phase of development. In addition, the earlier generated differentiated neurons provide extracellular signals such as cardiotropin-1 and Notch ligands, which alter the strength of JAK/STAT signal or DNA methylation in the late precursor cells [60, 61]. However, the identity of the key molecular switch that initiates an uncommitted cell to follow the astrocyte fate remains unknown.

Development of Oligodendrocyte Lineage Cells

In both the spinal cord and forebrain, the majority of oligodendrocyte lineage cells arise from discrete ventral domains in the VZ under the influence of Sonic hedgehog (Shh) (reviewed in [39]). In the spinal cord, the oligodendrogliogenic domain overlaps with the domain that generates motor neurons (pMN) and is coded by the basic helix-loop-helix transcription factors Olig1 and Olig2 [35–37]. The ventrally adjacent p3 domain, characterized by the expression of the homeodomain transcription factor Nkx2.2, also contributes to oligodendrocytes. NG2 and PDGFRα become detectable after these cells migrate out of the ventricular zone and begin to expand and occupy the entire spinal cord [34, 44]. In addition to these ventral sources, some OPCs also arise from the dorsal structures independently of Shh [62–64].

In the mouse forebrain, the earliest committed OPCs identified by PDGFRα expression appear in mid-gestation at embryonic day 12 (E12) in the median ganglionic eminence (MGE) and anterior entopeduncular region (AEP), and their appearance is dependent on the homeodomain transcription factor Nkx2.1, which is necessary for the correct expression of Shh [65, 66]. In the avian forebrain, cells in the AEP supply all the oligodendrocytes in the telencephalon [67]. In the mouse, a second wave of OPCs arises after E16.5 from the precursors that express Gsx2 in the lateral ganglionic eminence (LGE) and eventually replace the earlier generated OPCs [68]. A third wave of PDGFRα + cells appears mainly postnatally from dorsal Emx1+ cells and generates oligodendrocytes in the pallium including the neocortex and the corpus callosum.

From late embryonic stages throughout adulthood, the SVZ around the lateral ventricles provides a source for OPCs. Retroviral marking of perinatal SVZ showed that OPCs and oligodendrocytes in the corpus callosum and neocortex are generated from the neonatal SVZ [69]. In adult mice, retroviral marking of GFAP+ neural stem cells in the SVZ resulted in generation of OPCs cells in the corpus callosum in normal and demyelinated states [59, 70]. Besides the SVZ, local proliferation of NG2 cells also contributes to the maintenance of the oligodendrocyte lineage cells in the mature CNS.

In addition to the ventral sources of tangentially migrating cells in embryonic stages and the neocortical SVZ in postnatal rodents, radial glia have also been implicated as a source for oligodendrocytes [54, 71]. While it has been shown that the majority of oligodendrocyte lineage cells in the forebrain are generated from Gsx2+ cells in the embryonic ganglionic eminences and Emx1+ cells in the pallium, the relative contribution of radial glial cells and SVZ cells to the oligodendrocyte is not clear.

The Fate of OPCs

Developmental Fate of OPCs (NG2 Cells): Are They Multipotent Cells or Committed OPCs?

Search for Bipotential O-2A Progenitor Cells In Vivo

Since the identification of bipotential O-2A progenitor cells in vitro [20], numerous attempts were made to identify the astrocyte progeny of OPCs in vivo, but cells that coexpressed markers for OPCs and astrocytes could never be found in vivo. To circumvent the problem that the expression of A2B5 and other OPC antigens were downregulated before the appearance of astrocyte-specific antigens, isolated A2B5+ O-2A progenitor cells were transplanted into the brain, and the fate of the donor cells was examined to see whether the donor cells could give rise to astrocytes as well as oligodendrocytes in the host environment. These studies, however, failed to unequivocally demonstrate that astrocytes could be generated from the grafted A2B5+ cells, in addition to oligodendrocytes [72, 73]. Furthermore, retroviral lineage tracing of immature cells in the rat cerebral cortex revealed labeled clones that generated both neurons and glia but not astrocytes and oligodendrocytes [74], again failing to demonstrate the astrogliogenic fate of OPCs. Thus, the predominant view in the early 1990s was that type 2 astrocytes were an in vitro artifact and that OPCs are the committed progenitor cells of the oligodendrocyte lineage.

By the end of the 1990s, interest in the fate of OPCs was rekindled with the rediscovery of neural stem cells in the SVZ and the hippocampal subgranular zone of the adult CNS [75, 76]. This was fueled by the discovery that OPCs from the perinatal rat optic nerve could be reprogrammed to generate neurons after a prolonged period in culture, via a type 2 astrocyte-like stage [77]. The question of multipotency of OPCs became once again a subject of intense investigation.

BrdU Pulse-Chase Labeling to Follow the Fate of OPCs

One unique property of OPCs is that they continue to proliferate in the postnatal rodent CNS after other cell types have ceased to proliferate. Acute labeling of rats with the thymidine analog 5-bromo-2′-deoxyuridine (BrdU) revealed that 1.5 % of NG2 cells in the adult rat cerebral cortex incorporated BrdU [78]. Cumulative BrdU labeling after continuous exposure to BrdU resulted in labeling of 40–90 % of NG2 cells in the adult cerebral cortex and corpus callosum [79]. By immunolabeling for Ki-67, which is expressed in all active phases of the cell cycle including G1, S, G2, and M, but not in G0, it was shown that all of the Ki-67+ cells in nonneurogenic regions of the mature rodent and human forebrain were NG2+ [80, 81]. In the spinal cord of postnatal day 18 (P18) mice, more than 90 % of BrdU+ cells were found to be NG2+ [82]. The proportion of BrdU+ cells was highest at this developmental stage, which marks peak myelination, and dropped to 70 % in the adult spinal cord [78, 83]. BrdU+ cells that were NG2-negative were likely to include proliferating astrocyte precursors and microglia, as well as cells undergoing DNA repair. These findings are consistent with the earlier ultrastructural studies combined with ^{3}H-thymidine labeling, in which the majority of the proliferating cells were found to be "small glioblasts" or "spongioblastic precursor cells of oligodendrocytes," which likely correspond to NG2+ OPCs [84].

One can take advantage of this unique proliferative property of OPCs in the mature CNS and follow the fate of proliferated cells by performing pulse-chase labeling with BrdU, similar to what had been done by ultrastructural studies using ^{3}H-thymidine labeling [85]. When the phenotype of BrdU+ cells in the spinal cord was examined 1 day and 4 weeks after BrdU administration to adult rats, the percentage of BrdU+ cells that expressed NG2 declined, with a concomitant increase in the number of BrdU+ oligodendrocytes [83]. This indicated that the proliferated NG2+ OPCs had differentiated into oligodendrocytes over the course of 4 weeks. Similarly, BrdU pulse-chase labeling in the spinal cord after a single injection at P18 resulted in the disappearance of BrdU+ NG2+ cells accompanied by an increase in BrdU+ CC1+ oligodendrocytes in the white matter within 2 weeks [82]. In the Horner study, some S100β + astrocytes were observed after 4 weeks, but these experiments could not distinguish between astrocytes that had been generated from NG2-negative proliferated astrocyte progenitor cells or from NG2+ cells. In an earlier ultrastructural study combined with ^{3}H-thymidine pulse-chase labeling of the optic nerve, it was demonstrated that astrocytes are generated before the end of the first postnatal week, prior to the peak of oligodendrocyte generation, and it was suggested that astrocytes and oligodendrocytes are generated from distinct precursor cells [86]. However, the question of whether astrocytes are generated from NG2 cells could not be unequivocally resolved, as the percentage of BrdU+ cells that were NG2+ was less than 100 % immediately after labeling.

Genetic Fate Mapping in the Normal CNS

While pulse-chase labeling with BrdU provided some insight into the fate of proliferating OPCs, the existence of a minority of NG2-negative BrdU-incorporated cells posed a limitation to this approach, for one could never be certain whether a BrdU+ cell with a certain differentiated phenotype had been generated from a proliferated OPC or from another type of progenitor cell that had been in the S phase at the time of BrdU administration. To circumvent this problem, genetic fate-mapping tools were developed (Fig. 3.3) as an alternate approach to studying the fate of OPCs. Two lines of transgenic mice are used for these studies. In one line, the site-specific Cre recombinase is expressed under the control of

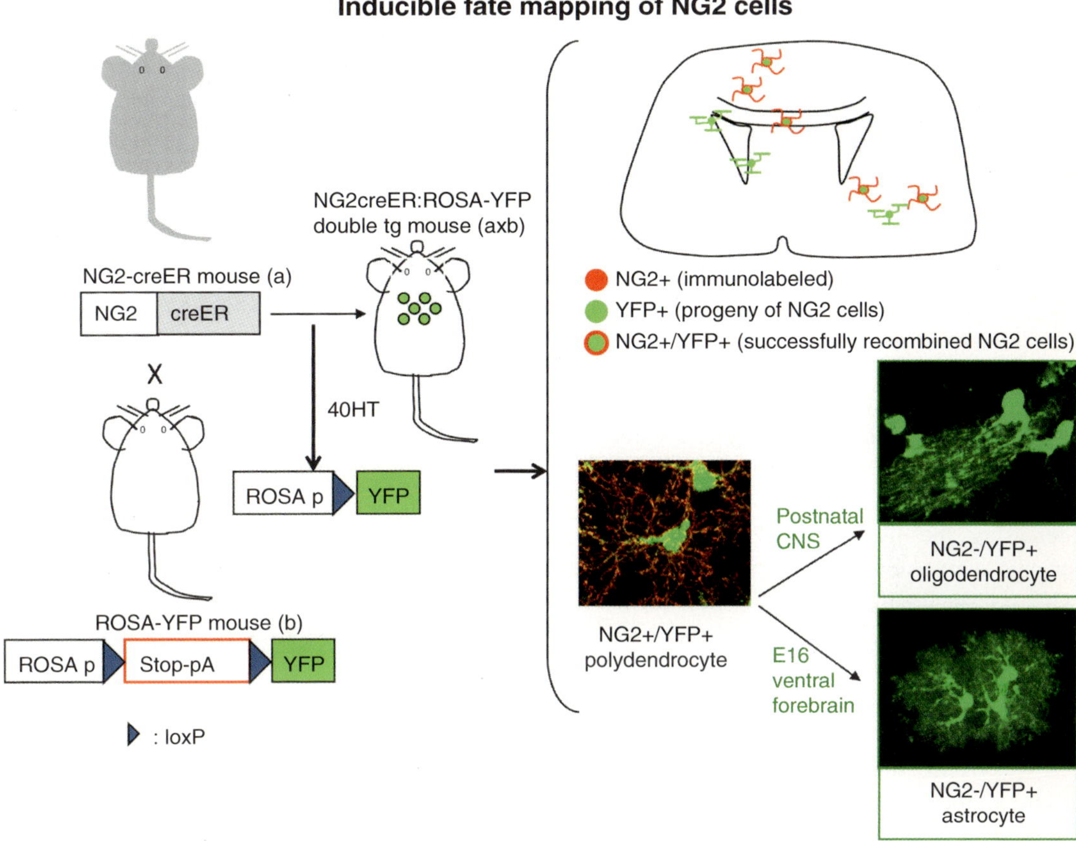

Fig. 3.3 Inducible cell fate mapping using NG2creER:ROSA-YFP mice. When NG2creER transgenic mice (*gray*) are crossed to the Cre reporter ROSA-YFP mice (*white*) and Cre is induced by 4-hydroxytamoxifen (4OHT), the fate of NG2 cells can be followed by YFP expression (*green*). In the ROSA-YFP reporter mouse, the expression of YFP is prevented by a transcriptional stop polyadenylation signal (Stop-pA) flanked by the Cre target recognition sequence loxP (*triangles*). When 4OHT is administered, CreER (fusion protein between Cre and mutated estrogen receptor) is activated in NG2 cells and is translocated into the nucleus, where it excises loxP-flanked polyadenylation signal and permanently activates the expression of YFP. CreER is expressed under the control of the NG2 promoter, while YPF is driven by the ubiquitous strong promoter of the ROSA26 locus [87]. Thus, CreER is only expressed in NG2 cells, but once YFP expression is activated, it will be expressed in all cell types, including NG2-negative progeny of NG2 cells. Thus, NG2+ YFP+ cells (*red* and *green* in the image on the left) represent cells in which Cre is active, while NG2-negative YFP+ cells (*green* but not *red* in the right two panels) represent the progeny of NG2 cells. *p* promoter, *pA* polyadenylation signal (Modified from Nishiyama [46], with permission. Copyright © 2007 SAGE Publications)

the promoter of a gene that is expressed in OPCs, such as NG2, PDGFR α, Olig2, and PLP1. These mice are crossed to any one of the available Cre reporter mouse lines, such as gtROSA26R [87], gtROSA-YFP [88], or Z/EG [89]. A Cre reporter mouse contains in its genome a cDNA encoding a reporter gene such as lacZ (gtROSA26) or a fluorescent protein such as yellow fluorescent protein (YFP) in gtROSA-YFP or Z/EG, following a transcriptional stop signal that is flanked by the Cre recognition sequence loxP. The entire cassette is placed under the regulation of a ubiquitous promoter which is active in most cells of the body. In double-transgenic mice, activation of the Cre recombinase in OPCs will cause excision of the stop sequence, thereby permanently activating transcription of the reporter gene. Since the expression of the reporter gene is driven by a ubiquitous promoter, the reporter will continue to be expressed as OPCs differentiate into other cell types and downregulate the expression of the OPC-specific gene driving the expression of Cre (Fig. 3.3).

Astrocyte Fate of NG2 Cells in Constitutive NG2-cre:ZEG Double-Transgenic Mice

The first such study of the fate of OPCs was performed using a constitutively active Cre driven under the regulatory elements of the NG2 (Cspg4) gene in a bacterial artificial chromosome (BAC) transgenic mouse line crossed to the Cre reporter Z/EG [90, 91]. BAC transgenesis allows one to generate a large transgenic cassette greater than 200 kb in size and includes the transgene in the context of the entire tissue-specific gene, thereby increasing the likelihood of attaining tissue-specific expression of the transgene, compared with transgenic mouse lines generated by the conventional approach of using a shorter promoter segment. The BAC approach has additional advantages over the gene knock-in approach in that both alleles of the endogenous gene will be left intact, and a higher level of expression can be achieved due to multiple copies of transgene inserted into the genome.

In NG2-cre:ZEG double-transgenic mice, EGFP expression was detected not only in NG2+ OPCs and oligodendrocytes, as expected, but also in a subpopulation of protoplasmic astrocytes in the ventral gray matter of the forebrain [90] and in the gray matter of the spinal cord [91]. In the ventral forebrain, EGFP+ astrocytes comprised more than 35 % of astrocytes that expressed S100β, and more than 40 % of the EGFP+ cells exhibited the astrocyte morphology, characterized by the highly branched bushy morphology typical of protoplasmic astrocytes. Thus, in the ventral gray matter of the forebrain, NG2 cell-derived astrocytes seemed to constitute a significant proportion of the resident astrocytes (Fig. 3.2). By contrast, in the neocortex, EGFP+ astrocytes constituted only 1 % of total astrocytes. A surprising outcome was that none of the astrocytes in white matter tracts throughout the CNS, including the optic nerve, corpus callosum, anterior commissure, cerebellum, and spinal cord, expressed EGFP (Fig. 3.2).

At first glance, the observation that both astrocytes as well as oligodendrocytes were EGFP+ and hence generated from NG2+ OPCs seemed to be consistent with the NG2+ cells being the in vivo equivalent of O-2A progenitor cells. However, the original discovery of O-2A progenitor cells had been made with cells isolated from the white matter, while the genetic fate-mapping studies failed to reveal an astrocyte fate of NG2+ OPCs in white matter. These studies could not determine whether a single NG2+ OPC could generate both oligodendrocytes and astrocytes or whether NG2+ OPCs comprised a heterogeneous population of astrocyte and oligodendrocyte precursor cells. Neither did these studies reveal whether NG2+ OPCs could generate astrocytes throughout the life of the mouse.

Astrocytes Are Generated from NG2+ OPCs in the Embryonic but Not Postnatal Brain

Studies from several labs ensued, in which the fate of OPCs in the adult brain was examined using tamoxifen-inducible Cre. In this system, the Cre recombinase is fused to the mutated ligand-binding domain of estrogen receptor (CreER^{T}), so that Cre is activated only when its ligand tamoxifen is present (Fig. 3.3). Different amino acid substitutions of the ER portion of CreER^{T} have been created that include CreER^{T}, CreER^{T2} [92], and CreER™ [93]. When a BAC transgenic mouse line expressing CreER^{T2} driven by regulatory elements of the Pdgfra gene (Pdgfra-CreER^{T2}) was crossed to gtROSA-YFP and Cre was activated in the adult, YFP was detected in NG2+ OPCs, oligodendrocytes, and a small number of neurons in the piriform cortex but not in astrocytes [94]. In another study, CreER™ was knocked into the Olig2 locus (Olig2-CreER™), and the mice were crossed to Z/EG or gtROSA26R Cre reporter mice [95]. When Cre was induced in adult mice, approximately 5 % of the reporter+ cells in the gray matter of the sensorimotor cortex expressed the astrocyte antigens GFAP and S100β, and this fraction increased only slightly to 6–7 % or 11 % after 2 or 6 months of survival time, respectively, after Cre induction. No reporter-expressing astrocytes were detected in the white matter. The interpretation of these results is somewhat complicated, as Olig2 is expressed more widely than PDGFRα and NG2, and it is possible that a low level of Olig2 that is expressed in some astrocytes or neural stem cells contributed to reporter+ astrocytes [96]. Since CreER™ was knocked into the endogenous Olig2 locus, one must also consider the possibility that loss of one allele of Olig2 has a subtle effect on the fate of OPCs. A small number of reporter+ astrocytes were also observed in PLP-creER^{T}:gtROSA-YFP mice after Cre induction in early postnatal mice [97], but the specificity of transgene expression in these mice is somewhat questionable [98].

In a more recent study, mice that were double transgenic for NG2-creER™ and gtROSA-YFP or Z/EG were used to directly compare the fate of NG2+ OPCs at different developmental stages. When Cre was induced at embryonic day 16.5 (E16.5), reporter+ astrocytes were detected in the gray matter of ventral forebrain. However, when Cre was induced at postnatal day 2 (P2) or in adult mice, none of the reporter+ cells exhibited astrocytic morphology or expressed astrocyte-specific antigens such as GFAP and aldehyde dehydrogenase 1L1 (Fig. 3.2) [99], which is consistent with the Rivers data that postnatal OPCs do not generate astrocytes. Additional support for the notion that NG2+ OPCs in the embryonic brain can generate astrocytes comes from the observation of "transitional cells" in the ventral forebrain of E18.5 NG2-cre:ZEG mice [90]. The "transitional cells," which were reporter+, expressed low levels of NG2 immunoreactivity in their distal processes but not in their cell bodies. Their processes appeared more branched and "bushy" than the typical NG2+ OPCs, and these cells also expressed glial glutamate aspartate transporter (GLAST), which is expressed by a subpopulation of astrocytes. Clusters of such "transitional cells" were found adjacent to the typical NG2+ GLAST-negative OPCs with fewer slender processes and more robust immunoreactivity for NG2. The "transitional cells" were found in the ventral parenchyma away from the germinal zone, which makes it unlikely that reporter+ astrocytes had been generated as a result of ectopic mis-expression of Cre in multipotent cells of the germinal zone.

It is not known why only NG2+ OPCs in the ventral forebrain but not in the postnatal brain or dorsal forebrain generate astrocytes. One attractive hypothesis is that NG2+ cells with astrogliogenic potential are generated from a different source than the oligodendrogliogenic NG2 cells. It has been shown that OPCs in the forebrain develop sequentially from three domains in the germinal zones defined by the expression of Nkx2.1, Gsx2, and Emx1 [68]. Thus, one could speculate that the Nkx2.1-expressing early OPCs but not the other later-born OPCs have the ability to generate astrocytes.

The low recombination efficiency in NG2-creER™:ZEG double-transgenic mice allowed clonal analysis of isolated EGFP+ cells after Cre induction in embryos. When Cre was induced at E16.5 and the phenotype for EGFP+ cells were analyzed at P14, clusters of EGFP+ astrocytes and clusters of EGFP+ oligodendrocyte lineage cells were both found in the ventral forebrain. However, each cluster of EGFP+ cells consisted exclusively of astrocytes or OPCs and oligodendrocytes, and none of the clusters of EGFP+ cells contained both astrocytes and oligodendrocyte lineage cells, suggesting that the diversification of astrocyte and oligodendrocyte fate occurs early and that OPCs divide symmetrically to produce either astrocytes or oligodendrocyte lineage cells [99]. Thus, NG2+ cells do not appear to be bipotential glial progenitor cells but are likely to represent a heterogeneous population in the embryonic ventral forebrain consisting of cells capable of differentiating into astrocytes or oligodendrocytes.

Genetic Fate Mapping in Injury Response

Several labs have used a similar Cre-loxP fate-mapping approach to determine whether NG2 cells in the mature CNS could generate reactive astrocytes in response to various types of injury. Using a neocortical stab wound model, Dimou and colleagues used Olig2-creER™ knock-in mice described above and showed that approximately 5 % of reporter+ cells were GFAP+ in the injured cortex 3 days after lesioning (3 dpl) and that the percentage did not increase but decreased slightly by 30 dpl [95, 100]. These observations differ significantly from a similar study by Tatsumi et al. [101] who used the same Olig2-creER™ mice crossed to ROSA-GAP43-EGFP reporter mice and found that the majority of reporter+ cells had acquired the bushy protoplasmic astrocyte morphology by 7 days after a cryoinjury in the cerebral cortex, and Olig2 was localized in the cytoplasm of these cells. One explanation for the discrepancy between the two studies could be that Olig2 transcription is upregulated in resident astrocytes or their precursor cells [29] during their reactive response to injury [102], rather than lineage progression from NG2 cells to astrocytes. Even a transient low level of expression of Cre in resident astrocytes will be sufficient to activate Cre-mediated recombination and permanently activate reporter cells. There is also the possibility that the ROSA-GAP43-EGFP reporter line either allowed more efficient recombination in Olig2+ cells or had a greater level of reporter expression in astrocytes, although this is unlikely, as both the ROSA-GAP43-EGFP and ROSA-YFP reporter lines are generated by inserting the reporter expression cassette into the same ROSA26 locus. It is also possible that cryoinjury alters the fate of Olig2-expressing cells in a different way than a simple stab wound.

In a neocortical stab wound created in NG2creER™:ZEG double-transgenic mice, it was shown that reporter+ cells that were GFAP+ appeared transiently at 10 dpl but mostly disappeared by 30 dpl [103]. A subpopulation of reporter+ GFAP+ cells that appeared at 10 dpl also expressed NG2 and displayed morphology that more closely resembled OPCs (polydendrocytes) rather than reactive astrocytes. The transient appearance of NG2+ GFAP+ positive cells at early time points after lesioning was also reported by Zhao et al. [104]. One interpretation of these observations is that NG2 cells attempt to undergo astrocyte differentiation by transiently upregulating GFAP expression, but the process of reprogramming is aborted before the cells become bona fide astrocytes, and the cells are either reverted back to the oligodendrocyte lineage or undergo cell death. In a spinal cord stab injury model, cells coexpressing NG2 and GFAP were only rarely detected, but the number of NG2+ GFAP+ cells was significantly increased when the activity of bone morphogenetic proteins (BMPs) was augmented by infusing neutralizing antibodies to the BMP inhibitor Noggin, suggesting that GFAP expression in the normal and injured CNS is repressed by endogenous Noggin [105]. Thus, upregulation of BMP in the lesion does not appear to be sufficient to counter the inhibitors and convert the fate of NG2 cells into astrocytes.

Several pieces of evidence suggest that GFAP expression can be upregulated in a small subpopulation of NG2 cells under normal conditions. GFAP mRNA has been detected by single cell RT-PCR from cells in the hippocampus with the electrophysiological characteristics of NG2 cells [106]. In transgenic mice that express EGFP under the human GFAP promoter [107], EGFP was detected in a subpopulation of NG2 cells [108]. These observations suggest that repression of GFAP transcription is released in NG2 cells under certain circumstances, leading to the presence of a low level of GFAP mRNA, which does not necessarily signify that they have become astrocytes. The converse case of NG2 mRNA in astrocytes, however, has never been reported, and NG2DsRedBAC transgenic mice or NG2YFP mice do not express DsRed in astrocytes [90, 109]. Further understanding of the molecular mechanisms that repress GFAP transcription in NG2 cells may shed light on the lineage relationship between astrocytes and oligodendrocytes and the evolutionary diversification of the two glial cell types.

Additional recent studies have used genetic fate mapping to examine whether reactive astrocytes are generated from NG2 cells. Zawadzka et al. [110] created an acute chemically induced demyelination in the spinal cord of adult Pdgfra-CreER^{T2}:ROSA-YFP mice and showed that approximately 3 % of the astrocytes around the rim of the lesion expressed YFP but that the majority of reactive astrocytes did not express YFP. They further used another Cre driver activated by the Fgfr3 gene, which is expressed in astrocyte precursor cells [28, 29] and demonstrated that 93 % of cells expressing GFAP around the lesion expressed the Cre reporter, thereby convincingly demonstrating that reactive astrocytes are derived from FGFR3+ precursors that are distinct from NG2 cells. Similarly, in experimental autoimmune encephalomyelitis (EAE) lesions created in Pdgfra-CreER^{T2}:ROSA-YFP mice, the majority of YFP+ cells in the inflammatory lesions were NG2+, and less than 3 % of the reporter+ cells were GFAP+ [111]. Furthermore, Kang et al. [112] showed that in a mouse model of amyotrophic lateral sclerosis (ALS) created in a new line of Pdgfra-creER^{T2} mice crossed to the Z/EG or ROSA-YFP reporter mice, there was no evidence of astrocyte differentiation of NG2 cells. Collectively, these observations suggest that NG2 cells are not a major source of reactive astrocytes in both acute and chronic injury.

The findings from the above genetic fate-mapping studies do not support the conclusions of several BrdU pulse-chase labeling experiments published over the past several years. The first of the series of studies using BrdU pulse-chase labeling showed a decrease in the percentage of the BrdU+ cells that were NG2+ and a concomitant increase in the percentage of BrdU+ cells that were GFAP+ between 2 and 6 days after stab wound injury [113]. Based on these observations, the author concluded that proliferated NG2 cells had differentiated into GFAP+ reactive astrocytes. Using a similar approach, other studies have also made similar conclusions in various lesion paradigms including spinal cord injury, stab wound, EAE, and a mouse model of amyotrophic lateral sclerosis (ALS) [104, 114–118]. However, in all of these studies, only a subpopulation of the proliferating cells expressed NG2 (33 % in [113]; up to 50 % in [104, 114]; 22 % in [118]; 55 % in [117]. Therefore, it is highly likely that the BrdU+ GFAP+ cells that appeared in the lesion in these studies had originated from resident astrocytes or their precursor cells that expressed neither NG2 nor GFAP. Since GFAP is not readily detectable in protoplasmic astrocytes in the rodent neocortex, the lack of detection of GFAP in BrdU+ cells cannot be used as a basis for concluding that the GFAP-negative cells are NG2 cells. Fate mapping of GFAP- astrocytes using Fgfr3-cre or Aldh1L1-cre would result in reporter expression in the majority of reactive astrocytes.

Many of the above BrdU pulse-chase labeling experiments noted cytoplasmic appearance of Olig2 during the course of injury response [101, 104, 115, 117, 118]. These studies describe cytoplasmic Olig2 localization found in BrdU+ GFAP+ cells and conclude that BrdU+ cells that retain Olig2 in the nucleus become oligodendrocytes while those that translocate Olig2 to the cytoplasm become GFAP+ astrocytes. In one aspect, this resembles the results of an in vitro study on neural stem cells in which blocking cytoplasmic translocation of Olig2 inhibited their differentiation into astrocytes in response to activation of AKT by CNTF [119, 120]. However, in other studies, Olig2 was never detected in the cytoplasm [103, 111]. It is expected that some Olig2 proteins must reside in the cytoplasm when it is being synthesized or degraded or when the cell is undergoing mitosis. While all NG2 cells express Olig2 [121], Olig2 expression has been detected in the germinal zones in cells that are neither NG2+ nor mature oligodendrocytes [59, 120, 122, 123], and the effects of downregulating Olig2 in those neural stem cells may be different from the effects of downregulating Olig2 in committed oligodendrocyte lineage cells. Further studies are necessary to clarify the significance and the role of Olig2 cytoplasmic translocation in astrocyte differentiation from OPCs and neural stem cells.

Genetic Approach to Study the Contribution of NG2 Cells to Glioma

Although the cellular origin of glial tumors is discussed in other chapters of this book, it is worth mentioning here a few observations in the context of the biology of NG2 cells and genetic fate mapping. Since NG2 cells constitute the major proliferative cell population in the mature CNS, they may be more likely to accumulate genetic mutations and undergo neoplastic transformation during the life of an individual. The importance of mutations in tumor suppressor genes such as p53 and neurofibromatosis 1 (NF1) in gliomagenesis has been well documented in human tumor tissue and animal models [124, 125]. However, it has not been clear whether neural stem cells or more committed glial cells undergo transformation [126].

A recent study used an application of the Cre-loxP-mediated fate mapping described above to demonstrate that OPCs rather than neural stem cells provide the cellular context that is necessary

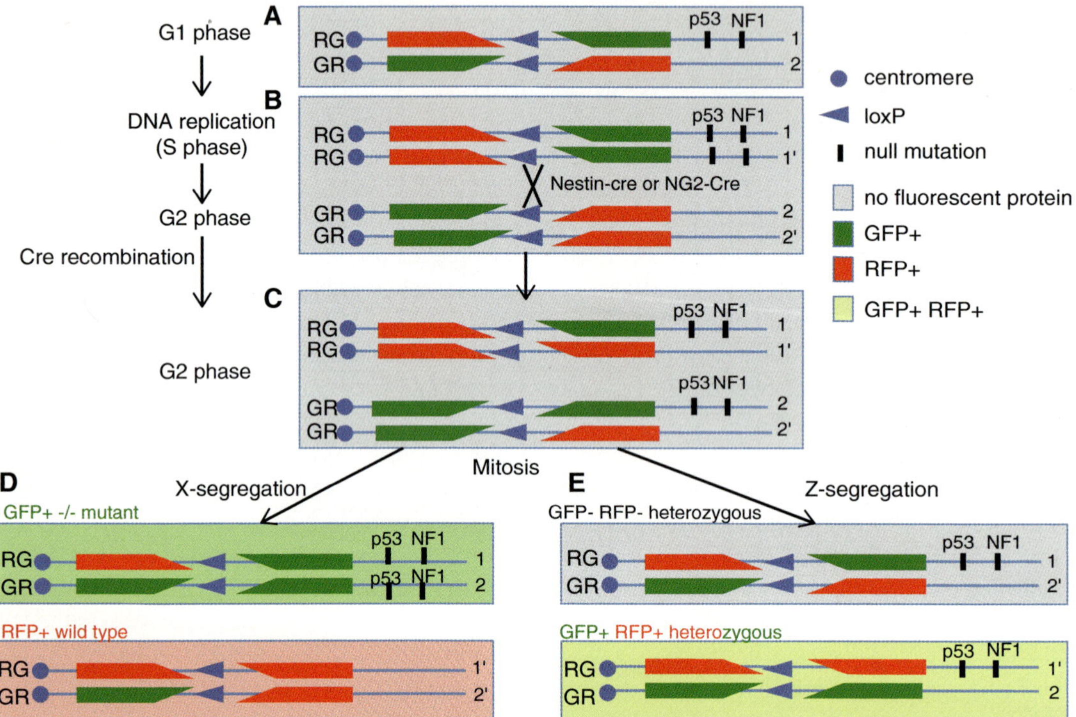

Fig. 3.4 MADM (mosaic analysis of double markers) applied to glioma study. This technique uses Cre-loxP-mediated mitotic recombination to generate a GFP+ cells with double null mutations in p53 and NF1 and sister wild-type red fluorescent protein (RFP)+ cell as well as heterozygous cells without color. The approach requires manipulation of two homologous alleles on chromosome 11. The first allele (1) contains N terminus of RFP and C terminus of GFP with a loxP site in between (RG). The second allele (2) contains N terminus of GFP and C terminus of RFP with a loxP site in between (GR). The N-terminal and C-terminal halves of RFP and GFP do not express functional protein. These mutations have been engineered into the Hipp11 locus, proximal to p53 and NF1 loci on chromosome 11 (centromere is indicated by *blue circle* on the *left*). In addition, null mutations in p53 and NF1 loci have been engineered into the RG expression cassette (*black bars*, **a**). During the G2 phase of mitosis, after DNA has undergone replication in S phase, homologous chromosomes are duplicated to generate 1 and 1′ from the RG cassette and 2 and 2′ from the GR cassette (**b**). When this mouse line is mated with Nestin-Cre or NG2-cre mouse, mitotic recombination occurs in neural stem cells or NG2 cells, respectively, and there is crossover between 1′ and 2 (**b** and **c**). At the end of mitosis, homologous chromosomes can undergo X-segregation (**d**) or Z-segregation (**e**). The latter generates two heterozygous daughter cells (p53+/–, NF1+/–) that are either GFP+ RFP+ (*yellow*) or colorless. X-segregation generates a wild-type RFP+ daughter cell and a double mutant GFP+ daughter cell. In this case, the progeny of cells with double p53–/– and NF1–/– mutations can be followed as GFP+ cells, and the proliferative behavior of the mutated GFP+ cells and the wild-type RFP+ progeny of the sister cells can be compared (Modified from Liu et al. [127], with permission)

for transformation into glioma [127]. In this study, the authors used a technique called MADM (mosaic analysis of double mutants) [128] (Fig. 3.4), which requires Cre-mediated mitotic recombination to express either GFP or red fluorescent protein (RFP) in each of the two daughter cells. In addition to expressing the fluorescent proteins, null mutations for p53 and NF1 were engineered in cis distal to the loxP site, so that only the GFP+ daughter cell carries the double mutation while the RFP+ daughter cell is wild type. Since mitotic recombination occurs at a low frequency of 10^{-4} to 10^{-5} [128], clonal analysis of the GFP+ progeny of the mutant cell can be performed and compared with the behavior of the progeny of its wild-type RFP+ sister cell (Fig. 3.4).

When mitotic recombination for p53 and NF1 double null mutation was targeted to neural stem cells by Nestin-Cre, GFP+ mutant cells began to proliferate, and after a lag period, tumors arose by 3 months of age. Immunolabeling of GFP+ tumor cells and transcriptome analyses of the tumor revealed that the neoplastic cells exhibited the characteristics of OPCs rather than neural stem cells with some properties that resembled neural stem cells, suggesting that even though mutations are created in neural stem cells, tumors arise from OPCs. This is consistent with the finding that NG2+ cells in glioblastoma represent a more proliferative tumorigenic cell population than NG2– cells [129]. Furthermore, tumors with identical characteristics arose when the double mutation was created in OPCs using NG2cre rather than in neural stem cells. These observations suggest that mutations need not occur in neural stem cells and that p53 and NF1 double null mutation occurring in OPCs is sufficient to cause neoplastic transformation in OPCs.

The notion that OPCs rather than neural stem cells can undergo neoplastic transformation has been suggested when unregulated proliferation of OPCs and subsequent tumor formation were observed by overexpressing PDGF in the brain. When PDGFB was delivered by retrovirus or the long form of PDGF A ($PDGFA_L$) was overexpressed by transgenesis, tumors consisting of NG2+ OPCs arose [130, 131]. Thus NG2+ OPCs, which are the major proliferative cell type in the mature CNS, are likely the source of some gliomas. These findings could be used to more effectively direct antineoplastic therapy specifically to OPCs.

Conclusions

The identification of bipotential O-2A glial progenitor cells by Raff and colleagues almost three decades ago has opened an exciting question of lineage plasticity of neural cells. After a long debate about whether OPCs or NG2 cells are bipotential glial progenitor cells or multipotent stem-like cells in vivo, recent genetic fate-mapping studies have provided a more direct answer to the question of the fate of NG2 cells. Most of the studies converge on the basic conclusion that the vast majority of NG2 cells and all of the NG2 cells in the postnatal CNS are precursors to oligodendrocytes. An exception to this rule is found in the gray matter of the embryonic ventral forebrain where some NG2 cells give rise to protoplasmic astrocytes. These observations provide the basis for future studies directed at elucidating the molecular mechanisms underlying fate restrictions, which could be used to reprogram proliferating OPCs to achieve greater myelin regeneration or convert a highly malignant glioma into a more quiescent cell type.

References

1. Baily P, Cushing H. A classification of the tumors of the glioma group on a histogenetic basis with a correlated study of prognosis. Philadelphia: J.B. Lippincott; 1926.
2. Virchow R. Die Cellularpathologie in ihrer Begrundung auf physiologische und pathologische Gewebelehre. Berlin: Verlag von August Hirschwald; 1858.
3. von Lenhossek M. Der feinere Bau des Nervensystems im Lichte neuester Forschungen. Berlin: Fischer. H. Kornfield; 1893.
4. Ramon y Cajal S. Contribucion al conocimiento de la neuroglia del cerebro humano. Trab Lab Invest Biol. 1913;11:255–315.
5. Del Rio-Hortega P. Tercera aportacion al conocimiento morfologico e interpretacion functional de la oligodendroglial. Mem Real Soc Exp Hist Nat. 1928;14:5–122.

6. Del Rio-Hortega P. Microglia. In: Penfield W, editor. Cytology and cellular pathology of the nervous system, vol. II. New York: Hafner; 1932. p. 483–534.
7. Bunge MB, Bunge RP, Pappas GD. Electron microscopic demonstration of connections between glia and the myelin sheath in the developing mammalian central nervous system. J Cell Biol. 1962;12:448–53.
8. Kimelberg HK. Functions of mature mammalian astrocytes: a current view. Neuroscientist. 2010; 16(1):79–106.
9. Paixao S, Klein R. Neuron-astrocyte communication and synaptic plasticity. Curr Opin Neurobiol. 2010;20(4):466–73.
10. Barker AJ, Ullian EM. Astrocytes and synaptic plasticity. Neuroscientist. 2010;16(1):40–50.
11. Robel S, Berninger B, Gotz M. The stem cell potential of glia: lessons from reactive gliosis. Nat Rev Neurosci. 2011;12(2):88–104.
12. Eng LF, Vanderhaeghen JJ, Bignami A, Gerstl B. An acidic protein isolated from fibrous astrocytes. Brain Res. 1971;28(2):351–4.
13. Kohler G, Milstein C. Continuous cultures of fused cells secreting antibody of predefined specificity. Nature. 1975;256(5517):495–7.
14. Eisenbarth GS, Walsh FS, Nirenberg M. Monoclonal antibody to a plasma membrane antigen of neurons. Proc Natl Acad Sci USA. 1979;76:4913–7.
15. Raff MC, Abney ER, Cohen J, Lindsay R, Noble M. Two types of astrocytes in cultures of developing rat white matter: differences in morphology, surface gangliosides, and growth characteristics. J Neurosci. 1983;3(6):1289–300.
16. Stallcup WB. The NG2 antigen, a putative lineage marker: immunofluorescent localization in primary cultures of rat brain. Dev Biol. 1981;83: 154–65.
17. Wilson SS, Baetge EE, Stallcup WB. Antisera specific for cell lines with mixed neuronal and glial properties. Dev Biol. 1981;83:146–53.
18. Raff MC, Abney ER, Miller RH. Two glial cell lineages diverge prenatally in rat optic nerve. Dev Biol. 1984;106(1):53–60.
19. Raff MC, Mirsky R, Fields KL, et al. Galactocerebroside is a specific cell-surface antigenic marker for oligodendrocytes in culture. Nature. 1978;274(5673):813–6.
20. Raff MC, Miller RH, Noble M. A glial progenitor cell that develops in vitro into an astrocyte or an oligodendrocyte depending on culture medium. Nature. 1983;303:390–6.
21. Ffrench-Constant C, Raff MC. Proliferating bipotential glial progenitor cells in adult rat optic nerve. Nature. 1986;319:499–502.
22. Shi J, Marinovich A, Barres BA. Purification and characterization of adult oligodendrocyte precursor cells from the rat optic nerve. J Neurosci. 1998;18:4627–36.
23. Wren D, Wolswijk G, Noble M. In vitro analysis of the origin and maintenance of O-2A^{adult} progenitor cells. J Cell Biol. 1992;116:167–76.
24. Miller RH, Raff MC. Fibrous and protoplasmic astrocytes are biochemical and developmentally distinct. J Neurosci. 1984;4:585–92.
25. Peters A, Palay SL, deF Webster H. The fine structure of the nervous system. 3rd ed. New York: Oxford; 1991.
26. Polito A, Reynolds R. NG2-expressing cells as oligodendrocyte progenitors in the normal and demyelinated adult central nervous system. J Anat. 2005;207(6):707–16.
27. Cahoy JD, Emery B, Kaushal A, et al. A transcriptome database for astrocytes, neurons, and oligodendrocytes: a new resource for understanding brain development and function. J Neurosci. 2008;28(1): 264–78.
28. Pringle NP, Yu WP, Howell M, Colvin JS, Ornitz DM, Richardson WD. Fgfr3 expression by astrocytes and their precursors: evidence that astrocytes and oligodendrocytes originate in distinct neuroepithelial domains. Development. 2003;130(1):93–102.
29. Young KM, Mitsumori T, Pringle N, Grist M, Kessaris N, Richardson WD. An Fgfr3-iCreER(T2) transgenic mouse line for studies of neural stem cells and astrocytes. Glia. 2010;58(8):943–53.
30. Raff MC, Lillien LE, Richardson WD, Burne JF, Noble MD. Platelet-derived growth factor from astrocytes drives the clock that times oligodendrocyte development in culture. Nature. 1988;333: 562–5.
31. Richardson WD, Pringle N, Mosley MJ, Westermark B, Dubois-Dalcq M. A role for platelet-derived growth factor in normal gliogenesis in the central nervous system. Cell. 1988;53:309–19.
32. Noble M, Murray K, Stroobant P, Waterfield MD, Riddle P. Platelet-derived growth factor promotes division and motility and inhibits premature differentiation of the oligodendrocyte/type-2 astrocyte progenitor cell. Nature. 1988;333:560–2.
33. Pringle NP, Mudhar HS, Collarini EJ, Richardson WD. PDGF receptors in the rat CNS: during late neurogenesis, PDGF alpha-receptor expression appears to be restricted to glial cells of the oligodendrocyte lineage. Development. 1992;115:535–51.
34. Pringle NP, Richardson WD. A singularity of PDGF alpha-receptor expression in the dorsoventral axis of the neural tube may define the origin of the oligodendrocyte lineage. Development. 1993;117:525–33.
35. Lu QR, Yuk D, Alberta JA, et al. Sonic hedgehog – regulated oligodendrocyte lineage genes encoding bHLH proteins in the mammalian central nervous system. Neuron. 2000;25(2):317–29.
36. Takebayashi H, Yoshida S, Sugimori M, et al. Dynamic expression of basic helix-loop-helix Olig family members: implication of Olig2 in neuron and oligodendrocyte differentiation and identification of a new member, Olig3. Mech Dev. 2000;99(1–2):143–8.
37. Zhou Q, Wang S, Anderson DJ. Identification of a novel family of oligodendrocyte lineage-specific basic helix-loop-helix transcription factors. Neuron. 2000;25(2):331–43.

38. Kuhlbrodt K, Herbarth B, Sock E, Hermans-Borgmeyer I, Wegner M. Sox10, a novel transcriptional modulator in glial cells. J Neurosci. 1998;18(1):237–50.
39. Kessaris N, Pringle N, Richardson WD. Specification of CNS glia from neural stem cells in the embryonic neuroepithelium. Philos Trans R Soc Lond B Biol Sci. 2008;363(1489):71–85.
40. Peru RL, Mandrycky N, Nait-Oumesmar B, Lu QR. Paving the axonal highway: from stem cells to myelin repair. Stem Cell Rev. 2008;4(4):304–18.
41. Stallcup WB, Beasley L. Bipotential glial precursor cells of the optic nerve express the NG2 proteoglycan. J Neurosci. 1987;7:2737–44.
42. Levine JM, Stallcup WB. Plasticity of developing cerebellar cells in vitro studied with antibodies against the NG2 antigen. J Neurosci. 1987;7:2721–31.
43. Nishiyama A, Lin XH, Giese N, Heldin CH, Stallcup WB. Interaction between NG2 proteoglycan and PDGF alpha-receptor on O2A progenitor cells is required for optimal response to PDGF. J Neurosci Res. 1996;43(3):315–30.
44. Nishiyama A, Lin XH, Giese N, Heldin CH, Stallcup WB. Co-localization of NG2 proteoglycan and PDGF alpha-receptor on O2A progenitor cells in the developing rat brain. J Neurosci Res. 1996;43(3):299–314.
45. LeVine SM, Goldman JE. Embryonic divergence of oligodendrocyte and astrocyte lineages in developing rat cerebrum. J Neurosci. 1988;8:3992–4006.
46. Nishiyama A. Polydendrocytes: NG2 cells with many roles in development and repair of the CNS. Neuroscientist. 2007;13(1):62–76.
47. Nishiyama A, Komitova M, Suzuki R, Zhu X. Polydendrocytes (NG2 cells): multifunctional cells with lineage plasticity. Nat Rev Neurosci. 2009;10(1):9–22.
48. Nishiyama A, Watanabe M, Yang Z, Bu J. Identity, distribution, and development of polydendrocytes: NG2-expressing glial cells. J Neurocytol. 2002;31(6–7):437–55.
49. Sim FJ, Lang JK, Waldau B, et al. Complementary patterns of gene expression by human oligodendrocyte progenitors and their environment predict determinants of progenitor maintenance and differentiation. Ann Neurol. 2006;59(5):763–79.
50. Levison SW, de Vellis J, Goldman JE. Astrocyte development. In: Rao M, Jacobson M, editors. Developmental neurobiology. 4th ed. New York: Plenum Publishers; 2005.
51. Voigt T. Development of glial cells in the cerebral wall of ferrets: direct tracing of their transformation from radial glia into astrocytes. J Comp Neurol. 1989;289(1):74–88.
52. Misson J, Takahashi T, Caviness Jr VS. Ontogeny of radial and other astroglial cells in murine cerebral cortex. Glia. 1991;4:138–48.
53. Hirano M, Goldman JE. Gliogenesis in rat spinal cord: evidence for origin of astrocytes and oligodendrocytes from radial precursors. J Neurosci Res. 1988;21(2–4):155–67.
54. Malatesta P, Hartfuss E, Gotz M. Isolation of radial glial cells by fluorescent-activated cell sorting reveals a neuronal lineage. Development. 2000;127(24):5253–63.
55. Noctor SC, Flint AC, Weissman TA, Dammerman RS, Kriegstein AR. Neurons derived from radial glial cells establish radial units in neocortex. Nature. 2001;409(6821):714–20.
56. Sudarov A, Turnbull RK, Kim EJ, Lebel-Potter M, Guillemot F, Joyner AL. Ascl1 genetics reveals insights into cerebellum local circuit assembly. J Neurosci. 2011;31(30):11055–69.
57. Levison SW, Goldman JE. Both oligodendrocytes and astrocytes develop from progenitors in the subventricular zone of postnatal rat forebrain. Neuron. 1993;10:201–12.
58. Doetsch F, Caille I, Lim DA, Garcia-Verdugo JM, Alvarez-Buylla A. Subventricular zone astrocytes are neural stem cells in the adult mammalian brain. Cell. 1999;97(6):703–16.
59. Menn B, Garcia-Verdugo JM, Yaschine C, Gonzalez-Perez O, Rowitch D, Alvarez-Buylla A. Origin of oligodendrocytes in the subventricular zone of the adult brain. J Neurosci. 2006;26(30):7907–18.
60. Namihira M, Kohyama J, Semi K, et al. Committed neuronal precursors confer astrocytic potential on residual neural precursor cells. Dev Cell. 2009;16(2):245–55.
61. Freeman MR. Specification and morphogenesis of astrocytes. Science. 2011;330(6005):774–8.
62. Cai J, Qi Y, Hu X, et al. Generation of oligodendrocyte precursor cells from mouse dorsal spinal cord independent of Nkx6 regulation and Shh signaling. Neuron. 2005;45(1):41–53.
63. Vallstedt A, Klos JM, Ericson J. Multiple dorsoventral origins of oligodendrocyte generation in the spinal cord and hindbrain. Neuron. 2005;45(1):55–67.
64. Fogarty M, Richardson WD, Kessaris N. A subset of oligodendrocytes generated from radial glia in the dorsal spinal cord. Development. 2005;132(8):1951–9.
65. Nery S, Wichterle H, Fishell G. Sonic hedgehog contributes to oligodendrocyte specification in the mammalian forebrain. Development. 2001;128(4):527–40.
66. Tekki-Kessaris N, Woodruff R, Hall AC, et al. Hedgehog-dependent oligodendrocyte lineage specification in the telencephalon. Development. 2001;128(13):2545–54.
67. Olivier C, Cobos I, Perez Villegas EM. Monofocal origin of telencephalic oligodendrocytes in the anterior entopeduncular area of the chick embryo. Development. 2001;128(10):1757–69.
68. Kessaris N, Fogarty M, Iannarelli P, Grist M, Wegner M, Richardson WD. Competing waves of oligodendrocytes in the forebrain and postnatal elimination of an embryonic lineage. Nat Neurosci. 2006;9(2):173–9.
69. Levison SW, Young GM, Goldman JE. Cycling cells in the adult rat neocortex preferentially generate oligodendroglia. J Neurosci Res. 1999;57(4):435–46.
70. Gonzalez-Perez O, Romero-Rodriguez R, Soriano-Navarro M, Garcia-Verdugo JM, Alvarez-Buylla A. Epidermal growth factor induces the progeny of

subventricular zone type B cells to migrate and differentiate into oligodendrocytes. Stem Cells. 2009;27(8):2032–43.
71. Ventura RE, Goldman JE. Dorsal radial glia generate olfactory bulb interneurons in the postnatal murine brain. J Neurosci. 2007;27(16):4297–302.
72. Espinosa de los Monteros A, Zhang M, de Vellis J. O2A progenitor cells transplanted into the neonatal rat brain develop into oligodendrocytes but not astrocytes. Proc Natl Acad Sci USA. 1993;90:50–4.
73. Groves AK, Barnett SC, Franklin RJM, et al. Repair of demyelinated lesions by transplantation of purified O-2A progenitor cells. Nature. 1993;362: 453–5.
74. Grove EA, Williams BP, Li D, Hajihosseini M, Friedrich A, Price J. Multiple restricted lineages in the embryonic rat cerebral cortex. Development. 1993;117:553–61.
75. Seki T, Arai Y. Highly polysialylated neural cell adhesion molecule (NCAM-H) is expressed by newly generated granule cells in the dentate gyrus of the adult rat. J Neurosci. 1993;13(6):2351–8.
76. Alvarez-Buylla A, Garcia-Verdugo JM. Neurogenesis in adult subventricular zone. J Neurosci. 2002;22(3): 629–34.
77. Kondo T, Raff M. Oligodendrocyte precursor cells reprogrammed to become multipotential CNS stem cells. Science. 2000;289(5485):1754–7.
78. Dawson MR, Levine JM, Reynolds R. NG2-expressing cells in the central nervous system: are they oligodendroglial progenitors? J Neurosci Res. 2000;61(5): 471–9.
79. Psachoulia K, Jamen F, Young KM, Richardson WD. Cell cycle dynamics of NG2 cells in the postnatal and ageing brain. Neuron Glia Biol. 2009;5(3–4):57–67.
80. Komitova M, Zhu X, Serwanski DR, Nishiyama A. NG2 cells are distinct from neurogenic cells in the postnatal mouse subventricular zone. J Comp Neurol. 2009;512(5):702–16.
81. Geha S, Pallud J, Junier MP, et al. NG2+/Olig2+ cells are the major cycle-related cell population of the adult human normal brain. Brain Pathol. 2010;20(2): 399–411.
82. Bu J, Banki A, Wu Q, Nishiyama A. Increased NG2(+) glial cell proliferation and oligodendrocyte generation in the hypomyelinating mutant shiverer. Glia. 2004;48(1):51–63.
83. Horner PJ, Power AE, Kempermann G, et al. Proliferation and differentiation of progenitor cells throughout the intact adult rat spinal cord. J Neurosci. 2000;20(6):2218–28.
84. Hommes OR, Leblond CP. Mitotic division of neuroglia in the normal adult rat. J Comp Neurol. 1967;129(3):269–78.
85. Adrian Jr EK, Walker BE. Incorporation of thymidine-H3 by cells in normal and injured mouse spinal cord. J Neuropathol Exp Neurol. 1962;21: 597–609.
86. Skoff RP. Gliogenesis in rat optic nerve: astrocytes are generated in a single wave before oligodendrocytes. Dev Biol. 1990;139:149–68.
87. Soriano P. Generalized lacZ expression with the ROSA26 Cre reporter strain. Nat Genet. 1999; 21(1):70–1.
88. Srinivas S, Watanabe T, Lin CS, et al. Cre reporter strains produced by targeted insertion of EYFP and ECFP into the ROSA26 locus. BMC Dev Biol. 2001;1:4.
89. Novak A, Guo C, Yang W, Nagy A, Lobe CG. Z/EG, a double reporter mouse line that expresses enhanced green fluorescent protein upon Cre-mediated excision. Genesis. 2000;28(3–4):147–55.
90. Zhu X, Bergles DE, Nishiyama A. NG2 cells generate both oligodendrocytes and gray matter astrocytes. Development. 2008;135(1):145–57.
91. Zhu X, Hill RA, Nishiyama A. NG2 cells generate oligodendrocytes and gray matter astrocytes in the spinal cord. Neuron Glia Biol. 2008;4(1):19–26.
92. Feil R, Wagner J, Metzger D, Chambon P. Regulation of Cre recombinase activity by mutated estrogen receptor ligand-binding domains. Biochem Biophys Res Commun. 1997;237(3):752–7.
93. Danielian PS, White R, Hoare SA, Fawell SE, Parker MG. Identification of residues in the estrogen receptor that confer differential sensitivity to estrogen and hydroxytamoxifen. Mol Endocrinol. 1993;7(2): 232–40.
94. Rivers LE, Young KM, Rizzi M, et al. PDGFRA/NG2 glia generate myelinating oligodendrocytes and piriform projection neurons in adult mice. Nat Neurosci. 2008;11(12):1392–401.
95. Dimou L, Simon C, Kirchhoff F, Takebayashi H, Gotz M. Progeny of Olig2-expressing progenitors in the gray and white matter of the adult mouse cerebral cortex. J Neurosci. 2008;28(41):10434–42.
96. Cai J, Chen Y, Cai WH, et al. A crucial role for Olig2 in white matter astrocyte development. Development. 2007;134(10):1887–99.
97. Guo F, Ma J, McCauley E, Bannerman P, Pleasure D. Early postnatal proteolipid promoter-expressing progenitors produce multilineage cells in vivo. J Neurosci. 2009;29(22):7256–70.
98. Miller MJ, Kangas CD, Macklin WB. Neuronal expression of the proteolipid protein gene in the medulla of the mouse. J Neurosci Res. 2009;87(13): 2842–53.
99. Zhu X, Hill RA, Dietrich D, Komitova M, Suzuki R, Nishiyama A. Age-dependent fate and lineage restriction of single NG2 cells. Development. 2011;138(4):745–53.
100. Simon C, Gotz M, Dimou L. Progenitors in the adult cerebral cortex: cell cycle properties and regulation by physiological stimuli and injury. Glia. 2010;59(6): 869–81.
101. Tatsumi K, Takebayashi H, Manabe T, et al. Genetic fate mapping of Olig2 progenitors in the injured

adult cerebral cortex reveals preferential differentiation into astrocytes. J Neurosci Res. 2008;86(16): 3494–502.

102. Chen Y, Miles DK, Hoang T, et al. The basic helix-loop-helix transcription factor olig2 is critical for reactive astrocyte proliferation after cortical injury. J Neurosci. 2008;28(43):10983–9.
103. Komitova M, Serwanski DR, Lu QR, Nishiyama A. NG2 cells are not a major source of reactive astrocytes after neocortical stab wound injury. Glia. 2011;59(5):800–9.
104. Zhao JW, Raha-Chowdhury R, Fawcett JW, Watts C. Astrocytes and oligodendrocytes can be generated from NG2+ progenitors after acute brain injury: intracellular localization of oligodendrocyte transcription factor 2 is associated with their fate choice. Eur J Neurosci. 2009;29(9):1853–69.
105. Hampton DW, Asher RA, Kondo T, Steeves JD, Ramer MS, Fawcett JW. A potential role for bone morphogenetic protein signalling in glial cell fate determination following adult central nervous system injury in vivo. Eur J Neurosci. 2007;26(11): 3024–35.
106. Schools GP, Zhou M, Kimelberg HK. Electrophysiologically "complex" glial cells freshly isolated from the hippocampus are immunopositive for the chondroitin sulfate proteoglycan NG2. J Neurosci Res. 2003;73(6):765–77.
107. Nolte C, Matyash M, Pivneva T, et al. GFAP promoter-controlled EGFP-expressing transgenic mice: a tool to visualize astrocytes and astrogliosis in living brain tissue. Glia. 2001;33(1):72–86.
108. Matthias K, Kirchhoff F, Seifert G, et al. Segregated expression of AMPA-type glutamate receptors and glutamate transporters defines distinct astrocyte populations in the mouse hippocampus. J Neurosci. 2003;23(5):1750–8.
109. Karram K, Goebbels S, Schwab M, et al. NG2-expressing cells in the nervous system revealed by the NG2-EYFP-knockin mouse. Genesis. 2008;46(12):743–57.
110. Zawadzka M, Rivers LE, Fancy SP. CNS-resident glial progenitor/stem cells produce Schwann cells as well as oligodendrocytes during repair of CNS demyelination. Cell Stem Cell. 2010;6(6):578–90.
111. Tripathi RB, Rivers LE, Young KM, Jamen F, Richardson WD. NG2 glia generate new oligodendrocytes but few astrocytes in a murine experimental autoimmune encephalomyelitis model of demyelinating disease. J Neurosci. 2010;30(48): 16383–90.
112. Kang SH, Fukaya M, Yang JK, Rothstein JD, Bergles DE. NG2+ CNS glial progenitors remain committed to the oligodendrocyte lineage in postnatal life and following neurodegeneration. Neuron. 2010;68(4): 668–81.
113. Alonso G. NG2 proteoglycan-expressing cells of the adult rat brain: possible involvement in the formation of glial scar astrocytes following stab wound. Glia. 2005;49(3):318–38.
114. Horky LL, Galimi F, Gage FH, Horner PJ. Fate of endogenous stem/progenitor cells following spinal cord injury. J Comp Neurol. 2006;498(4):525–38.
115. Cassiani-Ingoni R, Coksaygan T, Xue H, et al. Cytoplasmic translocation of Olig2 in adult glial progenitors marks the generation of reactive astrocytes following autoimmune inflammation. Exp Neurol. 2006;201(2):349–58.
116. Cassiani-Ingoni R, Muraro PA, Magnus T, et al. Disease progression after bone marrow transplantation in a model of multiple sclerosis is associated with chronic microglial and glial progenitor response. J Neuropathol Exp Neurol. 2007;66(7):637–49.
117. Magnus T, Carmen J, Deleon J, et al. Adult glial precursor proliferation in mutant SOD1G93A mice. Glia. 2008;56(2):200–8.
118. Magnus T, Coksaygan T, Korn T, et al. Evidence that nucleocytoplasmic Olig2 translocation mediates brain-injury-induced differentiation of glial precursors to astrocytes. J Neurosci Res. 2007;85(10): 2126–37.
119. Fukuda S, Kondo T, Takebayashi H, Taga T. Negative regulatory effect of an oligodendrocytic bHLH factor OLIG2 on the astrocytic differentiation pathway. Cell Death Differ. 2004;11(2):196–202.
120. Setoguchi T, Kondo T. Nuclear export of OLIG2 in neural stem cells is essential for ciliary neurotrophic factor-induced astrocyte differentiation. J Cell Biol. 2004;166(7):963–8.
121. Ligon KL, Kesari S, Kitada M, et al. Development of NG2 neural progenitor cells requires Olig gene function. Proc Natl Acad Sci USA. 2006;103(20): 7853–8.
122. Parras CM, Galli R, Britz O, et al. Mash1 specifies neurons and oligodendrocytes in the postnatal brain. EMBO J. 2004;23(22):4495–505.
123. Miyoshi G, Butt SJ, Takebayashi H, Fishell G. Physiologically distinct temporal cohorts of cortical interneurons arise from telencephalic Olig2-expressing precursors. J Neurosci. 2007;27(29): 7786–98.
124. Network TCGAR. Comprehensive genomic characterization defines human glioblastoma genes and core pathways. Nature. 2008;455(7216):1061–8.
125. Zhu Y, Guignard F, Zhao D, et al. Early inactivation of p53 tumor suppressor gene cooperating with NF1 loss induces malignant astrocytoma. Cancer Cell. 2005;8(2):119–30.
126. Stiles CD, Rowitch DH. Glioma stem cells: a midterm exam. Neuron. 2008;58(6):832–46.
127. Liu C, Sage JC, Miller MR, et al. Mosaic analysis with double markers reveals tumor cell of origin in glioma. Cell. 2011;146(2):209–21.
128. Zong H, Espinosa JS, Su HH, Muzumdar MD, Luo L. Mosaic analysis with double markers in mice. Cell. 2005;121(3):479–92.

129. Al-Mayhani MT, Grenfell R, Narita M, et al. NG2 expression in glioblastoma identifies an actively proliferating population with an aggressive molecular signature. Neuro Oncol. 2011;13(8):830–45.
130. Assanah M, Lochhead R, Ogden A, Bruce J, Goldman J, Canoll P. Glial progenitors in adult white matter are driven to form malignant gliomas by platelet-derived growth factor-expressing retroviruses. J Neurosci. 2006;26(25):6781–90.
131. Nazarenko I, Hedren A, Sjodin H, et al. Brain abnormalities and glioma-like lesions in mice overexpressing the long isoform of PDGF-A in astrocytic cells. PLoS One. 2011;6(4):e18303.

Stem Cells and Brain Cancer

4

Sara G.M. Piccirillo

Abstract

Cancer stem cells have been identified in human tumors. Initial findings in human leukemia suggested that tumors are organized hierarchically with a tumor-initiating rare cell population on the top of the hierarchy being responsible for tumor growth and metastasis.

In the last decade, these findings have been extended to several human cancers, and although several convincing results have been published, it is still controversial if cancer stem cells represent a rare, immutable sub-population perpetuating in the tumor or rather a functional state that many tumor cells can acquire.

Since the term "cancer stem cells" seems to imply that tumor cells derive from normal stem cells of the same tissue, the definition as "tumor-initiating cells" found a better consensus. The ability of being tumor-initiating cells is the most important feature of these cells, and this is evaluated by the formation of phenocopies of the original tumor in animal models.

Keywords

Central nervous system • Cancer stem cells • Tumorigenicity • Brain cancers • Glioblastoma

Introduction

Brain cancers are rare diseases but unfortunately they are responsible for 7 % of the years of life lost from cancer [1]. The commonest brain cancers are gliomas and among them the most aggressive form is the glioblastoma multiforme (GBM), which is characterized by a very poor prognosis [2] despite surgical resection, radiotherapy, and chemotherapy.

The infiltrative nature of GBM is one of the main features of this tumor. A complete resection

S.G.M. Piccirillo, Ph.D.
Department of Clinical Neurosciences, Cambridge Centre for Brain Repair, University of Cambridge, Forvie Site, Robinson Way, Cambridge, CB2 0PY, UK
e-mail: sp577@cam.ac.uk

C. Watts (ed.), *Emerging Concepts in Neuro-Oncology*,
DOI 10.1007/978-0-85729-458-6_4, © Springer-Verlag London 2013

Table 4.1 Summary of the cardinal features of somatic stem cells and tumor-initiating cells

Somatic stem cells	Tumor-initiating cells
Undifferentiated cells	Undifferentiated cells
Ability to self-maintain	Extensive ability to self-maintain
Ability to generate differentiated progeny	Ability to generate aberrant differentiated progeny
Flexible in differentiation fate choice	Flexible in differentiation fate choice
Capable of tissue homeostasis and repair	Tumor-initiating ability

of the tumor is not possible since GBM cells migrate from the tumor core and give rise to tumor recurrence [3]. In this regard, in the last decades, several cell and gene therapy approaches based on stem cells have been developed. These strategies imply the delivery of therapeutic molecules and the use of adult stem cells (in particular neural stem cells [NSCs]) as cellular vehicles since they show a marked tropism for tumor cells [4].

Although stem cell-based therapeutic approaches for human GBM require further investigation, they represent a valuable tool to design novel effective treatments for GBM patients.

Starting from 2001, a completely different perspective revolutionized the field of neuro-oncology. At that time, several groups demonstrated that cells carrying all the functional characteristics of stem cells constitute glioma and other brain tumors and that they are responsible for tumor growth. According to previous findings in other human cancers, these cells were called "cancer stem cells" (reviewed in [5]).

The field of "cancer stem cells," better defined as tumor-initiating cells (TICs), lies at the intersection of oncology and stem cell biology. This field developed from the striking parallels between tumor cells and somatic stem cells (Table 4.1). In the brain, the "no new neuron" dogma postulated by Santiago Ramon y Cajal in the beginning of the last century has been replaced by evidence of continuous neurogenesis that occurs even in adulthood. Based on this concept, it has been proposed that brain cancers can derive from abnormal proliferation of stem cells in the brain [6, 7].

In particular, in the newborn area of cancer stem cells, a lot of evidence came from human brain tumors, in particular from GBM and medulloblastoma (MDB) [5]. Similar to human leukemia, in 2004 it was originally proposed that GBM and MDB tumor-initiating cells can be identified by fluorescence-activated cell sorting (FACS) analysis using CD133 antigen [8]. This protein of unknown function has been used to identify NSCs, and it is mainly expressed on undifferentiated cell populations [9]. A subsequent characterization of CD133+ and CD133 – populations in GBM has then revealed that also the negative fraction is endowed with tumor-initiating activity [10–15], and it has been proposed that CD133+ can rather identify a pool of chemo- and radioresistant cell populations [16, 17] which secrete vascular endothelial growth factor (VEGF) and give rise to highly vascularized tumors in mouse models [18].

More recently, several other antigens have been proposed as a marker for a better identification of brain TICs, but unfortunately, there is no consensus on any of them. It has been therefore proposed to isolate these cells using side population (SP) [19, 20] or autofluorescent properties [21], but these attempts have not been yet adopted by the whole community working in this field and the use of SP has been recently challenged [22].

In the meanwhile, the existence of TICs has been exploited for therapeutic targeting, and their functional properties in vitro and in vivo have been recently complemented with genetics aimed at understanding their origin and clonal architecture in defined human tumors using tissue biopsies [23]. In this view, the challenge is to understand tumor development by evaluating genetic aberrations in a late stage of its growth. The use of primary tissue coming directly from the patients really restricts the possibilities of identifying driven mutations and the cell(s) of origin of the tumor.

Tracking tumor growth is only possible in genetically engineered mouse models [24]. Accordingly, a lot of effort has been put to develop transgenic or conditionally targeted gene technologies by combining oncogenes and tumor suppressors in different cell populations. By using mouse models, it has been shown that neurogenic regions of the brain such as the subventricular zone (SVZ) are more keen to tumor development after infusion of platelet-derived growth factor (PDGF) in that area [25, 26] and that tumors can be driven by tumor suppressor inactivation in neural stem/progenitor [27]. In support of all these findings, it has been demonstrated that p53 mutations preferentially occur in SVZ in mouse models [28]. This raises the question on whether TICs in the brain directly derive from NSCs which reside in neurogenic regions. Until now, the results coming from studies based on mouse models have indirectly indicated that this is the case but by using a transgenic cell-labeling system known as mosaic analysis with double markers (MADM) [29], Liu et al. have proposed that the cells of origin in GBM are oligodendrocyte precursor cells (OPCs). These findings have been reinforced by a recent study suggesting that a marker of OPCs, neuroglia-2 (NG2), identifies a highly proliferative cell subpopulation in human GBM [30]. Unfortunately, the use of mouse models for the study of brain cancers has been hampered by the lack of cell-type-specific promoters and the selection of a few of the whole set of mutations occurring in GBM [24].

Conversely, mouse models have proven indispensable in the identification of cells of origin in MDB. It has been demonstrated that two subtypes of MDB (either with constitutive hedgehog signaling or activating mutations in WNT pathway) have distinct cells of origin. In the first case, this has been identified as granule neuron precursor cells [31, 32], and in the latter, cells of the dorsal brainstem outside the cerebellum have been recently proposed as the cell of origin of this subtype [33].

Epidemiology and Classification of Brain Cancers

Each year in the United Kingdom, around 4,300 new cases of central nervous system (CNS) cancers are diagnosed, around 6 per 100,000 of general population.

Although brain cancers account for less than 2 % of all primary tumors, they are responsible for 7 % of the years of life lost from cancer before age 70 [1]. If the burden of disease is considered in terms of the average years of life lost per patient, brain cancers are one of the most lethal cancers with over 20 years of life lost [2]. The high rates of mortality make these rare cancers into the third leading cause of cancer-related death among economically active men between 15 and 54 years of age and the fourth leading cause of cancer-related death among economically active women between 15 and 34 years of age [34].

Gliomas constitute over 50 % of total primary CNS tumor cases and can be classified based on morphological criteria into astrocytic tumors, oligodendrogliomas, mixed oligoastrocytomas, and ependymomas [35].

GBM is classified as a neuroepithelial glial tumor and is considered to be the commonest type of primary brain tumor in adults. Median life expectancy in optimally managed patients is only 12–14 months [36].

Current clinical management of patients diagnosed with GBM involves a combination of surgery, radiotherapy, and chemotherapy. Radiotherapy has been the principle therapeutic modality since the late seventies [37]. Additional targeted chemotherapy is of only modest benefit and mainly in younger age group [36, 38]. The latest survival trends for patients with CNS malignancies have remained largely static with slight improvement in the last few years upon the introduction of temozolomide [36, 38]. This situation underlines the need for effective therapeutic treatments for patients with these cancers.

The Need for an In Vivo Model for Human Gliomas

The development of new markers, the identification of specific molecular targets, and the overall process of developing therapeutics for gliomas have been severely hampered by the lack of information as to the actual identity and nature of the normal cell type(s) that was hit by transformation and the consequent lack of animal models that may faithfully reproduce the occurrence growth, spreading, and recurrence of the human disease [24]. In fact, while animal brain tumor models have been widely used in experimental neuro-oncology, it is clear that no animal model which resembles human high-grade gliomas is available, to date.

Initial models of human GBMs were obtained using established tumor cell lines with specific genetic alterations or human tumor explants injected in rodents [39], and in the last years, new models were available by the generation of transgenic/knock-out mice that represent the common genetic alteration seen in human GBM (i.e., epidermal growth factor receptor (EGFR), Ink4a/Arf, Ras, Nf1, p53) [40]. Although by using these mouse models, it has become possible to understand which alterations/mutations support tumor growth, no model completely resembled the human pathology. Therefore, the development of effective therapies for GBM has been strongly affected [40].

It is a common belief that valid animal model for high-grade glioma should derive from glial-like cells, should be feasible to grow in vitro as stable cell lines and in vivo by serial transplantation, and should retain glioma-like growth features including infiltration, lack of encapsulation, migratory ability, neovascularization, and alteration of the blood–brain barrier. Furthermore, following tumor implantation, the survival time until death should be of sufficient duration so as to permit therapy [41].

In the last years, the establishment of a bona fide, reliable GBM model that can be used to assess and validate the efficacy of classical or innovative approaches for the diagnosis and cure of brain tumors has been obtained using TICs [7, 42]. These cells are extremely stable in culture and can produce tumors which resemble the original pathology (Fig. 4.1) to a much greater extent than any of the cell lines previously available. For instance, upon transplantation into the brain of nude mice, they give rise to lethal tumors, which display the general histology and typical infiltrating behavior of the parental neoplasia (Fig. 4.1). Such features have never been observed when using xenografts or allograft-based brain tumor models [7, 43].

Identification of Brain Tumor-Initiating Cells

Starting from the initial identification of cancer stem cells in nonsolid tumors, one decade ago the first experimental evidence of cells with stemlike features were reported in human GBM and then in human MDB, ependymoma, and neurocytoma (reviewed in [5]).

Up to date, the search for a marker of TICs in brain tumors has characterized the development of this field. In 2004, it was proposed that similar to the normal counterpart, stem cells in GBM and MDB could be identified by using CD133, a glycosylated 120-kDa protein with unknown function and originally identified in hematopoietic precursor cells [9]. It was suggested that the TICs reside in the CD133+ pool since only these cells were capable of initiate tumors in vivo when injected into immunosuppressed mice [8]. Recently, it has been reported by several groups that the CD133- cell population in GBM is tumorigenic [10–15], and other markers have been proposed (A2B5, CD15, L1CAM) [14, 44, 45]. Unfortunately, with regard to CD15 a subsequent direct comparative study has shown that CD15+ and CD15− cells in GBM are similar in proliferation and tumor-initiating activity [46].

Other strategies aimed at a definitive identification of brain TICs included functional assay based on neurosphere formation, side population, and cell autofluorescence. The first method found a lot of consensus and currently represents the best strategy to enrich in TICs and to develop in vivo models and therapeutic approaches based at targeting these cells both in vitro and in vivo [47]. More importantly, this method highlighted the similarities and differences between TICs and NSCs. Quite interestingly, it was found that similar to their normal counterpart, TICs are capable of growing in absence of serum and presence of epidermal growth factor (EGF) and fibroblast growth factor-2 (FGF-2); they can be expanded as

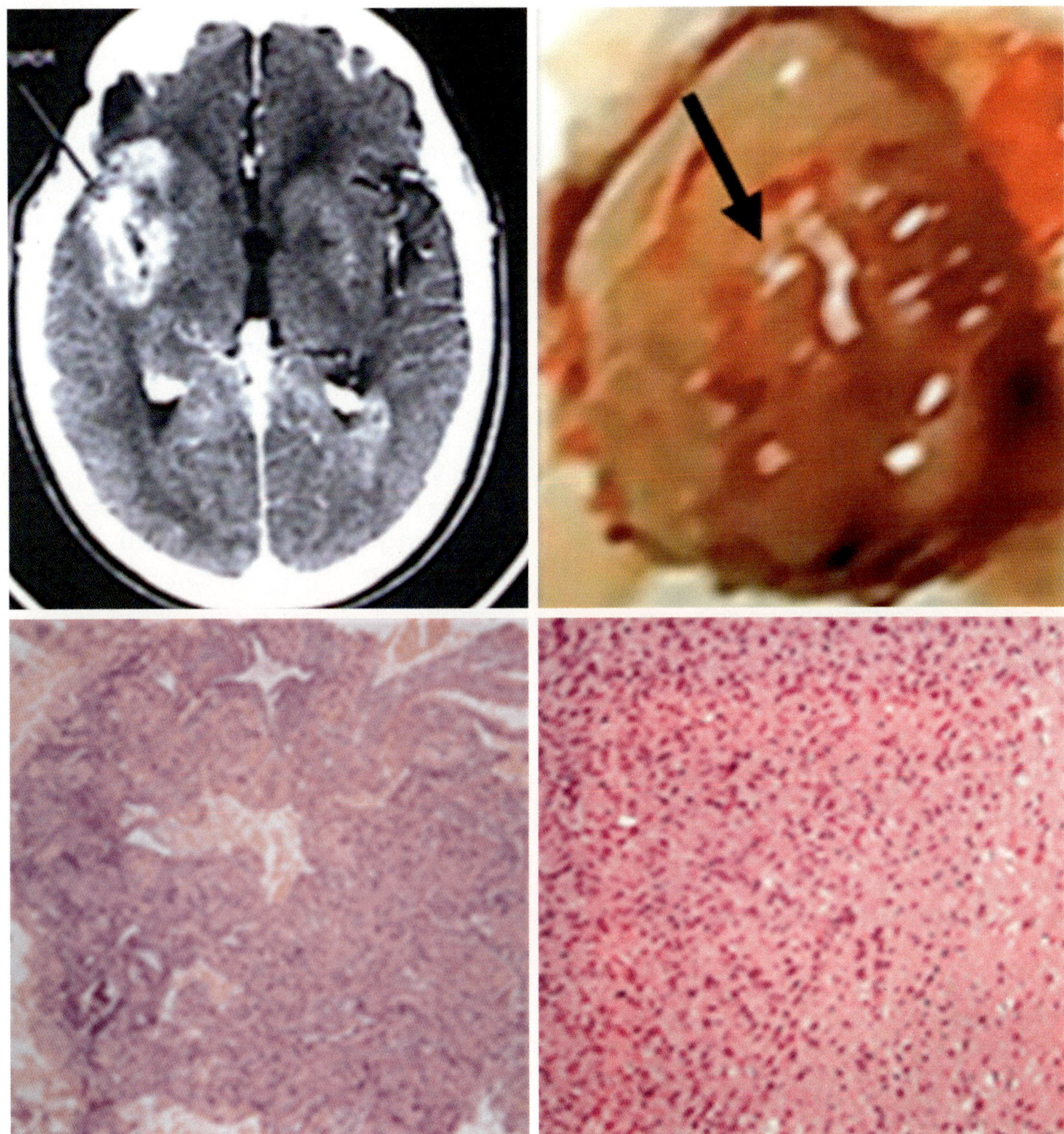

Fig. 4.1 *Left*, MRI of a GBM (*top*) and hematoxylin and eosin staining of the tumor (*bottom*, magnification 200×). *Right*, tumor formation in the brain of a Nod/Scid mouse is indicated by an *arrow* (*top*) and hematoxylin and eosin staining (*bottom*, magnification 200×) of note; tumor cells are capable of infiltrating the mouse brain

neurospheres (Fig. 4.2) and retain multipotency, that is the ability to generate all the three main neural lineages: neurons, astrocytes, and oligodendrocytes (Fig. 4.3) [7]. Conversely, it was reported that some TICs from GBM can generate mature progeny showing a coaberrant expression of markers of astrocytes and neurons [43], suggesting that these cells do not undergo a terminal differentiation. Another study reported that these cells can also be grown in absence of mitogens [48]. The above findings have been exploited to develop new therapeutic approaches targeting TICs in GBM and sparing NSCs.

Side population and cell autofluorescence were proposed more recently as a marker-independent alternative to all the previous approaches, and convincing results have been reported [19–21]. But very recently, a new study suggested

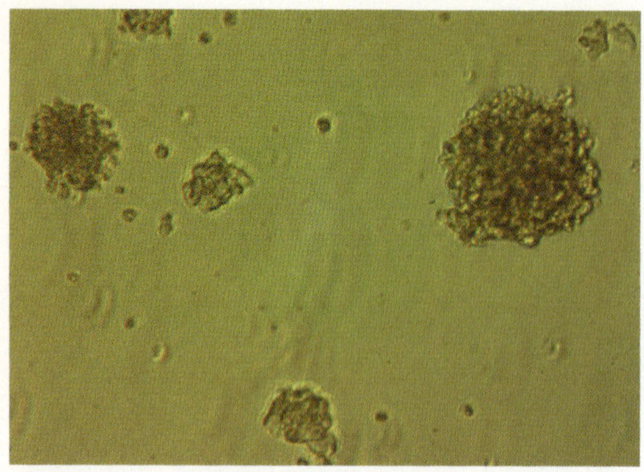

Fig. 4.2 Neurosphere formation of TICs isolated from human GBM. Cells can be extensively expanded in presence of mitogens (EGF and FGF-2) and absence of serum (magnification 100×)

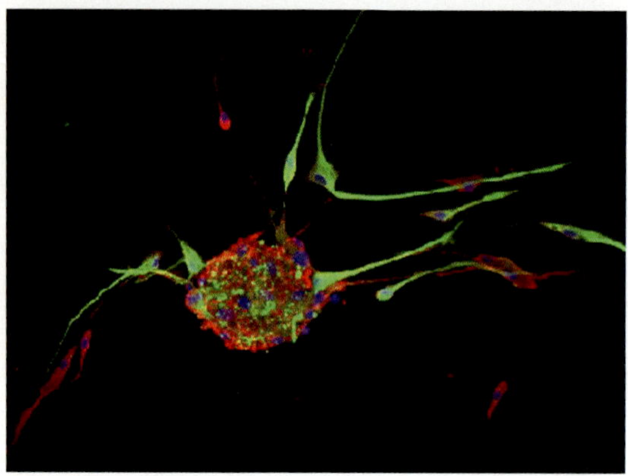

Fig. 4.3 Similar to NSCs, TICs from GBM are multipotent and can generate glial fibrillary acidic protein (GFAP)+cells (*green*, a marker of astrocytes) and neuron class III beta-tubulin (Tuj1)+cells (*red*, a marker of neurons). Cells have been counterstained with 4′-6-Diamidino-2-phenylindole (DAPI) (*blue*). Magnification 200×

that TICs do not reside in the SP both in GBM cell lines and primary GBM cells challenging the use of this method for enriching or isolating TICs [22].

The absence of a single marker or a combination of markers to definitely identify TICs in brain tumors can be ascribed to the cellular and molecular heterogeneity of tumor cells in GBM. In this view, recently it was demonstrated that distinct TIC populations exist in human GBM, suggesting that these cells can also be responsible for the high level of cell heterogeneity of GBM [23]. Although the cell(s) of origin in GBM has not been identified yet, in the last months some progress in this direction has been made. By using an elegant lineage tracing technique (MADM) in mice, Liu et al. demonstrated that only transformed NSCs giving rise to oligodendrocyte precursor cells (OPCs) developed in malignancies on the contrary, all the other neural lineages remained unaffected [29]. This suggests that TICs can derive from OPCs and not directly from NSCs, although the latter can represent the suitable cell substrate for the initiation of the malignant transformation [29].

Brain Tumor-Initiating Cells and Glioblastoma: Genetic Alterations Sustaining Tumor Growth and Progression

So far most of the focus and knowledge regarding TICs has been limited toward their identification, purification, and tumorigenicity, but there is a considerable need to explore and better understand the normal and perturbed regulatory pathways required for the maintenance and expansion of TICs to clarify which types of cells are the initially transformed cells and which signaling pathways are the driving transforming forces [7]. Considering that most of the tumors occur due to a dysregulation of epigenetic and genetic factors in particular cell types in a seed versus soil manner and that the tumor initiation and multistepped progression processes are hard to envisage for GBM, investigation of oncogenic and tumor suppressor pathways in TIC populations could provide new insights into the gliomagenesis, better prognosis prediction, and guidance for new treatment [7].

Common alterations in specific pathways that have been extensively studied and found in the majority of gliomas are:

1. TP53/p14^{ARF}: loss of or mutations in TP53 or alterations of a TP53 regulator (HDM2) are frequently observed in pediatric gliomas and low-grade and secondary high-grade adult gliomas [49].

   p14^{ARF} (p19^{ARF}in mouse) lies upstream of TP53 and negatively regulates HDM2 (MDM2 in mouse). $p14^{ARF}$ is frequently deleted or silenced in gliomas [50, 51]. TP53 is a master regulator of cell cycle arrest and apoptosis.

2. $p16^{INK4A}$/pRB: pRB restricts proliferation through the control of G1 phase cell cycle progression and is regulated by CDK4, which is blocked by CDK inhibitors $p16^{INK4A}$ and $p15^{INK4B}$. Loss of RB1 expression, amplification of CDK4, and the deletion or silencing of $p16^{INK4A}$ and $p15^{INK4B}$ occur in gliomas [50, 52].
3. Growth factor pathways: different growth factor pathways have been described in gliomas: the epidermal growth factor (EGF) [53], platelet-derived growth factor (PDGF) [54], insulin-like growth factor-1 (IGF1) [55–57], hepatocyte growth factor (HGF) [58], and vascular endothelial growth factor (VEGF) [59, 60]. Although this list might appear heterogeneous, the mentioned pathways share functional commonalities and converge onto three common downstream mediators: PI3K/AKT, RAS, and PKC. Importantly, the three intracellular points of convergence are in turn linked to the TP53/$p14^{ARF}$ and pRB pathways [61, 62].
4. c-MYC pathway: transcription factors of the MYC family mediate the effect of growth factors on cell cycle progression, mainly through activation of the pRB pathway. C-MYC is found overexpressed in many gliomas [63–65]. In several cases, the overexpression can be explained by an amplification of the respective genomic locus [63, 65], inactivation of MYC pathway antagonists, or prolongation of MYC half-life. Upregulation of c-Myc prevents G_1/G_0 cell cycle arrest in glioma cells, and MYC was identified as a central network component influencing many other downstream genes specifically expressed in glioma [65].

Interestingly, preliminary evidence suggests that several of the pathways perturbed in GBM also control the maintenance of NSCs through the regulation of self-renewal cell divisions, differentiation, and apoptosis [66]. For example, the Polycomb group protein BMI1 that influences the TP53/$p14^{ARF}$ and the $p16^{INK4A}$/pRB pathways by repressing $p16^{INK4A}$, $p19^{ARF}$i (human $p14^{ARF}$), and p21 [67, 68] is important for the self-renewal of embryonic and postnatal NSCs [68, 69] and was found to be expressed in cultured TICs from GBM [70]. Thus, it is important to understand pathway differences, which distinguish NSCs from TICs, and whose interference would block uncoupled tumorigenic growth while ideally not affecting NSCs.

To achieve this knowledge, architectures of critical pathways with respect to self-regulatory components and robustness need to be established. So far, targeted anticancer drugs typically interfere at upstream levels of signal transduction pathways, for example, at growth factor receptors, or at components which are only immediately downstream of receptor complexes. However, many therapies targeting such pathway components have not lead to a significant therapeutic benefit, and it has become clear that cancer cells often bypass the inhibition of upstream pathway components [71].

Further, even when the treatments developed prove to be of considerable clinical benefit [72–74], it has become obvious that TICs themselves remain resistant to these treatments [75–77].

New Therapeutic Approaches Targeting Brain Tumor-Initiating Cells

The original definition of TICs implies the ability of these cells to escape or to resist the conventional treatments aimed at blocking the tumor growth [78, 79].

According to this view, in human GBM, it was demonstrated that cells identified by CD133 were chemo- and radioresistant and that they are capable of activating the DNA repair machinery more efficiently than the corresponding negative fraction in response to ionizing radiation [16, 17].

In the same period, it was suggested a different approach aimed at exploiting one of the features of TICs, that is, to undergo differentiation as for their normal counterpart. In particular, it was proposed to inhibit the self-renewal of cancer stem cells by inducing a differentiation program with the bone morphogenetic proteins (BMPs) which are capable of inducing differentiation along the astrocytic lineage. This strategy was successful both in vitro and in vivo by using different experimental approaches: in vitro pretreatment of the cells with BMPs and in vivo delivery of BMPs in orthotopic injection of TICs. More importantly, the same findings were extended also to primary cells from human GBM [47].

Since GBM is a highly vascularized tumor, also antiangiogenic therapies aimed at perturbing the VEGF axis have been investigated. Initially, Bao et al. reported that cancer stem cells secrete VEGF and that they respond to the anti-VEGF antibody, bevacizumab, both in vitro and in vivo [18].

These findings were reinforced by the demonstration that brain TICs reside closely to endothelial cells and that their self-renewal and tumor-initiating ability can increase in presence of endothelial cells or blood vessels [80]. More recently it has been demonstrated not only that TICs are in proximity of endothelial cells in GBM, MDB, ependymoma, and oligodendroglioma [80] but that they can generate endothelial cells both in vitro and in vivo [81, 82].

In this perspective, antiangiogenic therapies should take into account the ability of TICs to transdifferentiate.

Furthermore, since it is well known that extrinsic interaction with the tumor microenvironment may promote therapeutic resistance [83], the next challenge for the development of new therapies will be to improve our understanding of CSCs' interaction with the surrounding microenvironment.

Future Perspectives

The CSC hypothesis is in continuous evolution. The last decade has been very important for the identification of these cells in different brain tumors and for the first description of new therapeutic approaches targeting these cells. What remains unclear, however, is to understand how these cells can be responsible for the high degree of cellular and molecular heterogeneity of brain tumors and how it is possible to identify the cell(s) of origin of these diseases. Although mouse model and human tumor tissue can partially answer these questions, the scenario seems to be much more complicated [83] and the absence of a reliable marker for a definitive identification of brain cancer stem cells really affect the development of the field in this direction.

The field has moved forward to a second phase where it seems important to investigate the molecular mechanisms and the driven mutations which lead to tumor initiation and formation [23]. This will complement studies aimed at characterizing the phenotype of cancer stem cells which will take into account the recent findings suggesting that cancer stem cells are not so rare and they might simply represent a surviving fraction of TICs evading the immune system in mouse models with partial (*nu/nu* and NOD/SCID) or more pronounced (NOD/SCID IL2R [gamma]^{null}) immune suppression.

Conclusions

Stem cell-based therapies for brain cancers have been extensively investigated. Different adult stem cells have been used for cell therapy or as vehicle for gene therapy. Some of these approaches are advanced and clinical studies are ongoing.

From a complete different perspective, the newborn area of cancer stem cells in brain cancers has invigorated the field of neuro-oncology and hopefully will provide new insights into the development of curative treatments. Similar to all the new born areas in scientific research, emerging concepts associated with cancer stem cells are rapidly evolving and challenged. Although this field will require extensive investigation for the next years, more than 450 original papers have been published in the last decade. It is therefore an exciting time for challenging devastating diseases such as brain cancer and to develop new therapeutic approaches that can improve survival and quality of life for patients.

References

1. Office for National Statistics. Cancer statistics registrations: registrations of cancer diagnosed in 2008, England. (PDF 544KB) Series MB1 no.39. London: National Statistics; 2010.
2. Burnet NG, Jefferies SJ, Benson RJ, Hunt DP, Treasure FP. Years of life lost (YLL) from cancer is an

important measure of population burden – and should be considered when allocating research funds. Br J Cancer. 2005;92(2):241–5.
3. Denysenko T, Gennero L, Roos MA. Glioblastoma cancer stem cells: heterogeneity, microenvironment and related therapeutic strategies. Cell Biochem Funct. 2010;28(5):343–51.
4. Tabatabai G, Wick W, Weller M. Stem cell-mediated gene therapies for malignant gliomas: a promising targeted therapeutic approach? Discov Med. 2011;11(61): 529–36.
5. Piccirillo SG, Binda E, Fiocco R, Vescovi AL, Shah K. Brain cancer stem cells. J Mol Med. 2009;87(11): 1087–95.
6. Huntly BJ, Gilliland DG. Leukaemia stem cells and the evolution of cancer-stem-cell research. Nat Rev Cancer. 2005;5(4):311–21.
7. Vescovi AL, Galli R, Reynolds BA. Brain tumour stem cells. Nat Rev Cancer. 2006;6(6):425–36.
8. Singh SK, Hawkins C, Clarke ID, et al. Identification of human brain tumour initiating cells. Nature. 2004;432(7015):396–401.
9. Miraglia S, Godfrey W, Yin AH, et al. A novel five-transmembrane hematopoietic stem cell antigen: isolation, characterization, and molecular cloning. Blood. 1997;90(12):5013–21.
10. Beier D, Hau P, Proescholdt M, et al. CD133(+) and CD133(−) glioblastoma-derived cancer stem cells show differential growth characteristics and molecular profiles. Cancer Res. 2007;67(9):4010–5.
11. Wang J, Sakariassen PO, Tsinkalovsky O, et al. CD133 negative glioma cells form tumors in nude rats and give rise to CD133 positive cells. Int J Cancer. 2008;122(4):761–8.
12. Gunther HS, Schmidt NO, Phillips HS, et al. Glioblastoma-derived stem cell-enriched cultures form distinct subgroups according to molecular and phenotypic criteria. Oncogene. 2008;27(20): 2897–909.
13. Joo KM, Kim SY, Jin X, et al. Clinical and biological implications of CD133-positive and CD133-negative cells in glioblastomas. Lab Invest. 2008;88(8): 808–15.
14. Ogden AT, Waziri AE, Lochhead RA. Identification of A2B5+CD133− tumor-initiating cells in adult human gliomas. Neurosurgery. 2008;62(2):505–14, discussion 514–5.
15. Yi JM, Tsai HC, Glockner SC, et al. Abnormal DNA methylation of CD133 in colorectal and glioblastoma tumors. Cancer Res. 2008;68(19):8094–103.
16. Liu G, Yuan X, Zeng Z, et al. Analysis of gene expression and chemoresistance of CD133+ cancer stem cells in glioblastoma. Mol Cancer. 2006;5:67.
17. Bao S, Wu Q, McLendon RE, et al. Glioma stem cells promote radioresistance by preferential activation of the DNA damage response. Nature. 2006;444(7120): 756–60.
18. Bao S, Wu Q, Sathornsumetee S, et al. Stem cell-like glioma cells promote tumor angiogenesis through vascular endothelial growth factor. Cancer Res. 2006;66(16):7843–8.
19. Harris MA, Yang H, Low BE, et al. Cancer stem cells are enriched in the side population cells in a mouse model of glioma. Cancer Res. 2008;68(24):10051–9.
20. Bleau AM, Huse JT, Holland EC. The ABCG2 resistance network of glioblastoma. Cell Cycle. 2009;8(18):2936–44.
21. Clement V, Marino D, Cudalbu C, et al. Marker-independent identification of glioma-initiating cells. Nat Methods. 2010;7(3):224–8.
22. Broadley KW, Hunn MK, Farrand KJ. Side population is not necessary or sufficient for a cancer stem cell phenotype in glioblastoma multiforme. Stem Cells. 2011;29(3):452–61.
23. Piccirillo SG, Combi R, Cajola L, et al. Distinct pools of cancer stem-like cells coexist within human glioblastomas and display different tumorigenicity and independent genomic evolution. Oncogene. 2009;28(15):1807–11.
24. Visvader JE. Cells of origin in cancer. Nature. 2011;469(7330):314–22.
25. Jackson EL, Garcia-Verdugo JM, Gil-Perotin S, et al. PDGFR alpha-positive B cells are neural stem cells in the adult SVZ that form glioma-like growths in response to increased PDGF signaling. Neuron. 2006;51(2):187–99.
26. Assanah MC, Bruce JN, Suzuki SO, Chen A, Goldman JE, Canoll P. PDGF stimulates the massive expansion of glial progenitors in the neonatal forebrain. Glia. 2009;57(16):1835–47.
27. Alcantara Llaguno S, Chen J, Kwon CH. Malignant astrocytomas originate from neural stem/progenitor cells in a somatic tumor suppressor mouse model. Cancer Cell. 2009;15(1):45–56.
28. Wang Y, Yang J, Zheng H, et al. Expression of mutant p53 proteins implicates a lineage relationship between neural stem cells and malignant astrocytic glioma in a murine model. Cancer Cell. 2009;15(6): 514–26.
29. Liu C, Sage JC, Miller MR. Mosaic analysis with double markers reveals tumor cell of origin in glioma. Cell. 2011;146(2):209–21.
30. Al-Mayhani MT, Grenfell R, Narita M. NG2 expression in glioblastoma identifies an actively proliferating population with an aggressive molecular signature. Neuro Oncol. 2011;13(8):830–45.
31. Schuller U, Heine VM, Mao J, et al. Acquisition of granule neuron precursor identity is a critical determinant of progenitor cell competence to form Shh-induced medulloblastoma. Cancer Cell. 2008;14(2):123–34.
32. Yang ZJ, Ellis T, Markant SL, et al. Medulloblastoma can be initiated by deletion of Patched in lineage-restricted progenitors or stem cells. Cancer Cell. 2008;14(2):135–45.
33. Gibson P, Tong Y, Robinson G. Subtypes of medulloblastoma have distinct developmental origins. Nature. 2010;468(7327):1095–9.
34. Kesari S, Schiff D, Henson JW, et al. Phase II study of temozolomide, thalidomide, and celecoxib for newly diagnosed glioblastoma in adults. Neuro Oncol. 2008;10(3):300–8.

35. Louis DN, Ohgaki H, Wiestler OD, et al. The 2007 WHO classification of tumours of the central nervous system. Acta Neuropathol. 2007;114(2): 97–109.
36. Stupp R, Mason WP, van den Bent MJ, et al. Radiotherapy plus concomitant and adjuvant temozolomide for glioblastoma. N Engl J Med. 2005; 352(10):987–96.
37. Walker MD, Alexander Jr E, Hunt WE. Evaluation of BCNU and/or radiotherapy in the treatment of anaplastic gliomas. A cooperative clinical trial. J Neurosurg. 1978;49(3):333–43.
38. Hegi ME, Diserens AC, Gorlia T, et al. MGMT gene silencing and benefit from temozolomide in glioblastoma. N Engl J Med. 2005;352(10):997–1003.
39. Pieper RO. Defined human cellular systems in the study of glioma development. Front Biosci. 2003;8:s 19–27.
40. Gutmann DH, Baker SJ, Giovannini M, Garbow J, Weiss W. Mouse models of human cancer consortium symposium on nervous system tumors. Cancer Res. 2003;63(11):3001–4.
41. Barth RF. Rat brain tumor models in experimental neuro-oncology: the 9L, C6, T9, F98, RG2 (D74), RT-2 and CNS-1 gliomas. J Neurooncol. 1998;36(1): 91–102.
42. Piccirillo SG, Vescovi AL. Brain tumour stem cells: possibilities of new therapeutic strategies. Expert Opin Biol Ther. 2007;7(8):1129–35.
43. Galli R, Binda E, Orfanelli U, et al. Isolation and characterization of tumorigenic, stem-like neural precursors from human glioblastoma. Cancer Res. 2004;64(19):7011–21.
44. Son MJ, Woolard K, Nam DH, Lee J, Fine HA. SSEA-1 is an enrichment marker for tumor-initiating cells in human glioblastoma. Cell Stem Cell. 2009;4(5):440–52.
45. Bao S, Wu Q, Li Z, et al. Targeting cancer stem cells through L1CAM suppresses glioma growth. Cancer Res. 2008;68(15):6043–8.
46. Patru C, Romao L, Varlet P. CD133, CD15/SSEA-1, CD34 or side populations do not resume tumor-initiating properties of long-term cultured cancer stem cells from human malignant glio-neuronal tumors. BMC Cancer. 2010;10:66.
47. Piccirillo SG, Reynolds BA, Zanetti N, et al. Bone morphogenetic proteins inhibit the tumorigenic potential of human brain tumour-initiating cells. Nature. 2006;444(7120):761–5.
48. Kelly JJ, Stechishin O, Chojnacki A, et al. Proliferation of human glioblastoma stem cells occurs independently of exogenous mitogens. Stem Cells. 2009;27(8):1722–33.
49. Zhu Y, Guignard F, Zhao D, et al. Early inactivation of p53 tumor suppressor gene cooperating with NF1 loss induces malignant astrocytoma. Cancer Cell. 2005;8(2):119–30.
50. Simon M, Koster G, Menon AG, Schramm J. Functional evidence for a role of combined CDKN2A (p16-p14(ARF))/CDKN2B (p15) gene inactivation in malignant gliomas. Acta Neuropathol. 1999;98(5): 444–52.
51. Ichimura K, Bolin MB, Goike HM, Schmidt EE, Moshref A, Collins VP. Deregulation of the p14ARF/MDM2/p53 pathway is a prerequisite for human astrocytic gliomas with G1-S transition control gene abnormalities. Cancer Res. 2000;60(2):417–24.
52. Buschges R, Weber RG, Actor B, Lichter P, Collins VP, Reifenberger G. Amplification and expression of cyclin D genes (CCND1, CCND2 and CCND3) in human malignant gliomas. Brain Pathol. 1999;9(3):435–42, discussion 432–3.
53. Nicholas MK, Lukas RV, Jafri NF, Faoro L, Salgia R. Epidermal growth factor receptor-mediated signal transduction in the development and therapy of gliomas. Clin Cancer Res. 2006;12(24):7261–70.
54. Feldkamp MM, Lau N, Guha A. Signal transduction pathways and their relevance in human astrocytomas. J Neurooncol. 1997;35(3):223–48.
55. Hirano H, Lopes MB, Laws Jr ER, et al. Insulin-like growth factor-1 content and pattern of expression correlates with histopathologic grade in diffusely infiltrating astrocytomas. Neuro Oncol. 1999;1(2): 109–19.
56. Chakravarti A, Loeffler JS, Dyson NJ. Insulin-like growth factor receptor I mediates resistance to anti-epidermal growth factor receptor therapy in primary human glioblastoma cells through continued activation of phosphoinositide 3-kinase signaling. Cancer Res. 2002;62(1):200–7.
57. Trojan J, Cloix JF, Ardourel MY, Chatel M, Anthony DD. Insulin-like growth factor type I biology and targeting in malignant gliomas. Neuroscience. 2007;145(3):795–811.
58. Abounader R, Laterra J. Scatter factor/hepatocyte growth factor in brain tumor growth and angiogenesis. Neuro Oncol. 2005;7(4):436–51.
59. Maity A, Pore N, Lee J, Solomon D, O'Rourke DM. Epidermal growth factor receptor transcriptionally up-regulates vascular endothelial growth factor expression in human glioblastoma cells via a pathway involving phosphatidylinositol 3′-kinase and distinct from that induced by hypoxia. Cancer Res. 2000;60(20):5879–86.
60. Pore N, Liu S, Haas-Kogan DA, O'Rourke DM, Maity A. PTEN mutation and epidermal growth factor receptor activation regulate vascular endothelial growth factor (VEGF) mRNA expression in human glioblastoma cells by transactivating the proximal VEGF promoter. Cancer Res. 2003;63(1):236–41.
61. Shen L, Glazer RI. Induction of apoptosis in glioblastoma cells by inhibition of protein kinase C and its association with the rapid accumulation of p53 and induction of the insulin-like growth factor-1-binding protein-3. Biochem Pharmacol. 1998;55(10):1711–9.
62. Narita Y, Nagane M, Mishima K, Huang HJ, Furnari FB, Cavenee WK. Mutant epidermal growth factor receptor signaling down-regulates p27 through activation of the phosphatidylinositol 3-kinase/Akt pathway in glioblastomas. Cancer Res. 2002;62(22):6764–9.

63. Trent J, Meltzer P, Rosenblum M, et al. Evidence for rearrangement, amplification, and expression of c-myc in a human glioblastoma. Proc Natl Acad Sci USA. 1986;83(2):470–3.
64. Engelhard HH, Butler AB, Bauer KD. Quantification of the c-myc oncoprotein in human glioblastoma cells and tumor tissue. J Neurosurg. 1989;71(2): 224–32.
65. Bredel M, Bredel C, Juric D, et al. Functional network analysis reveals extended gliomagenesis pathway maps and three novel MYC-interacting genes in human gliomas. Cancer Res. 2005;65(19):8679–89.
66. Pardal R, Clarke MF, Morrison SJ. Applying the principles of stem-cell biology to cancer. Nat Rev Cancer. 2003;3(12):895–902.
67. Bracken AP, Kleine-Kohlbrecher D, Dietrich N, et al. The Polycomb group proteins bind throughout the INK4A-ARF locus and are disassociated in senescent cells. Genes Dev. 2007;21(5):525–30.
68. Fasano CA, Dimos JT, Ivanova NB, Lowry N, Lemischka IR, Temple S. shRNA knockdown of Bmi-1 reveals a critical role for p21-Rb pathway in NSC self-renewal during development. Cell Stem Cell. 2007;1(1):87–99.
69. Molofsky AV, Pardal R, Iwashita T, Park IK, Clarke MF, Morrison SJ. Bmi-1 dependence distinguishes neural stem cell self-renewal from progenitor proliferation. Nature. 2003;425(6961):962–7.
70. Hemmati HD, Nakano I, Lazareff JA, et al. Cancerous stem cells can arise from pediatric brain tumors. Proc Natl Acad Sci USA. 2003;100(25):15178–83.
71. Piloto O, Wright M, Brown P, Kim KT, Levis M, Small D. Prolonged exposure to FLT3 inhibitors leads to resistance via activation of parallel signaling pathways. Blood. 2007;109(4):1643–52.
72. Kantarjian H, Sawyers C, Hochhaus A, et al. Hematologic and cytogenetic responses to imatinib mesylate in chronic myelogenous leukemia. N Engl J Med. 2002;346(9):645–52.
73. O'Brien SG, Guilhot F, Larson RA, et al. Imatinib compared with interferon and low-dose cytarabine for newly diagnosed chronic-phase chronic myeloid leukemia. N Engl J Med. 2003;348(11):994–1004.
74. Li Y, Zhu Z. FLT3 antibody-based therapeutics for leukemia therapy. Int J Hematol. 2005;82(2):108–14.
75. Michor F, Hughes TP, Iwasa Y, et al. Dynamics of chronic myeloid leukaemia. Nature. 2005;435(7046): 1267–70.
76. Copland M, Hamilton A, Elrick LJ, et al. Dasatinib (BMS-354825) targets an earlier progenitor population than imatinib in primary CML but does not eliminate the quiescent fraction. Blood. 2006;107(11):4532–9.
77. Ritchie E, Nichols G. Mechanisms of resistance to imatinib in CML patients: a paradigm for the advantages and pitfalls of molecularly targeted therapy. Curr Cancer Drug Targets. 2006;6(8):645–57.
78. Reya T, Morrison SJ, Clarke MF, Weissman IL. Stem cells, cancer, and cancer stem cells. Nature. 2001;414(6859):105–11.
79. Dean M, Fojo T, Bates S. Tumour stem cells and drug resistance. Nat Rev Cancer. 2005;5(4):275–84.
80. Calabrese C, Poppleton H, Kocak M, et al. A perivascular niche for brain tumor stem cells. Cancer Cell. 2007;11(1):69–82.
81. Ricci-Vitiani L, Pallini R, Biffoni M, et al. Tumour vascularization via endothelial differentiation of glioblastoma stem-like cells. Nature. 2010;468(7325):824–8.
82. Wang R, Chadalavada K, Wilshire J, et al. Glioblastoma stem-like cells give rise to tumour endothelium. Nature. 2010;468(7325):829–33.
83. Rosen JM, Jordan CT. The increasing complexity of the cancer stem cell paradigm. Science. 2009;324(5935):1670–3.

Part II

Models of Brain Cancer

5 In Vitro Models of Brain Cancer

David J. Ryan and Colin Watts

Abstract

Glioblastoma multiforme (GBM) is the most common primary brain malignancy in adults. Despite continuing advances in surgical treatment and combined chemoradiotherapy, little improvement in overall median survival has been seen. Therapeutic advances in neuro-oncology are likely to arise through the systematic dissection of the fascinating tumor biology that exists in GBM. If we are to tackle questions such as "What is the cell of origin in brain cancer?" and "How do these cells evade standard treatment methods and ultimately identify the Achilles' heel of this aggressive disease?" a scientific prerequisite is the availability of a robust and reliable in vitro model of glioma. What follows in this chapter is a discussion of the current state of knowledge in the generation of in vitro models of glioblastoma. Past and current models will be considered with their advantages and shortcomings highlighted. We will discuss the principles of in vitro cytotoxic assays and how translatable therapies emerge from this approach. We discuss the cell of mutation and cell of origin in GBM and how modeling oncogenic transformation can shed new light on this controversial topic. This chapter, we hope, will function as a timeline in the evolution of in vitro models of brain cancer. It illustrates how far we have come in our understanding of brain cancer but additionally highlights the barriers we face and must overcome to ensure that a cure remains within sight.

D.J. Ryan, MB, BCh, BAO, BMedSc, MRCS (Ireland)
Mouse Cancer Genetics,
Wellcome Trust Sanger Institute,
Cambridge, UK

C. Watts, MBBS, Ph.D. (Cantab), FRCS (SN) (✉)
Department of Clinical Neurosciences,
University of Cambridge,
Hills Road, Cambridge, CB2 0QQ, UK
e-mail: cw209@cam.ac.uk

C. Watts (ed.), *Emerging Concepts in Neuro-Oncology*,
DOI 10.1007/978-0-85729-458-6_5, © Springer-Verlag London 2013

Keywords
Glioblastoma • In vitro models • Cell derivation • Cell propagation • Neurospheres • Monolayers • Cytotoxic assays • Radioresistance

Introduction

Glioblastoma multiforme (GBM) is the most common primary brain malignancy in adults. With a median survival of 14 months from time of diagnosis, it has the second poorest survival of all cancers. With such a dismal outcome using the current standard of combined surgery and chemoradiotherapy, we have gladly returned to the bench in search of this tumor's Achilles' heel.

In order to study glioma in vitro, a scientific prerequisite is the availability of robust, reliable, and repeatable experimental protocols that facilitate efficient derivation of the glioma line from the primary tissue specimen. Following derivation, the cell line must be capable of sustained passage in vitro and, of paramount importance, must retain the characteristic cytogenetic profile that is representative of the parental tumor. What follows in this chapter is a discussion of the current state of knowledge in the generation of in vitro models of glioblastoma. Past and current models will be considered with their advantages and shortcomings highlighted. Concluding remarks will consider the future of cancer modeling in central nervous system (CNS) tumors.

In Vitro Culture of Glioblastoma

The Past: Use of Serum in Glioma Culture

Traditional methods of culturing animal cells including cancer cell lines relied on the use of serum culture conditions. Indeed, serum-containing media was used in glioma culture for many years. However, in the past two decades, its use came into question with the observation that the in vitro behavior of these tumor cells was grossly divergent from that of the parental tumor. An interesting early observation by Bigner et al. [1] was that in serum culture from derivation until establishment of a permanent glioblastoma line in vitro, the karyotype of these cells became markedly different from that of the primary tumor. Between passages 12–30, the original diploid tumor line became tetraploid and showed significant chromosomal rearrangements. Two conclusions for this finding were firstly it may represent selection and divergent clonal evolution in vitro. If so, then serious questions would have to be asked of the reliability of this method of in vitro modeling of glioma. What would be the impact of a doubling in the chromosomal number? Would this affect the in vitro biology that we observe? With such divergence from the parental tumor, how could we translate any observed therapeutic opportunities?

The second explanation for the karyotypal abnormalities was that it could represent chromosomal progression, which was in fact reproducing the natural history of the disease. Lacking the ability to serially biopsy glioma, a definitive explanation could not be offered.

Beyond the aforementioned chromosomal abnormalities, these traditional serum-cultured glioma lines often lacked a cardinal property of a tumor cell line, which is tumorgenicity in vivo. Furthermore, even when established in vivo, the histology was not comparable to that of the primary tumor as the tumors were noninfiltrative [2]. Understanding of the association between serum-cultured glioma and altered tumor biology was evident in a comparative in vivo tumorgenicity assessment study. In this experiment, the authors compared serum-cultured U-87 versus organotypic spheroids cultured directly from the primary tissue in vitro for 11–18 days prior to intracerebral injection. At histological examination, extensive white matter tract infiltration and cell migration was seen in the precultured organotypic spheroids. The U-87 glioma line, which was maintained in serum culture long term, did demonstrate tumor

Table 5.1 Biological similarity between normal neural stem cells and stemlike glioma cells

	Neural stem cells	Stemlike glioma cells
Neural stem cell markers (Nestin, SOX2, etc.)	Yes	Yes
Self-renewal	Yes	Yes
Proliferative potential	Yes	Yes
Multipotency	Yes	Yes
Migratory	Yes	Yes
Genomic alteration	No	Yes
Tumorgenicity	No	Yes

Observations that recognized the similarities between neural stem cells and stemlike glioma cells were significant findings which resulted in the development of serum-free derivation of glioma lines. Both of these cell types have the capacity for self-renewal, migration, and differentiation. Significant genomic alterations (TCGA) [6] and tumorgenicity distinguish the two cell types

formation following engraftment. However, the tumor was localized and compressive with no infiltrative margin being evident [3]. As you will see, it was not until the advent of serum-free derived culture for glioma that we came to realize just how poorly representative serum-cultured glioma was of the in vivo situation.

The Present: Serum-Free Glioma Culture

What finally led to the move away from serum? Astute observations that recognized the similarities between neural stem cells and glioma cells in situ were a defining moment in this transition. Both of these cell types have the capacity for self-renewal, migration, and differentiation [4, 5]. Of course what distinguishes these cells is that stem-like glioma cells are tumorgenic (Table 5.1) [6]. Assuming that there was this biological similarity between normal neural stem cells and stem-like glioma cells, the question centered on the difference in in vitro maintenance of these cell types. Glioma as we have seen was cultured in serum; however, neural stem cells were propagated and maintained in serum-free conditions as serum induces differentiation [7].

Culturing glioma in serum-free media in the presence of epidermal growth factor (EGF) and basic fibroblast growth factor (bFGF) produced an in vitro model whose genotype and phenotype was now more representative of the primary tumor [8]. Gliomas cultured in serum-free conditions in contrast to the standard serum conditions were self-renewing, multipotent, clonogenic, tumorigenic at all passages, and had gene expression profiles similar to the primary glioblastoma. Furthermore, in contrast to serum maintained lines whose in vivo tumorgenicity was nonrepresentative of the pathological process as previously described, serum-free cultured glioma produced tumors that were a phenocopy of the parental tumor.

Serum-free culture clearly represented a significant advance in neuro-oncology by providing an in vitro culture system that could maintain the unique biology of the individual patient's tumor. This created a tremendous opportunity to not only advance our knowledge of the intrinsic tumor biology but, by doing so, create novel personalized therapeutic strategies.

A third benefit of serum-free conditions is that it allowed us to retrospectively interpret the classic karyotype abnormalities that were seen in serum-cultured glioma. We know that the proliferation index in serum-free cultured glioma remains constant, but in serum culture, there is an initial period of limited growth followed by a plateau and subsequent exponential growth [8]. This observation in conjunction with the described work of Bigner and Fine allows reasonable speculation that this abnormal karyotype could be explained by in vitro selection and divergent clonal evolution producing a tumor that was no longer representative of the parental specimen.

Derivation and Propagation in Serum-Free Culture

Although serum-free culture was an improvement over traditional methods, it was not without its own shortcomings. Firstly, the efficiency of derivation and establishment of a stable line in these conditions was circa 50 % [9]. The derivation protocol relied on neurosphere formation in the presence of EGF and bFGF which was adapted from the work of Gritti et al. who derived

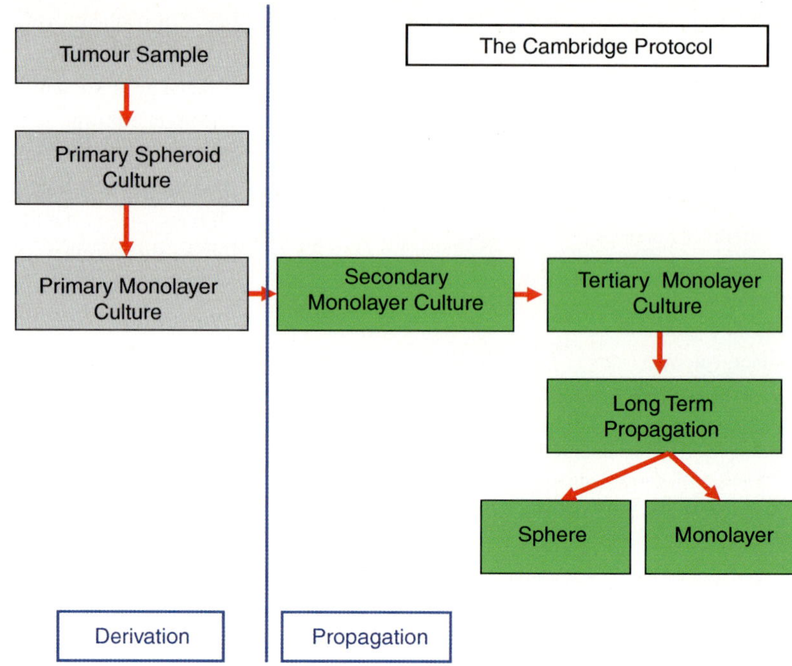

Fig. 5.1 The Cambridge protocol. The Cambridge protocol, by combining spheroid and monolayer culture, produced a 100 % derivation rate from fresh tumor samples. Propagation was successful in 92 % of samples. The established cell lines could be propagated and maintained long term in suspension or adherent culture. On comparative genomic hybridization (CGH) analysis, the derived lines preserved the characteristic cytogenetic profile of the parental tumor (Adapted and reproduced from Fael Al-Mayhani et al. [11]; with permission)

multipotent neural stem cells from murine brain in the presence of bFGF [10]. Unfortunately, in these conditions, the majority of cells did not survive beyond the first or second split.

A significant advance in serum-free derivation came with the refinement of the technique to include primary spheroid culture and primary monolayer culture in the derivation phase [11]. This modification, known as the Cambridge protocol (Fig. 5.1), produced successful derivation in 100 % of samples. Furthermore, on comparative genomic hybridization (CGH) analysis, the derived cell lines conserved the characteristic cytogenetic profile of the original tumor which included gains of chromosome 7 and monosomy of chromosome 10.

The Cambridge protocol was an improvement on previous methods and produced an in vitro model system that could be used to evaluate response to therapeutic strategies, specifically vascular endothelial growth factor (VEGF) receptor antagonism [12]. However, the failure of the derived lines to exhibit epidermal growth factor receptor (EGFR) amplification which was found in the primary tumor represented a shortcoming that was common with previous in vitro models [13]. This observation may reflect a diminished role for EGFR in cell survival in vitro, although an alternative explanation is that it is in fact a consequence of the in vitro effect on the cell line [11]. The Cambridge protocol combined spheroid and monolayer culture in the derivation phase in order to increase their efficiency. Following successful derivation, in the maintenance and propagation of glioma, which culture conditions should you use?

Monolayer Versus Suspension Culture

Growth of glioma in 2D culture is an attractive option due to ease of use, low cost, and allowance for high throughput therapeutic screening experiments. Suspension culture is time consuming and the slower proliferation rate of the cells in this culture condition removes the potential for high throughput science. However, are there differences in the biology of the tumor between these two culture conditions?

Recently, it has been suggested that the genomic profile of glioma cultured in monolayer is frequently deviant from that of the parental tumor, whereas primary spheroid culture produces a genetically more representative model system [14]. Examples of the genomic divergence include loss of chromosome 10 in the parental tumor, which was reconstituted in the primary cell line, and amplification of the EGFR locus in the parental tumor, which was lost in the primary cell line and preserved in spheroid culture conditions. The author's explanation for the divergence considered the faster proliferative index of glioma in 2D culture, which created a greater opportunity for mutation, clonal selection in the culture conditions, and subsequent divergent clonal evolution. The genomic profile in spheroid culture remained stable after 12 weeks in vitro. The author's conclusion was that spheroid culture is a genetically more representative model of glioblastoma.

The above suggests that in culture, extensive genomic instability may arise depending on the culture condition and produce de novo mutations. However, there is another body of evidence that suggests that once a tumor cell line has become established, the genomic profile remains stable over extended passaging. Jones et al. documented that a discordant CDKN2A/p16 status was observed between a primary bladder tumor and its derived tumor cell line. Fifty-four percent of cell lines had a p16 mutation versus 19 % of the primary tumors [15]. A similar observation was noted in malignant glioma, and Westphal et al. decided to study this entity by tracking the in vitro evolution of glioma lines from their establishment onward [16]. They found that at establishment of the line, there was a selection pressure which facilitated the outgrowth of p16 negative/mutated clones, but once the line had become established, p16 wild type or mutated, there was no change in the mutation status over 8 years of extended passaging.

Investigation of Therapeutic Strategies Using In Vitro Models of Glioblastoma

Chemoresistance

The cancer stem cell hypothesis states that within certain solid tumors and hematological malignancies, there exists only a small population of cells that are in fact tumorgenic [17]. Such cells entitled cancer stem cells with the properties of self-renewal, multipotency and in vivo tumorgenicity on serial transplantation have been isolated from glioblastoma [18]. Although the identification of markers that are pathognomic of the cancer stem cell phenotype remains controversial, we do know that cancer cells with stem-like properties possess inherent chemoresistance [19–21]. The mechanism for chemoresistance in this population has been attributed to the classical mechanism of adenosine triphosphate- (ATP-) mediated drug efflux transporters. A rather elegant experiment by Eramo et al. showed that fluorescent doxorubicin was not actively extruded from glioblastoma cells relative to a positive control lung carcinoma cell line [22]. The uptake was similar to the control cell line, but nuclear compartmentalization of the drug was observed in the glioma cell supporting the hypothesis that mechanistic evasion of apoptosis played a crucial role in chemoresistance. With the properties described above, it becomes clear why the cancer stem cell is an attractive therapeutic target, being likely responsible for treatment failure and disease progression/recurrence.

Principles of In Vitro Cytotoxic Assays

So, prior to testing your cytotoxic compound of choice, consideration must be given to the culture condition that you use. Despite the potential shortcomings of in vitro culture and possible de novo genomic divergence from the parental tumor, whether these observations have an effect on response to therapy is largely unknown. Work on 9L gliosarcoma suggests that use of neurosphere/suspension culture rather than monolayer selects for a more chemoresistant, aggressive tumor subpopulation [23]. Glioma cells growing as neurospheres were self-renewing, multipotent, expressed SOX2 and formed larger, and more aggressive tumors than the equivalent cells propagated in monolayer. Anti-apoptotic drug resistance genes were also more highly expressed in the neurosphere population. An interesting finding, however, was a twofold larger luciferase signal observed in the monolayer condition. An explanation for this is that more luciferase-negative cells were selected for in the neurosphere group due to lower end organ concentration of the chemical selection compound in suspension culture. If this is the case, we have to question the merits of the neurosphere assay for cytotoxic experiments. 2D adherent culture has the advantage of ensuring uniformity of delivery of not only factors conducive to growth but also extrinsically delivered compounds in chemical screens [24].

So what are the basic pharmacokinetic principles that underlie in vitro cytotoxic drug testing? Let us consider this question in a reverse fashion. When treating patients, drug prescribing can present some greatly challenging clinical dilemmas. These dilemmas arise due to intra-patient and inter-patient differences in pharmacokinetics of absorption, distribution, metabolism, and excretion of the pharmaceutical compound. Certainly, the complexities of these physiological processes are not reproduced in current in vitro models, however, what the in vitro model does provide is a platform to understand the biological response in standardized conditions [25]. 2D culture allows even drug dispersal bypassing the in vivo problems of absorption and distribution that facilitates observation of biological effect at the target cell and furthermore permits knowledge to be attained of dose–response relationships. Taking the information gained using this model back to the in vivo encounter provides a conceptual framework to understand the in vivo response. Using our standardized in vitro cytotoxic assay, we can infer that in order to produce a cytotoxic effect on a specific target cell, we need dose X at the level of the cell in order to elicit maximal pharmacodynamics. Is treatment failure a result of unique chemoresistant properties in the cancer cell or rather as a result of inadequate active drug delivery to the target site or most likely a combination of both?

Failure of conventional chemotherapy to date due to one or both of the above factors has resulted in increased interest in local therapies for malignant glioma where a defined chemotherapeutic dose can be delivered to the target site at the time of surgery. Specifically, Gliadel wafers which are carmustine impregnated wafers have resulted in improved survival without an increased incidence of adverse events over placebo wafers when used for primary disease therapy, as concluded from a Cochrane review [26]. The GALA5 trial which combines fluorescence guided surgical resection with gliadel wafer insertion is currently recruiting [27].

Glioblastoma and Radioresistance

Again, borrowing from the cancer stem cell hypothesis, it is the same subpopulation with self-renewing characteristics that are felt to possess inherent radioresistance [28]. In an interesting study by Tamura et al., patients with malignant glioma were treated using combined gamma knife surgery and external beam irradiation. Examination of tumor histological samples pre- and post-adjuvant therapy showed that the percentage of CD133 positive cells was much higher in the posttreatment tumor material [29]. Although we cannot make any causal inference from this study, what we should conclude is that there is higher proportion of cells bearing a marker of stemlike behavior in the cellular fraction surviving radiotherapy. What is required in order to prove that the surviving fraction is a cancer stem cell? As we have previously described, the properties of self-renewal with the clonogenic assay being the gold standard, multipotency, and tumorgenicity must be present.

When investigating radioresistance in glioblastoma, one readily sees that the cancer stem cell does not exist in a vacuum. Much of the observed biological behavior is a product of the microenvironment. In a recent review article, Chalmers et al. discuss the "microenvironment stem cell unit" where the role of the microenvironment in regulating stem cell biology and glioma cell radioresistance is outlined [30]. Factors including endothelial cells, extracellular matrix, nitric oxide, and oxygen concentration were discussed. When considering and designing our in vitro model for studying glioblastoma, have we adequately considered the role of oxygen concentration and its effect on tumor biology?

Oxygen Concentration and In Vitro Glioma Culture

It is well-known that many cancers operate at a hypoxic level. The role of hypoxia in mediating radioresistance was first described by Schwarz in 1909 [31]. Differential radiosensitivity is thought to be accounted for by the different fates of free radicals depending on the oxygen concentration. Oxygen stabilizes free radicals making DNA damage more likely to occur. In the absence of oxygen, DNA damage is less likely to occur as free radicals are more likely to react with H^+ ions, reverting to their original form [30, 32]. Indeed, this knowledge of intratumoral hypoxia-mediated radioresistance has led to a registered clinical trial where hyperbaric hyperoxygenation with radiotherapy and temozolomide is being tested in adults with newly diagnosed glioblastoma [33].

Beyond the clinical observation of radioresistance, what does hypoxia do to the tumor biology? Li et al. found that in response to hypoxia in glioblastoma, the hypoxia inducible factor transcription network was preferentially activated [34]. Specifically; cancer stem cells activated HIF2A which was required for the maintenance of stem cell identity and regulation of the angiogenic switch. Furthermore, expression of HIF2A correlated with poor overall patient survival. With studies such as this suggesting that hypoxia selects for the cancer stem cell population, use of hypoxic chambers for cell culture may have a role to play. Alternatively, cells can be treated by 100 or 200 μM hypoxia-mimic chemical deferrioxamine mesylate [34].

Success of In Vitro Models of Radioresistance in Glioblastoma

The holy grail of in vitro models is one that furthers our understanding of the tumor biology in malignant glioma, and by doing so exposes an Achilles' heel, one potentially translatable to the clinic. The Chek1/2 story in modulating radioresistance proved to be one such example.

Chek1 is a checkpoint kinase in the cell cycle regulating the transition from G2/M. As seen in the figure (Fig. 5.2) below in response to ionizing radiation, DNA damage is sensed through the ATR complex which phosphorylates Chek1 [35]. Activation of Chek1 phosphorylates CDC25C, a protein phosphatase inhibiting its function. As CDC25C cannot dephosphorylate CDC2, it remains in its inactive state preventing cell cycle re-entry and mitosis. Chek2 is activated through ATM and, through a similar mechanism of action, prevents G1/S cell cycle progression. The purposes of these checkpoints are to allow DNA damage to be repaired through nonhomologous end joining and homologous recombination [36]. A cell that had increased ability to repair damaged DNA in response to ionizing radiation would be less likely to undergo apoptosis and by exhibiting such radioresistance, if a cancer cell, would lead to treatment failure.

Bao et al., as previously mentioned, identified the cancer stem cell population as the subpopulation within the tumor mass that possessed inherent

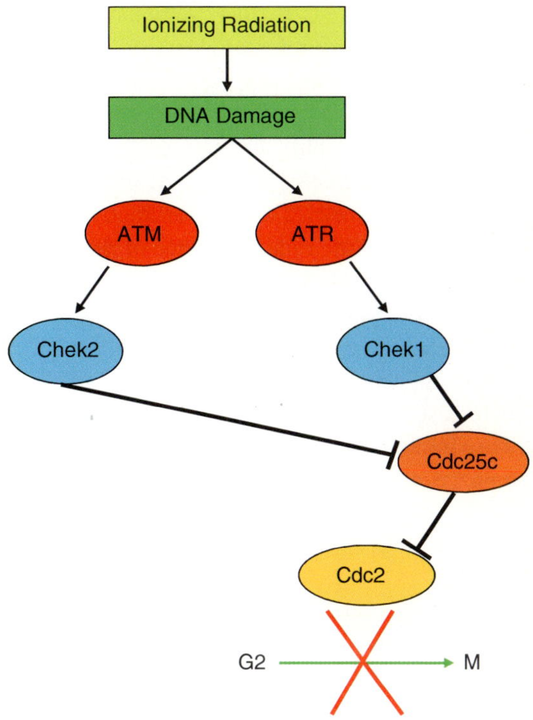

Fig. 5.2 Chek1/2 mechanism of induced cell cycle arrest. Chek1 is a checkpoint kinase in the cell cycle regulating the transition from G2/M. In response to ionizing radiation, DNA damage is sensed through the ataxia telangiectasia-related gene product (*ATR*) complex which phosphorylates Chek1. Activation of Chek1 phosphorylates CDC25C, a protein phosphatase inhibiting its function. As CDC25C cannot dephosphorylate CDC2, it remains in its inactive state preventing cell cycle re-entry and mitosis. Chek2 is activated through ataxia telangiectasia-mutated gene product (*ATM*) and through a similar mechanism prevents cell cycle re-entry until the DNA damage has been repaired through nonhomologous end joining or homologous recombination

radioresistance [28]. In their publication, they demonstrated how preferential activation of the Chek1/2 checkpoint kinases allowed cell cycle arrest and DNA damage repair to occur preventing the cell from undergoing apoptosis. Furthermore, to test this mechanism, they showed that inhibiting Chek1/2 using debromohymenialdisine disrupted the radioresistance of these cancer stem cells in vitro and in vivo.

Clearly, a conclusion from their work was that checkpoint kinase inhibitors may have a role to play in the clinic through their modulation of radioresistance. LY2606368, a checkpoint 1 kinase inhibitor (Chek1), is currently in a phase 1 clinical trial of patients with advanced/metastatic cancer including colorectal, ovarian, and non-small-cell lung carcinoma [37].

Cell of Mutation and Cell of Origin: Modeling Oncogenic Transformation in Glioblastoma

Generating cell lines from the primary tumor specimen, although unquestionably useful for understanding the established tumor biology, does not however permit knowledge to be attained of the cell of origin in glioblastoma. With the isolation of putative cancer stem cells from glioblastoma with the properties of self-renewal, multipotent differentiation into astroglial and neuronal lineages, the neural stem cell was proclaimed the cell of origin [18]. There has been much controversy surrounding this claim as it is well established that the cell of origin is not necessarily a stem cell [38]. Making the distinction between the cell of origin and the cancer stem cell is fundamentally important. By understanding this dichotomous yet not mutually exclusive relationship, one can appreciate that any somatic cell following sustaining a critical oncogenic hit can be the cell of origin. This cell can subsequently produce daughter cells with stemlike ability that sustains growth of the tumor.

The hierarchical model of cancer as seen in Fig. 5.3 depicts the cancer stem cell at the apex of the pyramid. This cell has the capacity for symmetric/asymmetric cell division producing either two additional cancer stem cells, a cancer stem cell and a differentiated progeny, or two differentiated progeny. This model maintains that only the cancer stem cell has the capacity for sustaining growth of the tumor. Human cancers that appear to follow this hierarchical arrangement include leukemia and some solid cancers including glioblastoma [17, 19]. In contrast to this is the example of malignant melanoma with a stochastic arrangement of propagation where any cell appears to have the capacity to sustain tumor growth [39]. So how do we attempt to track down the cell or origin in vitro/vivo?

Hypothesizing that the neural stem cell was the cell of origin in malignant glioma, Zhu et al. succeeded in producing high-grade glioma in a mouse model by inactivating p53 and NF1 in mouse neural stem cells [40]. In support of their hypothesis, they found that the earliest evidence of tumor formation was seen in the subventricular zone, a region known to be a neural stem cell niche. However, additional studies have suggested that glia including astrocytes and oligodendrocyte precursor cells may in fact be the cell of origin [41]. In this study, Lindberg et al. generated a Ctv-a mouse where tumor formation was confined to myelinating oligodendrocyte progenitor cells (OPCs) expressing 2′,3′-cyclic nucleotide 3′-phosphodiesterase. They demonstrated that platelet derived growth factor beta (PDGF-B) transfer to OPCs could induce gliomas with an incidence of 33 %. The tumors produced resembled WHO grade II oligodendroglioma based on their similarities in histopathology and expression of cellular markers. Supporting the latter hypothesis, Jane Visvader in her review article discussed how the neural stem cell does not have to be the cell of origin as the phenotype of a neural stem cell may in fact be a product of the transformation process [38].

A rather elegant experimental design entitled mosaic analysis with double markers (MADM) has recently offered some exciting insights into the cell of origin in glioma [42]. In this publication, the authors produced concurrent p53/NF1 mutations sporadically in mouse neural stem cells. Through MADM-based lineage tracing, they observed aberrant growth in the oligodendrocyte precursor cell but not in the neural stem

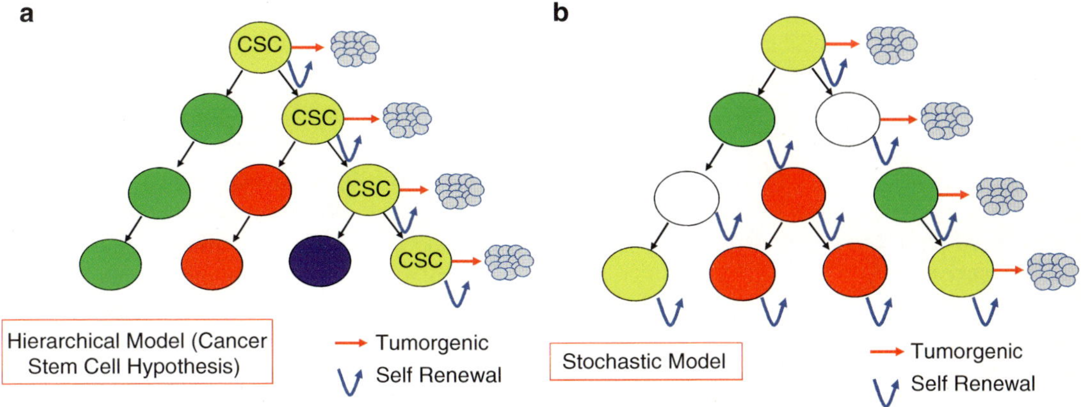

Fig. 5.3 Hierarchical versus stochastic model of cancer. (**a**) In the hierarchical model, known as the cancer stem cell hypothesis, it is only the cancer stem cells (*CSCs*), a small fraction of the total tumor population, that have the capacity to undergo self-renewal and that are tumorgenic. According to this hypothesis, if we do not eradicate the cancer stem cells, disease recurrence and progression is a certainty. (**b**) According to the stochastic model of cancer, any tumor cell can self-renew and has tumorgenic potential, so our therapeutic strategy should be directed to achieving complete surgical resection/disease eradication

cell or any other differentiated subpopulation. Analysis of formed tumors showed a transcriptome profile with OPC features. Introduction of the same mutations into the OPC resulted in consistent gliomagenesis. Based on these findings, the authors propose that the neural stem cell may reflect the cell of mutation, but the OPC is the cell of origin in malignant glioma. Consistent with this, analysis of fresh clinical GBM samples has confirmed the widespread expression of the OPC marker Neuroglia 2 (NG2) [43].

These studies have offered crucial insights into the development of malignant glioma; however, many controversies still remain unresolved. An additional fundamental point to remember is that these studies are assessing the potential for oncogenesis in mice. To what extent are these models representative of the equivalent human phenotype?

In an excellent review entitled "Comparative biology of mouse versus human cells: modeling human cancer in mice," Robert Weinberg discussed the differences in mouse and human biology in order to ensure that the scientific community continues to question their findings in mouse models and to strive to produce more biologically representative model systems [44]. Some of the key differences he discussed which are applicable to modeling cancer formation include, firstly, humans are 30 times larger than a mouse and live 30 times longer with 10^5 more cell divisions in a lifetime. Yet, 30 % of mice have cancer at the end of their second year, while 30 % of people have cancer by the ninth decade. The obvious conclusion to make is that humans have a lower cancer susceptibility which is a composite of multiple protective mechanisms at the level of the cell. Secondly, the spectrum of cancer from which mice suffer is different from that of humans being predominantly lymphomas and sarcomas, whereas humans suffer predominantly from epithelial cancers. Thirdly, mouse cells in vitro after extended passaging become immortalized, whereas human cells undergo replicative senescence due to loss of and failure to de-repress telomerase expression. So, what is the future of bottom up cancer modeling?

The ability to produce oncogenic transformation in glioblastoma using a human cell line would represent a significant advance in cancer modeling. This biological platform would allow investigation of driver and cooperative mutations and their effect in gliomagenesis. Fully characterized human fetal neural stem cells would be one potential cell type to consider for this purpose. With the advancement in cellular reprogramming and the use of non-integrational methods of induced pluripotent stem cell production (iPS), we may have another useful system to consider in our biological armamentarium in the very near future.

Future of In Vitro Modeling of Glioblastoma

As we have seen, in vitro models of glioblastoma have evolved tremendously over the past 20 years. The move from serum to serum-free culture conditions alone produced a disease model that was more representative of the parental tumor. Improvements in derivation and propagation resulted in cell lines that retained the characteristic cytogenetic profile of the parental glioblastoma. This created great excitement at the potential to apply chemical screens in vitro with the hope of uncovering new therapeutic targets. However, these improved systems are not without their problems. Choice of monolayer versus suspension culture can produce markedly different results in the same experiment. Whether suspension culture is superior in maintaining genomic stability long term is controversial. Whether the use of neurosphere culture is less appropriate than monolayer for chemical screens is questionable. All of these issues aside, as we mentioned in the radioresistance discussion, cancer cells do not exist in a vacuum. How representative is a tissue culture plate of the in vivo microenvironment? As cancer cells are influenced by cell-cell contact, endothelial cells, extracellular matrix, fluctuations in oxygen concentration, local ph and so on, the future must focus on the creation of in vitro models that incorporate a representative microenvironment. How close are we to this?

A fascinating publication by Vickerman et al. described the production of a novel microfluidics platform that allowed 3D real-time imaging of capillary morphogenesis in vitro [45]. Such systems could be applied to the investigation of tumor angiogenesis. A simpler, cost effective platform for studying the interaction between glioma and its microenvironment is the tissue explant. In this model, primary tissue is dissected to produce small slices (1 mm^3) which are placed on a porous fibronectin-coated membrane in a tissue insert. When the method was described for fetal nervous tissue, sprouting was demonstrated in vitro and long-term synaptic potentiation could be observed [46]. This organotypic culture system preserves cytoarchitecture, tumor stroma, and blood vessels in vitro offering a unique 3D culture platform that is truly representative of the tumor biology [47].

Alternatively, one can consider the use of 3D bioscaffolds. Biodegradable polymers have been used to direct differentiation of human embryonic stem cells [48]. Mouse neural stem cells remain viable, proliferate, and differentiate in 3D in vitro self-assembling peptide scaffolds [49]. The underlying premise for all these technological designs are that they will render the in vitro model more representative of the in vivo environment. If successful, then having such a robust, reliable, and repeatable experimental protocol would hopefully bring the bench slightly closer to the bedside, representing a major step forward in our battle against this presently incurable disease.

References

1. Bigner SH, Mark J, Bigner DD. Chromosomal progression of malignant human gliomas from biopsy to establishment as permanent lines in vitro. Cancer Genet Cytogenet. 1987;24(1):163–76.
2. Mahesparan R, Read TA, Lund-Johansen M, Skaftnesmo KO, Bjerkvig R, Engebraaten O. Expression of extracellular matrix components in a highly infiltrative in vivo glioma model. Acta Neuropathol. 2003;105(1):49–57.
3. Engebraaten O, Hjortland GO, Hirschberg H, Fodstad O. Growth of precultured human glioma specimens in nude rat brain. J Neurosurg. 1999;90(1):125–32.
4. Sanai N, Alvarez-Buylla A, Berger MS. Neural stem cells and the origin of gliomas. N Engl J Med. 2005;353(8):811–22.
5. Reya T, Morrison SJ, Clarke MF, Weissman IL. Stem cells, cancer, and cancer stem cells. Nature. 2001; 414(6859):105–11.
6. Cancer Genome Atlas Research Network. Comprehensive genomic characterization defines human glioblastoma genes and core pathways. Nature. 2008;455(7216):1061–8.
7. Reynolds BA, Tetzlaff W, Weiss S. A multipotent EGF-responsive striatal embryonic progenitor cell

produces neurons and astrocytes. J Neurosci. 1992;12(11):4565–74.
8. Lee J, Kotliarova S, Kotliarov Y, Li A, Su Q, Donin NM, Pastorino S, Purow BW, Christopher N, Zhang W, Park JK, Fine HA. Tumor stem cells derived from glioblastomas cultured in bFGF and EGF more closely mirror the phenotype and genotype of primary tumors than do serum-cultured cell lines. Cancer Cell. 2006;9(5):391–403.
9. Günther HS, Schmidt NO, Phillips HS, Kemming D, Kharbanda S, Soriano R, Modrusan Z, Meissner H, Westphal M, Lamszus K. Glioblastoma-derived stem cell-enriched cultures form distinct subgroups according to molecular and phenotypic criteria. Oncogene. 2008;27(20):2897–909.
10. Gritti A, Parati EA, Cova L, Frolichsthal P, Galli R, Wanke E, Faravelli L, Morassutti DJ, Roisen F, Nickel DD, Vescovi AL. Multipotential stem cells from the adult mouse brain proliferate and self-renew in response to basic fibroblast growth factor. J Neurosci. 1996;16(3):1091–100.
11. Fael Al-Mayhani TM, Ball SL, Zhao JW, Fawcett J, Ichimura K, Collins PV, Watts C. An efficient method for derivation and propagation of glioblastoma cell lines that conserves the molecular profile of their original tumours. J Neurosci Methods. 2009;176(2): 192–9.
12. Kenney-Herbert EM, Ball SL, Al-Mayhani TM, Watts C. Glioblastoma cell lines derived under serum-free conditions can be used as an in vitro model system to evaluate therapeutic response. Cancer Lett. 2011; 305(1):50–7.
13. Bigner SH, Mark J, Bigner DD. Cytogenetics of human brain tumors. Cancer Genet Cytogenet. 1990;47(2):141–54.
14. De Witt Hamer PC, Van Tilborg AA, Eijk PP, Sminia P, Troost D, Van Noorden CJ, Ylstra B, Leenstra S. The genomic profile of human malignant glioma is altered early in primary cell culture and preserved in spheroids. Oncogene. 2008;27(14):2091–6.
15. Spruck III CH, Gonzalez-Zulueta M, Shibata A, Shimoneau AR, Lin M-F, Gonzales F, Tsai YC, Jones PA. p16 gene in uncultured tumours. Nature. 1994;370:183–4.
16. Hartmann C, Kluwe L, Lücke M, Westphal M. The rate of homozygous CDKN2A/p16 deletions in glioma cell lines and in primary tumors. Int J Oncol. 1999;15(5):975–82.
17. Bonnet D, Dick JE. Human acute myeloid leukemia is organized as a hierarchy that originates from a primitive hematopoietic cell. Nat Med. 1997;3:730–7.
18. Galli R, Binda E, Orfanelli U, Cipelletti B, Gritti A, De Vitis S, Fiocco R, Foroni C, Dimeco F, Vescovi A. Isolation and characterization of tumorigenic, stem-like neural precursors from human glioblastoma. Cancer Res. 2004;64(19):7011–21.
19. Singh SK, Hawkins C, Clarke ID, Squire JA, Bayani J, Hide T, Henkelman RM, Cusimano MD, Dirks PB. Identification of human brain tumour initiating cells. Nature. 2004;432(7015):396–401.
20. Wang J, Sakariassen PØ, Tsinkalovsky O, Immervoll H, Bøe SO, Svendsen A, Prestegarden L, Røsland G, Thorsen F, Stuhr L, Molven A, Bjerkvig R, Enger PØ. CD133 negative glioma cells form tumors in nude rats and give rise to CD133 positive cells. Int J Cancer. 2008;122(4):761–8.
21. Ma S, Lee TK, Zheng BJ, Chan KW, Guan XY. CD133+ HCC cancer stem cells confer chemoresistance by preferential expression of the Akt/PKB survival pathway. Oncogene. 2008;27(12):1749–58.
22. Eramo A, Ricci-Vitiani L, Zeuner A, Pallini R, Lotti F, Sette G, Pilozzi E, Larocca LM, Peschle C, De Maria R. Chemotherapy resistance of glioblastoma stem cells. Cell Death Differ. 2006;13(7): 1238–41.
23. Ghods AJ, Irvin D, Liu G, Yuan X, Abdulkadir IR, Tunici P, Konda B, Wachsmann-Hogiu S, Black KL, Yu JS. Spheres isolated from 9L gliosarcoma rat cell line possess chemoresistant and aggressive cancer stem-like cells. Stem Cells. 2007;25(7):1645–53.
24. Pollard SM, Yoshikawa K, Clarke ID, Danovi D, Stricker S, Russell R, Bayani J, Head R, Lee M, Bernstein M, Squire JA, Smith A, Dirks P. Glioma stem cell lines expanded in adherent culture have tumor-specific phenotypes and are suitable for chemical and genetic screens. Cell Stem Cell. 2009;4(6):568–80.
25. Ekwall B, Silano V, Paganuzzi-Stammati A, Zucco F. Chap. 7: Toxicity tests with mammalian cell cultures. In: Short-term toxicity tests for non-genotoxic effects. SCOPE. Chichester: Wiley; 1990.
26. Hart MG, Grant R, Garside R, Rogers G, Somerville M, Stein K. Chemotherapy wafers for high grade glioma. Cochrane Database Syst Rev. 2011;3: CD007294. doi:10.1002/14651858.CD007294.pub2.
27. Watts C. CRUK/10/009: GALA-5 TRIAL: an evaluation of the tolerability and feasibility of combining 5-Amino-Levulinic Acid (5-ALA) with carmustine wafers (Gliadel) in the surgical management of primary glioblastoma. http://science.cancerresearchuk.org/research/who-and-what-we-fund/browse-by-location/cambridge/university-of-cambridge/grants/colin-watts-11869-cruk-10-009-gala-5-trial-an-evaluation.
28. Bao S, Wu Q, McLendon RE, Hao Y, Shi Q, Hjelmeland AB, Dewhirst MW, Bigner DD, Rich JN. Glioma stem cells promote radioresistance by preferential activation of the DNA damage response. Nature. 2006;444(7120):756–60.
29. Tamura K, Aoyagi M, Wakimoto H, Ando N, Nariai T, Yamamoto M, Ohno K. Accumulation of CD133-positive glioma cells after high-dose irradiation by Gamma Knife surgery plus external beam radiation. J Neurosurg. 2010;113(2):310–8.
30. Mannino M, Chalmers AJ. Radioresistance of glioma stem cells: intrinsic characteristic or property of the 'microenvironment-stem cell unit'? Mol Oncol. 2011;5(4):374–86.
31. Schwarz G. UeberDesensibilisierunggegenröntgen- und radiumstrahlen. MunchenerMedizinischeWoche nschrift. 1909;24:1–2.

32. Horsman MR, Overgaard J. The oxygen effect and tumour microenvironment. In: Steel G, editor. Basic clinical radiobiology. London: Arnold; 2002. p. 158–68.
33. Duic JP. NCT00936052. Phase II pilot trial of hyperbaric hyperoxygenation in conjunction with radiotherapy and temozolomide in adults with newly diagnosed glioblastomas. http://clinicaltrials.gov/ct2/show/NCT00936052.
34. Li Z, Bao S, Wu Q, Wang H, Eyler C, Sathornsumetee S, Shi Q, Cao Y, Lathia J, McLendon RE, Hjelmeland AB, Rich JN. Hypoxia-inducible factors regulate tumorigenic capacity of glioma stem cells. Cancer Cell. 2009;15(6):501–13.
35. Guo Z, Kumagai A, Wang SX, Dunphy WG. Requirement for Atr in phosphorylation of Chk1 and cell cycle regulation in response to DNA replication blocks and UV-damaged DNA in Xenopus egg extracts. Genes Dev. 2000;14(21):2745–56.
36. Sørensen CS, Hansen LT, Dziegielewski J, Syljuåsen RG, Lundin C, Bartek J, Helleday T. The cell-cycle checkpoint kinase Chk1 is required for mammalian homologous recombination repair. Nat Cell Biol. 2005;7(2):195–201.
37. NCT01115790. A phase 1 study of LY2606368 in patients with advanced cancer. www.clinicaltrials.gov.
38. Visvader JE. Cells of origin in cancer. Nature. 2011;469(7330):314–22.
39. Quintana E, Shackleton M, Sabel MS, Fullen DR, Johnson TM, Morrison SJ. Efficient tumour formation by single human melanoma cells. Nature. 2008;456(7222):593–8.
40. Zhu Y, Guignard F, Zhao D, Liu L, Burns DK, Mason RP, Messing A, Parada LF. Early inactivation of p53 tumor suppressor gene cooperating with NF1 loss induces malignant astrocytoma. Cancer Cell. 2005;8(2):119–30.
41. Lindberg N, Kastemar M, Olofsson T, Smits A, Uhrbom L. Oligodendrocyte progenitor cells can act as cell of origin for experimental glioma. Oncogene. 2009;28(23):2266–75.
42. Liu C, Sage JC, Miller MR, Verhaak RG, Hippenmeyer S, Vogel H, Foreman O, Bronson RT, Nishiyama A, Luo L, Zong H. Mosaic analysis with double markers reveals tumor cell of origin in glioma. Cell. 2011;146(2):209–21.
43. Al-Mayhani MT, Grenfell R, Narita M, Piccirillo S, Kenney-Herbert E, Fawcett JW, et al. NG2 expression in glioblastoma identifies an actively proliferating population with an aggressive molecular signature. Neuro Oncol. 2011;13(8):830–45.
44. Rangarajan A, Weinberg RA. Opinion: comparative biology of mouse versus human cells: modelling human cancer in mice. Nat Rev Cancer. 2003;3(12):952–9.
45. Vickerman V, Blundo J, Chung S, Kamm R. Design, fabrication and implementation of a novel multiparameter control microfluidic platform for three-dimensional cell culture and real-time imaging. Lab Chip. 2008;8(9):1468–77.
46. Stoppini L, Buchs PA, Muller D. A simple method for organotypic cultures of nervous tissue. J Neurosci Methods. 1991;37(2):173–82.
47. Hovinga KE, Shimizu F, Wang R, Panagiotakos G, Van Der Heijden M, Moayedpardazi H, Correia AS, Soulet D, Major T, Menon J, Tabar V. Inhibition of notch signaling in glioblastoma targets cancer stem cells via an endothelial cell intermediate. Stem Cells. 2010;28(6):1019–29.
48. Levenberg S, Huang NF, Lavik E, Rogers AB, Itskovitz-Eldor J, Langer R. Differentiation of human embryonic stem cells on three-dimensional polymer scaffolds. Proc Natl Acad Sci USA. 2003;100(22):12741–6.
49. Cunha C, Panseri S, Villa O, Silva D, Gelain F. 3D culture of adult mouse neural stem cells within functionalized self-assembling peptide scaffolds. Int J Nanomedicine. 2011;6:943–55.

Mouse Models of Glioma Pathogenesis: History and State of the Art

6

Sebastian Brandner

Abstract

In this chapter, we describe the evolution of experimental models for brain tumors. Model systems were established as early as the 1920s, when chemical carcinogenesis was used to elicit malignant neoplasms in various tissues or organs, including the central nervous system. A more systematic study of different carcinogens, with a detailed histological analysis, followed in the 1950s and 1960s. At the same time, retroviral carcinogenesis was used as an alternative approach, and refined virus delivery resulted in more realistic models for gliomas. Brain tumors resulting from these approaches were carefully characterized and resembled high-grade gliomas, including oligodendroglial tumors and glioblastomas. The models were limited in that a cell of origin could not be formally demonstrated, but the localization of the lesions suggested that the ventricular zone may have been the origin of some of the tumors. 20 years later, an entirely different approach, the transgenic expression of oncogenic (virus- derived) gene sequences, started a new era of cancer research. These technologies were soon followed by "straight" gene knockout models, and in the mid 1990s, the more refined conditional (Cre-Lox) gene knockout system, which was modified in various ways to allow tissue specific, temporally controlled expression of oncogenes or inactivation of tumor suppressor genes. For the first time, these models led to an in-depth understanding of the mechanisms of brain tumor pathogenesis and the identification of the cells giving rise to intrinsic brain tumors.

S. Brandner, M.D., FRCPath
Division of Neuropathology, UCL Institute of Neurology,
Queen Square, London, WC1N 3BG, UK
e-mail: s.brandner@ucl.ac.uk

C. Watts (ed.), *Emerging Concepts in Neuro-Oncology*,
DOI 10.1007/978-0-85729-458-6_6, © Springer-Verlag London 2013

Keywords
Chemical carcinogenesis • Viral carcinogenesis • Point mutations • Transgenic mouse • Gene knockout • Conditional gene inactivation • cre recombinase • loxP site • Brain tumor • Glioma • Glioblastoma • Oligodendroglioma • Rous sarcoma virus • Ethylnitrosourea • SV40 • Retinoblastoma gene • p53 tumor suppressor gene • PTEN tumor suppressor gene • INK4a-ARF locus

Introduction

As with any other experimental model of a human disease, mouse models for gliomas essentially aim at addressing two major issues: (1) to mimic a phenotypically similar pathology in rodents and (2) provide a model system that can be used as platform for treatments and other interventions. The development of glioma mouse models over the last 6 decades always aimed at mimicking a human counterpart, in particular astrocytomas (including the most malignant from glioblastoma) and oligodendroglial tumors. A secondary aim was the development of a system that can be used to test therapeutic options. The mouse model systems that were used and that are described here reflect the available technologies at the times of their generation.

Two fundamentally different approaches have been used to model the pathogenesis of gliomas: One model system aims at generating *de novo* tumors that arise from the host brain by various methods (Fig. 6.1), while the second uses the rodent brain as a vehicle for propagating xenografted glioma cell lines, tumor particles, or more refined glioma cultures. In this chapter, we will focus only on model systems of intrinsic brain tumors, that is, the various forms of allografts or xenografts will not be discussed.

Why do we need mouse models of human intrinsic brain tumors? It was recognized early on, and reiterated through decades of experimental neuro-oncology, that there are “three major objectives: (1) to develop experimental models with a reproducible high rate of incidence, and to compare them with each other and with naturally occurring human brain tumors; (2) to explore the etiology and pathogenesis of experimental neurogenic tumors as a basis for a better understanding of the still unknown causes and unresolved pathomorphogenesis of brain tumors in man; and (3) to utilize experimental tumors as models for testing chemotherapeutic agents and prophylactic measures” [1].

Natural Models of Brain Tumors: Spontaneous Mutations

While in principle an ideal reflection of spontaneous brain tumors in humans, the infrequent occurrence of spontaneous brain tumors in laboratory animals makes this model rather unattractive and unpractical. The most valuable information on the occurrence of brain tumors in various animal species came from studies in which animals are allowed to live out their lifespan. Since nearly all lifetime studies are confined to mice and rats, the incidences in these two animal species are probably the most reliable figures available from all animal species studied so far. In mice, brain neoplasms are exceedingly rare with the exception of two strains and their crosses, the VM and BRVR lines [2] with incidences of 1.6 and 1.1 %, respectively [3] (all abbreviations see Table 6.1). Astrocytomas constitute nearly 100 % of the intracranial neoplasms in these two strains. In all other mouse strains studied so far, the incidence does not exceed 0.3 % (Table 6.2). In comparison, rats having a higher incidence are much more common than in mice (Wistar AF rat: 7.1 %). Therefore, these animal models never became popular in experimental neuro-oncology [66].

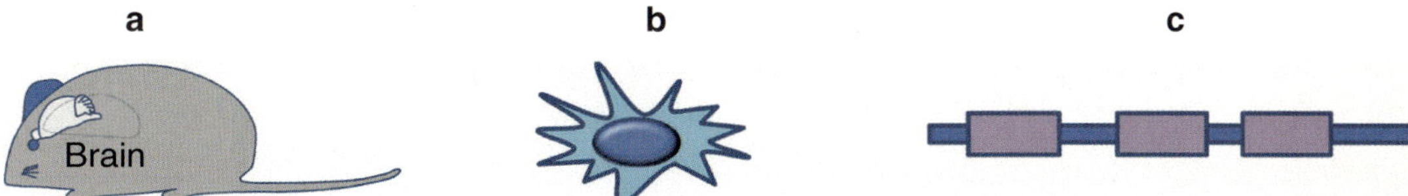

Fig. 6.1 Levels of modification in model systems for brain tumors. The highest level is the model organism. In the diagrams of the subsequent figures, the model organism symbol indicated how the model is being altered. The next level is the cellular level, with an indication of the cellular changes. Finally, the changes on the genomic level are indicated, in particular the corresponding genetic modifications in the conditional knockout mouse models. (**a**) Model organism: Mostly the current models are mice, but rats, hamsters, and other rodents have been used in the past. In this figure, the brain is highlighted as the primary target of neuro-oncology research. (**b**) Events on cellular level: In the past, it was not possible to direct mutagenesis to specific cells. Once this was possible by transgenic modeling, astrocytes or neural precursor cells were used as target for oncogene expression or tumor suppressor gene inactivation. (**c**) Events or modifications on genomic level: The symbol that is consistently used here is a simplified genomic structure with three exons, for example, representing a tumor suppressor gene or a random genomic sequence, depending on the context. Modifications are subsequently made to this structure

Carcinogen-Mediated Induction of Brain Tumors

First attempts to generate tumors in rodents *in vivo* were made by chemical carcinogenesis (Figs. 6.1 and 6.2). The observation that spontaneous brain tumors are practically nonexistent in common laboratory animals [66] rendered positive results immediately significant. Experiments date back to the 1930s and 1940s. Chemical carcinogenesis was either induced locally or systemically (Fig. 6.2). Local induction was achieved by implanting pellets of a carcinogen (such as methylcholanthrene, dibenzanthracene, and benzpyrene suspended in carriers such as lanolin, paraffin oil, Vaseline, or lard) intracranially. Following early, largely unsuccessful studies in guinea pigs, rabbits, and rats, mice were used for larger, systematic studies [39, 40]. Initially, the carcinogenic substances (methylcholanthrene, benzpyrene) were intracranially implanted as crystals or pellets [41]. Tumors developed with a latency of 10 months and corresponded histologically to oligodendroglioma, glioblastoma multiforme, medulloblastoma, as well as "unclassified gliomas" and soft tissue tumors (at the time classified as meningeal sarcomas). The efficacy of tumor induction was remarkable, reaching nearly 50 % [42, 43] (Table 6.2), half of which were of glial or at least intrinsic origin. Importantly, the site of implantation was an important factor for the type of tumor produced. Medulloblastomas followed implantation into the cerebellum, ependymomas into the ventricular wall, polar spongioblastomas (an entity that is much less commonly diagnosed nowadays [67]) into the brainstem, and oligodendrogliomas into the white matter of the hemispheres. Subsequently, this approach become rather fashionable and was used in several laboratories to generate intracranial neoplasms. Interestingly, generation of brain tumors was often favored by scientists over other types of tumors. In the 1950s and 1960s, novel carcinogenic compounds were developed and subsequently applied systemically ("resorptive carcinogens"). After systemic application *in vivo*, these compounds are rapidly decomposed, mainly in the liver [44]. Metabolism of the compounds by hydroxylation and demethylation requires enzymatic activation [68]. The carcinogenic action is not restricted to the liver but, depending on the chemical structure of the given compound, includes various organs such as the esophagus, kidney, bladder, lung, or nasal cavity. Most commonly N-nitroso compounds were used [69], specifically methylnitrosourea (MNU), dimethylnitrosourea (DMNU), trimethylnitrosourea (TMNU), and ethylnitrosourea (ENU). Most researchers used inbred rat strains for their experiments, and detailed neuropathological

Table 6.1 List of abbreviations and technical terms

Abbreviation/acronym		Explanation
VM	Inbred as "5 M" from Moredun Inst. stock and name later changed to conform with nomenclature rules	Congenic mouse strains VM/Dk and VM-Sincs7/Dk differ at the Sinc gene, which controls the incubation period of scrapie in mice
GFAP	Glial fibrillary acidic protein	Intermediate filament, predominantly, but not exclusively expressed by astrocytes. Also expressed by B-Type stem cells in the SVZ and transiently expressed by neural progenitors during development
hGFAP	Human GFAP (promoter)	See above
Nestin	Neuroectodermal stem cell marker	Intermediate filament, expressed by neural progenitor cells during development but also in SVZ progenitor cells
C neu	Oncogene, named from its derivation from a glioblastoma line, a "neural tumor"	The proto-oncogene Neu is more commonly known as HER-2 (human epidermal growth factor receptor 2) or as erbB-2 receptor tyrosine-protein kinase. For transgenic mouse models, see [4]
SV40	Simian (vacuolating) virus 40	The early region of the SV40 chromosome contains both the large T antigen and small T-antigen coding sequences. These gene products are required to transform cells in culture [5] and are oncogenic in hamsters and other rodents
v-src	Viral Sarc (oma)	v-src, the transforming gene of the Rous sarcoma virus (RSV), encodes pp60v-src which is nearly identical to pp60c-src, but lacks the carboxy-terminal region comprising Tyr-527 which switches on its kinase upon phosphorylation and is therefore constitutively active as a tyrosine kinase [6]
c-src		Src (pp60c-src) is an intracellular tyrosine kinase expressed ubiquitously in mammalian cells, with highest levels encountered in brain and platelets [7]
RCAS	Replication competent ALV [avian leukosis virus] with splice acceptor	Used to generate viral vectors and for a transgenic expression system; see also (RCAS)/TVA system [8, 9]
Nf1	Neurofibromatosis 1 gene	Individuals afflicted with neurofibromatosis type 1 (NF1) are predisposed to malignant astrocytoma in the brain with a greater than fivefold increased incidence throughout their lives (reference mouse model: [10])
Cre	"Cause of recombination"	Originally identified in the P1 phage, a gene product causing recombination of loxP sites in the phage P1 [11–13]. The Cre-loxP system is now widely used in conditional knock-out mouse models [14].
PTEN	Phosphatase and tensin homologue located on chromosome 10q23	PTEN is an antagonist of PI3Kinase. PTEN loss leads to a hyperphosphorylation of Akt to pAkt [15–17]
APC	Adenomatosis polyposis coli	Tumor suppressor gene and protein. Its loss (truncation) leads to constitutive activation of the wnt signaling pathway [18–20]
Gtv-a	GFAP-tv-a transgenic mouse	Transgenic mouse model that expresses TVA, the avian cell surface receptor for the retrovirus ALV-A (avian leucosis virus—subgroup A), under the control of the GFAP promoter [21, 22]
Ntv-a	Nestin-tv-a transgenic mouse	Transgenic mouse model that expresses TVA, the avian cell surface receptor for the retrovirus ALV-A (avian leucosis virus—subgroup A), under the control of the nestin promoter [21, 22]

Abbreviation/acronym		Explanation
EGFRvIII	Epidermal growth factor receptor with a truncation of the vIII domain	
INK4a and INK4a-ARF	Inhibitor of CDK4 Arf: alternate reading frame	The unique INK4A/ARF locus at chromosome 9p21 encodes two distinct proteins that link the pRB and p53 tumor suppressor pathways. p16INK4A is an inhibitor of the cell cycle, capable of inducing arrest in G1 phase. p14/p19ARF can induce both G1 and G2 arrest due to its stabilizing effects on the p53 transcription factor. The frequent mutation or deletion of INK4A/ARF in human tumors and the occurrence of tumors in the murine knockout models have identified both p16 and ARF as bona fide tumor suppressors [23–25]
EGFR	Epidermal growth factor receptor	Amplification of the EGFR is one of the most common genetic events in glioblastoma pathogenesis. Several mouse models have used constitutively active mutants to recapitulate glioblastoma histogenesis in mice [22, 26, 27]
CDK4	Cyclin-dependent kinase 4	The division cycle of eukaryotic cells is regulated by a family of protein kinases known as cyclin-dependent kinases. CDK4 and D type kinases have been implicated in the control of cell proliferation during the G1 Phase. The p16 protein binds to CDK4 and inhibits its catalytic function [25]
RSV	Rous sarcoma virus	After the original isolation of the agent responsible for the chicken myxosarcoma, a number of strains of RSV with different biological properties were described [28]. RSV proved effective in the nervous system of certain animals [29]. v-src is the transforming gene of RSV
p53	p53 is a tumor suppressor protein that in humans is encoded by the TP53 gene (protein with molecular weight of 53kDA)	Loss of genetic material on the short arm of chromosome 17 is observed in approximately 40 % of human astrocytomas and in approximately 30 % of cases of glioblastoma multiforme. The p53 gene, located on the short arm of chromosome 17, is frequently mutated in these tumors [30–32]
Rb	Retinoblastoma tumor suppressor gene	The retinoblastoma gene can be considered a model for a class of recessive human cancer genes that have a “suppressor” or “regulatory” function [33, 34]
CNS	Central nervous system	The central nervous system encompasses the brain and the spinal cord
SVZ	Subventricular zone	Located beneath the ependymal cells in the lateral ventricles of the brain, it contains neurogenic cells, that is, undifferentiated cells that are capable of self-renewal (“stem cells”) and their progeny [35, 36]. Mutations in this compartment are thought to be the origin of brain tumors [37, 38]
MBP	Myelin basic protein	Major constituent of the myelin sheath of oligodendrocytes and Schwann cells, the myelin forming cells of the CNS and the peripheral nervous system, respectively

Table 6.2 Modeling of intrinsic brain tumor pathogenesis in experimental animals

Technology	Species	Description	Resulting tumors	Approximate Tumor Incidence	References
Spontaneous brain tumors	Mouse	Two mouse strains and their crosses, the VM and BRVR spontaneously develop brain tumors	Astrocytomas	1–2 %	[2, 3]
Chemical carcinogenesis, intracranial implantation: methylcholanthrene, dibenzanthracene, benzpyrene	Rabbit, rat, guinea pig, mouse	Tumors developed with a latency of 10 months and corresponded histologically to as well as "unclassified gliomas" and soft tissue tumors (then classified as meningeal sarcomas). The efficacy of tumor induction was remarkable, reaching nearly 50 half of which were of glial or at least intrinsic origin	Oligodendroglioma, Glioblastoma multiforme, Medulloblastoma,	25 %	[39–43]
Ethylnitrosourea (ENU) mutagenesis, systemic ("resorptive carcinogens")	Rat	After systemic application *in vivo*, resorptive carcinogens are metabolized in the liver [44]. Rats were most commonly used	Glioblastomas, astrocytomas, and oligodendrogliomas	25 %	[1, 45]
		Transplacental administration. Offspring develops tumors. Frequent, often multiple tumors of the brain, the spinal cord, and the cranial and peripheral nerves	Glioblastomas, astrocytomas, and oligodendrogliomas.	90 %	
Virus-induced primary brain tumors: intracranial injection of oncogenic virus	Dog, rodents, including mice	Oncogenic viruses most commonly Rous Sarcoma virus, the active kinase of which is v-src.	Gliomas, choroid plexus papillomas, neuroblastomas	Up to 80 %	[28, 46–49]
Transgenic expression of a viral oncogene	Mouse	Metallothionein promoter; SV40 large T antigen,	Malignant choroid plexus tumors	26 %	[50, 51]
	Mouse	Myelin basic protein (MBP) promoter c-neu oncogene expression	Poorly differentiated neural tumors (PNET)	n.d.	[4]
	Mouse	GFAP promoter Vsrc expression	Astrocytomas, peripheral tumors	Up to 30 %	[52, 53]
Transgenic expression of a virus receptor (Tv-a) and gene transfer using a viral vector(ALV/RCAS) ("Tv-a+(ALV/RCAS) system") in combination with conditional gene knockout (Cre-loxP system)	Mouse	Induction of gliomas by RCAS vectors in tv-a transgenic mice: Transgenic mice express a virus receptor (t-va), under the control of the GFAP or nestin promoter. The activating oncogene (EGFR, CDK4, Akt, K-Ras) are delivered to the receptor expressing cells by a suitable viral vector (RCAS)			[22]
		Gtv-a (GFAP-Tv-a)+EGFR*; INK4a–/–	Gliomas	30 %	
		Ntv-a (nestin-Tv-a)+EGFR*; INK4a–/–;	Gliomas	50 %	
		Ntv-a (nestin-Tv-a)+EGFR*; INK4a–/–; cdk4			
		Ntv-a (nestin-Tv-a)+EGFR*; INK4a–/–; cdk4; bFGF			
Tv-a+(ALV/RCAS) system		Ntv-a (nestin-Tv-a)+Akt; -Kras	Glioblastoma	25 %	[54]

Tv-a + (ALV/RCAS) system + tumor suppressor knockout	Mouse	Ntv-a ARF–/–; Akt; -Kras	Glioblastoma	65 %	[55]
		Ntv-a Ink4a–/–; Akt + K-Ras	Glioblastoma	45 %	
		Gtv-a ARF–/–; Akt; -K-Ras	Glioblastoma	80 %	
		Gtv-a Ink4a–/–; Akt + K-Ras	Glioblastoma	5 %	
Tv-a + (ALV/RCAS) system+conditional knockout (Cre-loxP system)	Mouse	Ntv-a (nestin-Tv-a) $INK4a^{Lox/Lox}$; +MEK; + Cre (Cre induces recombination of INK4A in nestin-expressing cells)	Anaplastic Astrocytomas	60 %	[56]
	Mouse	Ntv-a (nestin-Tv-a) $INK4a^{Lox/Lox}$; +MEK; +Akt All cells remain INK4A wild type	Glioblastoma Oligoastrocytomas	60 %	
Tv-a + (ALV/RCAS) system	Mouse	Ntv-a (nestin-Tv-a) + Akt; + $KRas^{G12D}$ + Tet Tet-responsive element (TRE) inserted upstream of the K-Ras gene; the Akt or K-RAs become activated upon doxycycline injection	Glioblastoma	46 %	[57]
Tv-a + (ALV/RCAS) system + conditional knockout (Cre-loxP system)	Mouse	Ntv-a (nestin-Tv-a) $INK4a^{Lox/Lox}$; $KRas^{G12D}$; cre (Cre induces recombination of INK4A in nestin-expressing cells)	High-grade gliomas	100 %	[58]
		Ntv-a (nestin-Tv-a) $INK4a^{Lox/Lox}$; $BRAF^{V600E}$; cre (Cre induces recombination of INK4A in nestin-expressing cells)	Poorly differentiated tumors	~40 %	
		Ntv-a (nestin-Tv-a); $KRas^{G12D}$ + Akt	Invasive high-grade gliomas/ glioblastoma; angiogenesis	~50 %	
		Ntv-a (nestin-Tv-a); $BRAF^{V600E}$ + Akt	Highly pleomorphic glioblastomas	~50 %	
Heterozygous of tumor suppressor gens	Mouse	$NF1^{+/-}$; $p53^{+/-}$ tumor incidence depends on specific genetic backgrounds of mice	Low-grade astrocytomas, anaplastic astrocytomas, and glioblastomas	55–75 %	[59]
Inducible transgenic expression of oncogenic viral sequence, Heterozygous tumor suppressor genes	Mouse	GFAP; $SV40T_{121}$ ^{loxP}; $PTEN^{+/-}$GFAP; Lox $SV40T_{121}$ ^{loxP}; $p53^{+/-}$transgenic expression of T121 (a truncated SV40 T antigen) under the GFAP promoter Cre-mediated activation of $SV40T_{121}$, (removal of flanking loxP sites using an β actin-cre deleter mouse) resulting in inactivation of pRb, p107, and p130 in GFAP-expressing cells; in a PTEN or p53 heterozygous background PTEN but not p53 heterozygousity accelerate tumor development	Multifocal anaplastic astrocytomas and astrocyte dysplasia	100 %	[60]
Inducible transgenic expression of oncogenic viral sequence, conditional tumor suppressor gene	Mouse	GFAP; $SV40T_{121}$ ^{loxP}; $PTEN^{loxP/loxP}$ Recombination of PTEN by Adenovirus-mediated cre delivery to the brain (frontal cortex)—incidence not given, but referred to previous publication [60]	Multifocal anaplastic astrocytomas	100 %*	[61]

(continued)

Table 6.2 (continued)

Technology	Species	Description	Resulting tumors	Approximate Tumor Incidence	References
Tumor induction by transgenic Cre-mediated tumor suppressor gene inactivation	Mouse	hGFAP-Cre; $p53^{loxP/loxP}$; $PTEN^{lox/+}$ Inactivation of one PTEN allele and both p53 alleles by cre-mediated recombination in GFAP-expressing cells. GFAP is expressed during development, leading to a recombination already in newborn astrocytes and neural progenitors	Anaplastic astrocytomas, glioblastomas	~70 %	[62]
		hGFAP-Cre; $p53^{loxP/loxP}$ Inactivation of both p53 alleles by cre-mediated recombination in GFAP-expressing cells (i.e., astrocytes, neural progenitors, and SVZ stem cells)	Anaplastic astrocytomas	~20 %	
Tumor induction by transgenic Cre-mediated tumor suppressor gene inactivation	Mouse	hGFAP-Cre; NF1 $^{loxP/loxP}$; $p53^{-/-}$Inactivation of bothNF1 alleles by cre-mediated recombination in GFAP-expressing cells (i.e., astrocytes, neural progenitors and SVZ stem cells), in a p53 null background	Low-grade astrocytomas (30 %), anaplastic astrocytomas (60 %), and glioblastoma (15 %)	100 %	[10, 63]
		hGFAP-Cre; NF1 $^{loxP/loxP}$; $p53^{-/-}$Inactivation of bothNF1 alleles by cre-mediated recombination in GFAP-expressing cells (i.e., astrocytes, neural progenitors, and SVZ stem cells), in a $p53^{+/-}$ (heterozygous) background	Anaplastic Astrocytomas	5 %	
		hGFAP-Cre; NF1 $^{lox/lox}$; $p53^{loxP/-}$inactivation of bothNF1 and p53 alleles by cre-mediated recombination in GFAP-expressing cells (i.e., astrocytes, neural progenitors, and SVZ stem cells), in a p53 (heterozygous) background	Anaplastic astrocytomas (35 %) and glioblastoma (65 %)	100 %	
		Nestin-cre ER(T2); $NF1^{loxP/+}$; $p53^{loxP/loxP}$; $PTEN^{loxP/+}$ Nestin-cre ER(T2); $NF1^{loxP/loxP}$; $p53^{loxP/loxP}$; cre recombinase-modified estrogen receptor ligand-binding domain fusion protein (cre-ERT2) under the control of the nestin promoter/enhancer. Tamoxifen administration induces nuclear transfer of the cre-ERT2 protein in nestin-expressing cells, where it mediates loxP-dependent recombination. E13 embryos and adult mice were tamoxifen treated and both developed tumors Tumors seen throughout most of the brain parenchyma	Anaplastic astrocytomas (35 %) and glioblastoma (65 %)	93 %	[64]
		GFAP-CreERTM $p53^{loxP/loxP}$; $PTEN^{loxP/oxP}$ Widespread tumor development due to global activation of cre	Anaplastic astrocytomas and glioblastoma	100 %	[65]
		GFAP-CreERTM $p53^{loxP/loxP}$; $Rb^{loxP/oxP}$ Widespread tumor development due to global activation of cre	High-grade gliomas	68 %	

Tumor induction by adenoviral Cre-mediated tumor suppressor gene inactivation	$NF1^{loxP/loxP}$; $p53^{loxP/loxP}$ $NF1^{loxP/+}$; $p53^{loxP/loxP}$; $PTEN^{loxP/+}$ $NF1^{loxP/loxP}$; $p53^{loxP/-}$ Recombination in newborn or adult mice following Adenovirus injection into the SVZ	Anaplastic astrocytomas and Glioblastoma	100 %	[64]
	$p53^{loxP/loxP}$; $PTEN^{loxP/oxP}$ Recombination in stem/progenitor cells following adenovirus injection into the SVZ of adult mice	Anaplastic oligoastrocytomas	25 %	[37]
	$Rb^{loxP/loxP}$; $p53^{loxP/loxP}$; $PTEN^{loxP/oxP}$ Recombination in stem/progenitor cells following adenovirus injection into the SVZ of adult mice	Primitive neuroectodermal tumors (PNET)	40 %	
	$Rb^{loxP/loxP}$; $p53^{loxP/loxP}$; Recombination in stem/progenitor cells following adenovirus injection into the SVZ of adult mice	Primitive neuroectodermal tumors (PNET)	25 %	

This table systematically summarizes the model systems discussed in the text. Not all models develop gliomas, but they are important to understand the history, development, and refinement of model systems of intrinsic brain tumors

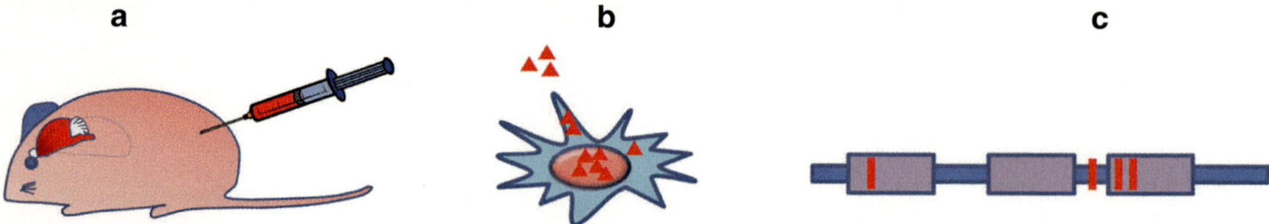

Fig. 6.2 Chemical systemic carcinogenesis. (**a**). Organism: Hamster, rats, mice, and others. Organotropic action depends on a combination of species, their genetic background, and the carcinogen. In early models, the carcinogen was injected or implanted directly into the brain, while in later models the carcinogen was systemically applied. (**b**) Carcinogen is resorbed by the cell and acts in the nucleus. The cell type that gave rise to tumors in these model systems models was probably an astrocyte progenitor. (**c**) Point mutations occur in the entire genome

analysis [1, 45, 70] showed tumors of the brain, the spinal cord, and the cranial and peripheral nerves. An important observation was the detection of a considerable number of "microtumors" in the brain which were not detectable grossly [1] and sometimes concurrent with contralateral macrotumors. The vast majority of these tumors corresponded to glioblastomas, astrocytomas, and oligodendrogliomas. It is noted that generally the pleomorphism of these tumors was considerably higher than in the human counterparts [45]. Another important observation, in particular in the view of many current hypotheses of the subventricular zone as a potential origin of brain tumors, is the occurrence of oligodendroglial tumors in the SVZ or in the ventricle (30 %) and in the white matter (45 %), while only a minority was located in the gray matter (15 %) (Table 6.2). Another important observation was the presence of pure oligodendrogliomas in the early phase, while more advanced forms showed a progressive admixture of astrocytic elements [45].

Virus-Induced Primary Brain Tumors

In 1962, the oncogenic potential of human adenovirus 12 in Syrian hamsters was discovered [71], but only extrinsic tumors were found [72]. A variety of strains of the Rous sarcoma virus (RSV), discovered in 1911, were subsequently derived and were found to be effective in the nervous system of dogs and to a lesser extent in rodents, including mice (Fig. 6.3): Gliomas and choroid plexus papillomas were generated by intracerebral inoculation of RSV in hamsters [28], dogs [73], and rabbits [74]. In mice, cerebellar medulloepitheliomas [46], forebrain gliomas [47], or neuroblastomas were described. Later, intracerebral inoculation of SV40 into newborn hamsters (reviewed in [48]) resulted in choroid plexus carcinomas.

Brain Tumors in Genetically Modified Mice

Following two decades of increasing refinement of virus injection, selections of virus strains, and characterizations of tumors arising in mice and other rodents that had received viruses, a new era of experimental neuro-oncology started with the avenue of transgenic mice. Henceforth, almost all experimental models were established in mice, and the use of hamsters, rats, and other species rapidly lost its relevance. Too great were the advantages and the potential of genetically modified mice, first transgenic mice, later knockout mice, and finally conditional knockout mice, in combination with cre-expressing transgenic mice or with cre-expressing expression vectors, for example, adenoviruses. Further modification of the technology allowed the cre-inducible expression or ablation of transgenes.

The technology used for the approaches will be described within each paragraph.

Transgenic Expression of Oncogenes

The first transgenic model of a brain tumor (albeit not yet a glioma) was engineered by expressing

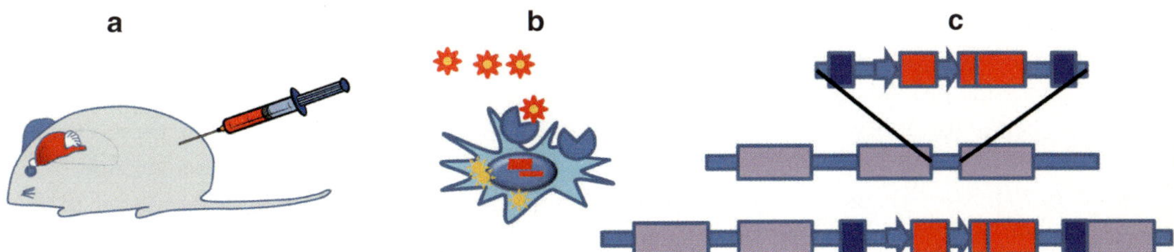

Fig. 6.3 Viral carcinogenesis. (**a**) Organism: Hamster, rats, mice, and others. Organotropic action depends on a combination of species, their genetic background, and the virus strain. (**b**) The virus docks onto cells using a (naturally existing) receptor and replicates using the cell's machinery. Retroviruses integrate into the genome. They transform the cell. (**c**) Genomic integration of a retrovirus. Retroviruses disrupt and potentially activate endogenous genes. In the case of RSV, the oncogenic v-src kinase is expressed

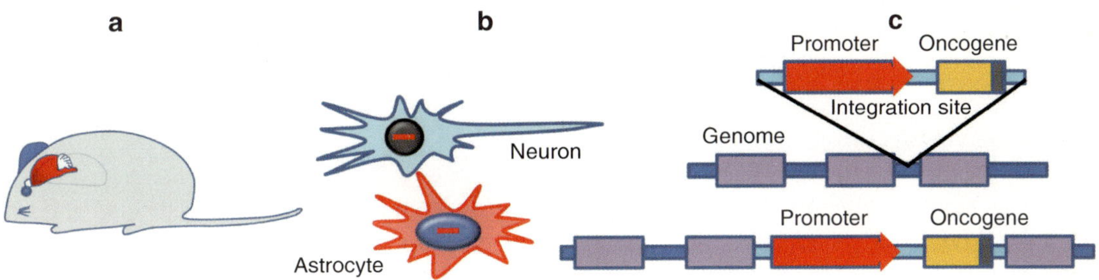

Fig. 6.4 Transgenic expression of an oncogene. (**a**) Organism: Mice. Organotropic or cell-specific action depends on the promoter that drives the expression of the oncogenic protein. (**b**) The transgene is integrated in every single cell of the organism. In this example it is illustrated that all cells (i.e., including astrocytes and neurones) carry the transgene, but only those cells in which the promoter is activated (e.g., GFAP in astrocytes) actually express the transgene (*red cell*). (**c**). Transgenes randomly integrate into the genome. Their expression mostly depends on the promoter (*Red arrow*), while the effect depends on the oncogenic signal (*yellow*). The transgene's expression pattern can further vary depending on the integration site

the SV40 early region coding for the large and small T antigens, under the control of a metallothionein promoter/enhancer [50] (Fig. 6.4) (Table 6.2). Three variations of the construct were injected into fertilized eggs, ensuring that all cells express the transgene, and resulted in the formation of malignant choroid plexus tumors [50, 51]. Several years earlier, Jaenisch and Mintz [75] had attempted to induce tumors by injecting SV40 DNA into the blastocoel cavity of mouse embryos, but never detected tumors, presumably due to an incomplete mosaicism of the resulting animals. An important question emerged from these experiments: Why is the SV40 T antigen active in epithelial cells of the choroid plexus, but not in other cells of the CNS? The plasmids were injected into fertilized eggs, and in transgenic mice, every cell carries identical, integrated copies of the SV-MK or SV-MGH genes (Fig. 6.4). Yet, in many different transgenic mice, each with a presumed random integration site, the primary site, of oncogenesis is the choroid plexus. This suggests that the site of integration is not important. Thus, development of brain tumors is not a consequence of T antigen being present in all cells, with the choroid plexus being the most sensitive target tissue. Instead, it appears that the choroid plexus is more permissive for T-antigen gene activation. It is possible that some mechanism inactivates the injected SV40 genes during early development and then an infrequent event activates the gene in a few cells during later development.

For a number of years, the expression of the SV40 large T antigen under the control of various promoters remained the main approach to generate brain tumors. The SV40 large T antigen binds and suppresses the protein products of the tumor suppressor genes Rb and p53 (and other members of the "pocket protein" family, p107 and p130) [76–79]. An elegant follow-up study expressed truncation mutants of SV40, with defective p53 suppression in the context of a p53 null mutation to tease out the role of the tumor suppressor p53. This model demonstrated that p53 is not required for the initiation but for the progression of choroid plexus tumors [80].

A step toward generating a mouse model for intrinsic brain tumors was the generation of a model where the large SV40 T antigen was expressed under the control of the Moloney murine sarcoma virus (MSV) enhancer and the SV 40 promoter. These mice developed uniform midline brain neoplasms with features of primitive neuroectodermal tumors [81].

A more refined approach was used in subsequent models: Instead of using promoters with no tissue specificity, the next generation of mouse models expressed potent oncogenes under the control of a cell- or tissue-specific promoter, aiming at the expression in cell types that were thought to represent the likely origin of brain tumors (Fig. 6.4): GFAP for astrocytomas and oligodendrocytes promoters for the generation of oligodendrogliomas. For example, aiming to develop a mouse model for oligodendrogliomas, the c-neu oncogene (tyrosine kinase) was expressed under the control of a myelin basic protein promoter, directing the expression to mature oligodendrocytes. With hindsight the oligodendrocyte was not an ideal cell type to induce oncogenesis, but at the time it was assumed that these cells may be the progenitor of oligodendrogliomas. This explains why MBP/c neu transgenic mice developed primitive neuroectodermal tumors with large bizarre cells, expressing GFAP and neurofilaments [4] (Table 6.2). Similarly, transgenic mice expressing the SV40 large T antigen under control of the GFAP promoter exhibited an early and lethal proliferation of cells of the periventricular subependymal zone of the immature brain, associated with strong expression of the transgene. The tumor cells showed a uniform (undifferentiated) cellular morphology and diffuse invasion and secondary structuring around neurons and blood vessels. Both these features were also clearly evident on transplantation of early passage cultures into the brains of mature nude mice. The localization of the microneoplasms around the SVZ suggests that the GFAP promoter initiated a neoplastic transformation in the germinative matrix in newborns, but there were also numerous dysplastic mature astrocytes scattered in the hemispheres. This suggests that the SV40 LT antigen predominantly transformed immature cells in the developing brain but not mature astrocytes [82]. A similar model was generated by expressing the viral oncogene v-src under control of the GFAP promoter. Src ($pp60^{c-src}$) is an intracellular tyrosine kinase expressed ubiquitously in mammalian cells, with highest levels encountered in brain and platelets [7]. v-src, the transforming gene of the Rous sarcoma virus (RSV), encodes $pp60^{v-src}$ which lacks the carboxy-terminal region comprising Tyr-527 (which switches off the kinase activity), resulting $pp60^{v-src}$ to be constitutively active [6]. 20 % of GFAP v-src mice developed multifocal astrocytomas [52] at 4 weeks of age. Similar to GFAP SV40 transgenic mice, the oncogenic signal is targeted to all cells expressing GFAP, which includes mature astrocytes [52] (Fig 6.2a), (Table 6.2). Additional deletion of the tumor suppressor gene p53 however did not increase the incidence or accelerate tumor growth [53]. While these models reliably generated astrocytomas and recapitulated some features, such as angiogenesis [83], they did in fact not formally address the question of the cell of origin of brain tumors for two reasons: First, the GFAP promoter is expressed in a variety of progenitor cells during CNS development and in the adult brain in stem cells as well as in mature astrocytes. This expression pattern obscures the identification of a cell of origin. Second, the "forced" expression of a transgene in a given cell type bypasses a natural preference of a cell type to undergo transformation. Therefore, this approach was soon replaced by more refined models using more advanced technologies.

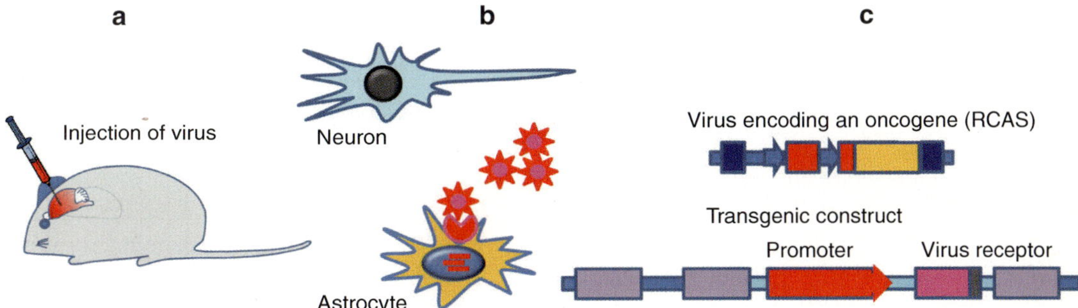

Fig. 6.5 Transgenic expression of a virus receptor and administration of an oncogene-expressing virus. (**a**) Organism: Mice. Organotropic or cell-specific action depends on the promoter that drives the expression of the receptor. Oncogenic action depends on the engineering of the virus. (**b**) The transgene encodes for a virus receptor (expressed on the surface of the desired cell), to match the genetically engineered virus expressing the oncogene of choice. The infected cell is shown in *yellow*. Cells which do not express the receptor will not be infected (neuron in this example). (**c**) Transgenes randomly integrate into the genome. Their expression pattern can vary depending on the integration site. The upper construct represents the viral expression vector, the *yellow* gene being the oncogene. The transgene expressing the virus receptor (*pink*) is integrated in the mouse genome but is active only in cells in which the transgene promoter (*red*) is active, for example, astrocytes (GFAP-*tv-a*)

Transgenic Expression of a Virus Receptor in Combination with a Genetically Engineered Virus Expressing Activated Oncogenes

In the late 1990s, a series of mouse models were generated in the laboratory of Harold Varmus. At the same time, significant progress had been made to identify genetic lesions in high-grade gliomas, which in turn allowed a more rational approach to mouse models. The most significant genetic lesions that were known at that time were p53 mutations, INK4a-ARF loss (resulting in loss of regulation of the downstream targets), Rb and p53, CDK4 amplification, EGFR amplification, and less commonly EGFRvIII mutations. Consequently the group engineered a flexible model system, which allowed for the (temporally and spatially controlled) delivery of a variety of oncogenes into a desired target cell population. Essentially, their system expressed a virus receptor in a specific cell type of the CNS (e.g., in astrocytes (GFAP) or in neural stem cells (nestin)) (Fig. 6.5). An elegant aspect and advantage over transgenic expression of an oncogene is the bypass of the developmental period. The gene transfer was accomplished by a genetically engineered retrovirus which delivers a gene of interest (e.g., FGF, EGFR, or CDK4) into the cell expressing the virus receptor.

Transgenic mice were engineered to express TVA, the avian cell surface receptor for the retrovirus ALV-A (avian leucosis virus—subgroup A), under the control of the GFAP promoter. These transgenic mice express the receptor, for example, in astrocytes, which renders them susceptible to ALV-A virus. The virus was genetically engineered to express multiple genes, including those enabling histological identification [84] (Table 6.2), (Fig. 6.5).

Using this system, a proof of principle (in vitro) experiment confirmed that *ex vivo* cultured mouse astrocytes expressing GFAP-*tv-a* can be infected with ALV expressing basic fibroblast growth factor. These cells did grow *in vitro*, showed increased proliferation and migration, and formed small clusters of transformed astrocytes following injection into a host mouse brain, but failed to form tumors [84]. Instead, a more aggressive phenotype was observed in cultures expressing CDK4 in astrocytes: In this setting, transformed astrocytes grew rapidly and became immortalized, similar to astrocytes with a deletion of the INK4A locus [21].

Having demonstrated the functionality *in vitro*, a follow-up study demonstrated the generation of gliomas *in vivo:* Transgenic mice expressing the *tv-a* receptor under control of the GFAP (*Gtv-a*) or the nestin promoter (*Ntv-a*) were infected with the virus expressing a constitutively active, mutant form of human EGFR with deletions of intra- and extracellular sequences (termed EGFR* by the authors). While *Gtv-a* and *Ntv-a* mice, transduced with EGFR*, did not develop any tumors, backcrossing into an INK4a heterozygous or INK4a null background increased the tumor incidence to nearly 50 % (Table 6.2). Further transduction with cdk4- and bFGF-expressing viruses increased the incidence to 10 % in the INK4a wild-type background. Instead, ablation of p53 had no significant effect.

It was concluded that the frequency of gliomagenesis is higher after infection of *Ntv-a* mice than of *Gtv-a* mice, probably because cells earlier in the glial lineage may be more susceptible to transformation than terminally differentiated astrocytes. Essentially, the constitutively active form of the EGF receptor can cooperate with mutations that disrupt the G1 cell cycle arrest pathways to induce lesions with some similarities to gliomas [22].

With the incremental discovery of new pathways involved in gliomagenesis, such as the Ras or the PTEN/Akt pathways, the system was further extended to test the role of these pathways in tumor initiation or progression in the CNS. Using the same receptor expressing mice (*Ntv-a* and *Gtv-a*) but a virus expressing the G12D mutant form of K-Ras or Akt was injected, respectively. Neither Ras or Akt alone could elicit a brain tumor in *Ntv-a* or *Gtv-a* mice, but only the combination of Ras and Akt in the context of nestin-Tv-a (*NTv-a*) mice resulted in malignant gliomas in 25 % of the animals [54] (Table 6.2).

This model was slightly modified and refined by separately backcrossing the GFAP (*Gtv-a*) or nestin-Tv-a (*Ntv-a*) mice into either the p16$^{Ink4a-/-}$ or the p19$^{ARF-/-}$ background. As before, mice were injected with Ras- and Akt-expressing vectors. Transduction of nestin-expressing cells was generally more effective than targeting GFAP-expressing cells, and the combination of Ras and Akt was more effective than expression of Akt or Ras alone. The highest rate of glioblastomas (83 %) was seen in *Gtv-a* mice transduced with Akt and K-Ras, in an $^{ARF-/-}$ background [55].

Using the same model system, Holmen and coworkers [57] injected *Ntv-a* mice with Ras, Akt, and tet on/off vectors. The tet on/off vector allows for a doxycycline-inducible activation of Ras and Akt. Prior to doxycycline administration, mice do not develop any tumors, but switching on the tet system (25 or 45 days duration) activates Ras and Akt gene expression activation, resulting in a glioblastoma incidence of nearly 50 %. More recently, the same group used the same system to express activated MEK (MAPK/extracellular signal-regulated kinase (ERK) kinase (MEK), a RAF effector), to induce tumors in vivo in the context of activated Akt or INK4a/Arf loss, showing that indeed activated MEK cooperates with INK4a/Arf loss or Akt activation to induce gliomas *in vivo* [56]. Similarly, activated (V600E mutant) BRAF again in the context of INK4a loss results in a high frequency of malignant gliomas, but also in poorly differentiated intrinsic tumors [56]. In human brain tumors, BRAF V600E mutations are seen in 5 % of pilocytic astrocytomas, 25 % of gangliogliomas, and nearly 70 % of pleomorphic xanthoastrocytomas [85].

Although the authors claim that these data imply a central role of these pathways in gliomagenesis, it should be considered that expression of strong oncogenic signal into any progenitor cell is likely to elicit a neoplastic transformation. It may be argued that this model, despite producing high-grade gliomas, still does not resolve essential questions about the histogenesis and the formal pathogenesis of brain tumors.

The Era of Conditional Gene Inactivation: A New Milestone Toward the Understanding of Brain Tumor Pathogenesis

In the mid-1990s, a new technology of *in vivo* gene inactivation revolutionized the way of disease modeling. Until then, inactivation of genes was done by replacing one or more functional exons in the open reading frame of a gene (null mutation). Because the null mutation is carried in the germ line of the mutant animals, it will exert its effects from the onset of animal development.

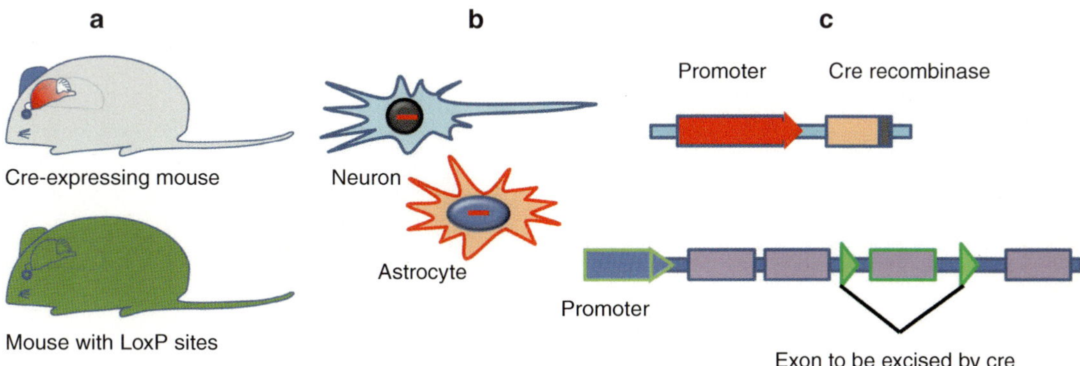

Fig. 6.6 Cre-loxP system (I). (**a**) Transgenic expression of the cre recombinase. Organotropic or cell specific action depends on the promoter that drives the expression of cre. Oncogenic action depends on the recombination of the target genes, which are flanked by loxP sites. loxP sites are in all cells of the body (*green* mouse). The cre mouse is generated as a separate line, where cre is expressed in an organ or cell of interest, e.g., brain. Cre and loxP mice must be crossed to generate a compound mutant. (**b**) The cre transgene is integrated in every single cell of the organism but is only expressed by those cells that activate the transgene promoter (in this example in astrocytes but not in neurons). (**c**) The upper construct is the transgene expressing cre recombinase (*orange*) under a cell-specific promoter (*red*). In a compound mutant mouse, cre recombinase recognizes pairs of loxP sites (*green triangles*) and forms a loop between them, hereby excising the DNA stretch between them. One loxP site remains in the genome (see Fig. 6.7). In glioma mouse models, often NF1, p53, or PTEN were used as target genes carrying loxP sites

This often results in early embryonic lethality, for example, in mice that were "knockout" for Rb [33], APC [18, 86–88], or PTEN [15, 16, 89]. Obviously, such an early embryonic lethality precluded further studies of the role of these tumor suppressor genes on the context of living organisms, requiring complementary *in vitro* studies, or the use of heterozygous or chimeric animals [18, 33, 86–89]. The breakthrough technology that could address these biological questions allowed for an inducible gene targeting in mice (Fig. 6.6) [14, 90, 91]. In this approach, the gene of interest (usually one or several exons) is engineered to carry a 32-base pair loxP recognition sequence on either side (i.e., 5′ and 3′) of the region to be excised (green triangles in the gene scheme in Fig. 6.6c). The excision of the loxP sites is accomplished by the action of the enzyme cre recombinase (Fig. 6.6c), which forms and excises a loop between the two loxP sites. The result is the removal of the sequence flanked by the two sites, and one single loxP site remains in the genome (Fig. 6.6c).

Several possibilities exist to achieve the expression of cre recombinase in the desired tissue. The most popular approach is the generation of a separate mouse line expressing cre recombinase under the control of a cell- or tissue-specific promoter (Fig. 6.6a). This cre-expressing mouse line is crossed with a mouse line carrying loxP sites (floxed) (Fig. 6.6a) in order to achieve a cell- or region-specific inactivation of the floxed gene (Fig. 6.7). For example, in order to achieve a functional inactivation of the PTEN gene in cells expressing GFAP (i.e., astrocytes, but also stem and progenitor cells), a GFAP-cre-expressing mouse is crossed with a PTEN conditional mouse to generate GFAP-cre and $PTEN^{loxP/loxP}$ genotype to achieve PTEN-negative astrocytes. However, there is an important caveat, which is essential for the correct interpretation of the phenotype: Cre-mediated recombination begins as soon as the cre transgene becomes activated. If this happens during development (as it is the case with the GFAP promoter), cre is transiently expressed in many neural precursor/progenitor cells, which results in a permanent deletion of the target gene, regardless of the fate of their progeny. For example, the GFAP-cre and $PTEN^{loxP/loxP}$ mouse mutant shows a widespread recombination not

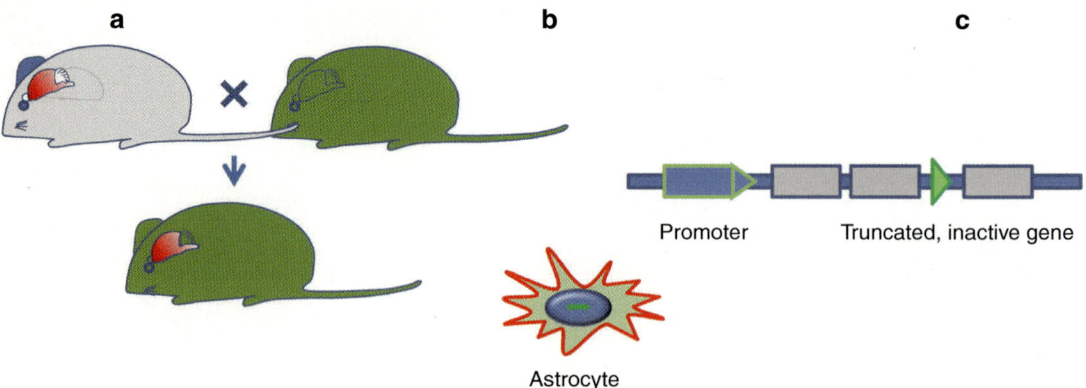

Fig. 6.7 Cre-loxP system (II). (**a**) The double mutants will have the gene of interest recombined in those cells which are expressing cre or which have previously expressed cre, for example, during development. Transient cre expression permanently excises the target sequence, leading to a deletion in all progeny. The *green* mouse is the floxed mouse, the *gray* mouse is the cre line, and the mouse below is the compound mutant, where every cell contains the cre and the floxed gene, but only those cells actually expressing cre will show recombination. (**b**) The target cells in which recombination took place (see also Fig. 6.6). (**c**) Target gene, in which the sequences between the loxP sites have been excised. The triangle indicates a remaining loxP site

only in mature astrocytes but has also a severe developmental phenotype with enlarged brains due to a disturbance of the neural precursor migration due to the effects of PTEN loss in neural stem and progenitor cells [92, 93]. An even more severe phenotype with embryonic lethality is observed in nestin-cre and $PTEN^{loxP/loxP}$ mice due to the widespread deletion of PTEN in a wide range of neural progenitors [94]. To circumvent this lethality, several techniques have been used: (1) a more restricted expression of cre, for example, using region-restricted promoters [95]; (2) a topical application of cre recombinase, for example, an adenovirus vector [96–99] and (3) an inducible cre transgene, that is, a mouse which expresses the cre transgene under the control of a cell-specific promoter, which has to be activated using the tamoxifen (cre ER(T) system, e.g., described in [100]).

Using the conditional gene inactivation technology, several studies demonstrated that high-grade intrinsic brain tumors of glial phenotype can be generated in mouse models where tumor suppressor genes are inactivated by conditional, cre-mediated gene expression in astrocytes, neural progenitor, and neural stem cells. Except for the mouse models from the vanDyke Group, where multifocal astrocytomas developed following inactivation of several Rb family members (Rb, p107 and p130), and of PTEN [60, 61] (Table 6.2), other models have in common an inactivation of the tumor suppressor gene p53 always in combination with other tumor suppressor genes, such as *Nf1*, *Pten,* and Rb [10, 37, 62–65] (Table 6.2). The mode of conditional gene inactivation in these models included adenovirus-mediated cre expression selective in the stem cell compartment of the SVZ [37], constitutive expressing in all stem progenitor cells using a GFAP-cre transgenic mouse line [10, 63, 64], in which the GFAP transgene constitutively activates Cre in all astrocytes, and stem cells from development through adult life (Fig. 6.8). A more refined, temporally controlled cre transgenic approach was used in a study where *Rb*, *p53,* and *PTEN* was recombined by a tamoxifen inducible cre expression. This model showed a recombination of parenchymal astrocyte as well as SVZ stem/progenitor cells [65]. Interestingly, the tumor phenotypes generated in this model were almost exclusively high-grade gliomas, in contrast to the findings of Jacques et al. [37], where *Rb/p53* and *Rb/Pten/p53* recombination in stem/progenitor cells resulted in a high proportion of primitive neuroectodermal tumors.

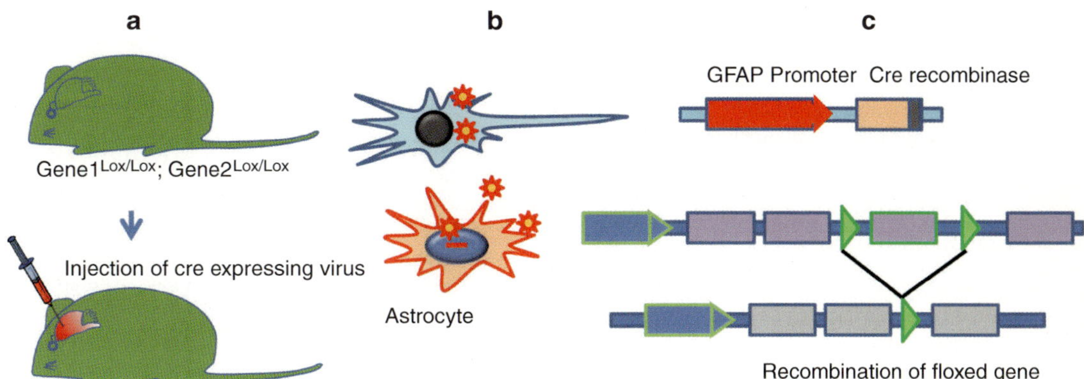

Fig. 6.8 Cre-loxP system (III). (**a**) Mice with loxP sites flanking tumor suppressor genes of interest (e.g., $p53^{Lox/Lox}$, $PTEN^{Lox/Lox}$) are injected with an adenovirus expressing cre recombinase. The virus is topically applied to minimize spread. (**b**) Cells that are infected with the adenovirus will undergo recombination. For example, an adeno-GFAP-cre virus can infect several cell types but will express cre only in GFAP-expressing astrocytes and stem cells. The neuron will be infected, but will not recombine the floxed genes. (**c**) Viral expression of cre (within viral construct). Below is the target gene, in which the sequences between the loxP sites have been recombined (excised)

These mouse models show striking similarities to human brain tumors and a change of the approach of modeling the human disease: In the history of modeling brain tumors, several key aims were pursued (see introduction) of which several important milestones have been addressed. This significant achievement was only possible in that research on the human disease mirrored the development of increasingly refined model systems and the stringency of the scientific question which drove these developments:

1. Generation of a tumor that is morphologically similar to a human counterpart. This was in principle already achieved in the carcinogenesis models in the 1960s, which showed striking similarities to human astrocytomas and oligodendrogliomas [1]. The limitation of these models was the unknown (and multiple) genetic lesions mechanism by which the tumors were induced (Fig. 6.2) and the scarcity of knowledge of molecular pathways involved in tumorigenesis.
2. Identification of the cell(s) of tumor origin. Addressing this issue is less straightforward as expression of a potent oncogene, an approach that had been pursued with transgenic models in the 1990s, results in tumors arising from cells that are forced to express the oncogene [4, 52], but did not necessarily address the question of the cell or origin. Recent work has narrowed down potential candidates [37].
3. Remodel genetic pathways to understand gliomagenesis in humans. This approach was successfully addressed following the identification of key glioma pathways and the availability of suitable conditional mouse models [62], including the comparison of genomic profiles of human and murine tumors, which has recently been achieved [65].

Conclusion

More than 6 decades of research in experimental neuro-oncology have resulted in the development and significant refinement of brain tumor models, in particular of gliomas. The current models allow for a better understanding of the cell of origin, the mechanisms leading to their malignant transformation, and the correlation of the experimental phenotype with the human counterpart. Further refinement of the model systems is now essential to extend their use to develop and test targeted cancer therapies.

References

1. Wechsler W, Kleihues P, Matsumot S, Zulch KJ, Ivankovi S, Preussma R. Pathology of experimental neurogenic tumors chemically induced during prenatal and postnatal life. Ann N Y Acad Sci. 1969; 159(A2):360.
2. Swenberg JA. Neoplasms of the nervous system. In: Foster H, Small J, JG F, editors. The mouse in biomedical research. New York: Academic; 1982. p. 529–37.
3. Fraser H, Mcconnell I. Experimental brain tumors. Lancet. 1975;1(7897):44.
4. Hayes C, Kelly D, Murayama S, Komiyama A, Suzuki K, Popko B. Expression of the neu oncogene under the transcriptional control of the myelin basic-protein gene in transgenic mice – generation of transformed glial-cells. J Neurosci Res. 1992;31(1):175–87.
5. Eibl RH, Kleihues P, Jat PS, Wiestler OD. A model for primitive neuroectodermal tumors in transgenic neural transplants harboring the SV40 large T antigen. Am J Pathol. 1994;144(3):556–64.
6. Cooper JA, Gould KL, Cartwright CA, Hunter T. Tyr527 is phosphorylated in Pp60c-Src – implications for regulation. Science. 1986;231(4744):1431–4.
7. Brugge J, Cotton P, Lustig A, Yonemoto W, Lipsich L, Coussens P, et al. Characterization of the altered form of the C-Src gene-product in neuronal cells. Genes Dev. 1987;1(3):287–96.
8. Greenhouse JJ, Petropoulos CJ, Crittenden LB, Hughes SH. Helper-independent retrovirus vectors with Rous-associated virus type-O long terminal repeats. J Virol. 1988;62(12):4809–12.
9. Hughes SH, Greenhouse JJ, Petropoulos CJ, Sutrave P. Adapter plasmids simplify the insertion of foreign DNA into helper-independent retroviral vectors. J Virol. 1987;61(10):3004–12.
10. Zhu Y, Guignard F, Zhao D, Liu L, Burns DK, Mason RP, et al. Early inactivation of p53 tumor suppressor gene cooperating with NF1 loss induces malignant astrocytoma. Cancer Cell. 2005;8(2):119–30.
11. Abremski K, Hoess R. Bacteriophage P1 site-specific recombination. Purification and properties of the cre recombinase protein. J Biol Chem. 1984;259(3): 1509–14.
12. Sternberg N, Hamilton D. Bacteriophage P1 site-specific recombination. I. Recombination between loxP sites. J Mol Biol. 1981;150(4):467–86.
13. Sternberg N, Hamilton D, Hoess R. Bacteriophage P1 site-specific recombination. II. Recombination between loxP and the bacterial chromosome. J Mol Biol. 1981;150(4):487–507.
14. Kuhn R, Schwenk F, Aguet M, Rajewsky K. Inducible gene targeting in mice. Science. 1995;269(5229): 1427–9.
15. Di Cristofano A, Pesce B, Cordon-Cardo C, Pandolfi PP. Pten is essential for embryonic development and tumour suppression. Nat Genet. 1998;19(4):348–55.
16. Stambolic V, Suzuki A, de la Pompa JL, Brothers GM, Mirtsos C, Sasaki T, et al. Negative regulation of PKB/Akt-dependent cell survival by the tumor suppressor PTEN. Cell. 1998;95(1):29–39.
17. Li J, Yen C, Liaw D, Podsypanina K, Bose S, Wang SI, et al. PTEN, a putative protein tyrosine phosphatase gene mutated in human brain, breast, and prostate cancer. Science. 1997;275(5308):1943–7.
18. Su LK, Kinzler KW, Vogelstein B, Preisinger AC, Moser AR, Luongo C, et al. Multiple intestinal neoplasia caused by a mutation in the murine homolog of the Apc gene. Science. 1992;256(5057):668–70.
19. Ashton-Rickardt PG, Dunlop MG, Nakamura Y, Morris RG, Purdie CA, Steel CM, et al. High frequency of APC loss in sporadic colorectal carcinoma due to breaks clustered in 5q21-22. Oncogene. 1989;4(10):1169–74.
20. Ichii S, Horii A, Nakatsuru S, Furuyama J, Utsunomiya J, Nakamura Y. Inactivation of both APC alleles in an early stage of colon adenomas in a patient with familial adenomatous polyposis (FAP). Hum Mol Genet. 1992;1(6):387–90.
21. Holland EC, Hively WP, Gallo V, Varmus HE. Modeling mutations in the G1 arrest pathway in human gliomas: overexpression of CDK4 but not loss of INK4a-ARF induces hyperploidy in cultured mouse astrocytes. Genes Dev. 1998;12(23):3644–9.
22. Holland EC, Hively WP, DePinho RA, Varmus HE. A constitutively active epidermal growth factor receptor cooperates with disruption of G1 cell-cycle arrest pathways to induce glioma-like lesions in mice. Genes Dev. 1998;12(23):3675–85.
23. Bates S, Phillips AC, Clark PA, Stott F, Peters G, Ludwig RL, et al. p14ARF links the tumour suppressors RB and p53 [letter]. Nature. 1998;395(6698): 124–5.
24. Palmero I, Pantoja C, Serrano M. p19ARF links the tumour suppressor p53 to Ras. Nature. 1998;395(6698):125–6.
25. Serrano M, Hannon GJ, Beach D. A new regulatory motif in cell-cycle control causing specific-inhibition of cyclin-D/Cdk4. Nature. 1993;366(6456):704–7.
26. von Deimling A, von Ammon K, Schoenfeld D, Wiestler OD, Seizinger BR, Louis DN. Subsets of glioblastoma multiforme defined by molecular genetic analysis. Brain Pathol. 1993;3(1):19–26.
27. Watanabe K, Tachibana O, Sata K, Yonekawa Y, Kleihues P, Ohgaki H. Overexpression of the EGF receptor and p53 mutations are mutually exclusive in the evolution of primary and secondary glioblastomas. Brain Pathol. 1996;6(3):217–23. discussion 23–4.
28. Rabotti GF, Raine WA, Sellers RL. Brain tumors (gliomas) induced in hamsters by Bryans strain of Rous sarcoma virus. Science. 1965;147(3657):505.
29. Mennel HD. Significance of experimental models in neurooncology. Neurosurg Rev. 1980;3(2):129–37.
30. Malkin D, Li FP, Strong LC, Fraumeni Jr JF, Nelson CE, Kim DH, et al. Germ line p53 mutations in a

familial syndrome of breast cancer, sarcomas, and other neoplasms [see comments]. Science. 1990;250(4985):1233–8.
31. Ohgaki H, Eibl RH, Schwab M, Reichel MB, Mariani L, Gehring M, et al. Mutations of the p53 tumor suppressor gene in neoplasms of the human nervous system. Mol Carcinog. 1993;8(2):74–80.
32. Levine AJ. p53, the cellular gatekeeper for growth and division. Cell. 1997;88(3):323–31.
33. Jacks T, Fazeli A, Schmitt EM, Bronson RT, Goodell MA, Weinberg RA. Effects of an Rb mutation in the mouse. Nature. 1992;359(6393):295–300.
34. Murphree AL, Benedict WF. Retinoblastoma: clues to human oncogenesis. Science. 1984;223(4640): 1028–33.
35. Alvarez Buylla A, Lois C. Neuronal stem cells in the brain of adult vertebrates. Stem Cells. 1995;13(3):263–72.
36. Doetsch F, Caille I, Lim DA, Garcia-Verdugo JM, Alvarez-Buylla A. Subventricular zone astrocytes are neural stem cells in the adult mammalian brain. Cell. 1999;97(6):703–16.
37. Jacques TS, Swales A, Brzozowski MJ, Henriquez NV, Linehan JM, Mirzadeh Z, et al. Combinations of genetic mutations in the adult neural stem cell compartment determine brain tumour phenotypes. EMBO J. 2010;29(1):222–35.
38. Jackson EL, Garcia-Verdugo JM, Gil-Perotin S, Roy M, Quinones-Hinojosa A, VandenBerg S, et al. PDGFR alpha-positive B cells are neural stem cells in the adult SVZ that form glioma-like growths in response to increased PDGF signaling. Neuron. 2006;51(2):187–99.
39. Peers JH. The response of the central nervous system to the application of carcinogenic hydrocarbons: I. Dibenzanthracene. Am J Pathol. 1939;15(2): 261–272.3.
40. Peers JH. The response of the central nervous system to the application of carcinogenic hydrocarbons: II. Methylcholanthrene. Am J Pathol. 1940;16(6): 799–816.5.
41. Seligman AM, Shear M. Studies in carcinogenesis. VIII. Experimental production of brain tumors in mice with methylcholanthrene. Am J Cancer. 1939;37:364–95.
42. Zimmerman HM, Arnold H. Experimental brain tumors: II. Tumors produced with benzpyrene. Am J Pathol. 1943;19(6):939–55.
43. Zimmerman HM, Arnold H. Experimental brain tumors. I. Tumors produced with methylcholanthrene. Cancer Res. 1941;1(12):919.
44. Magee PN, Barnes JM. The production of malignant primary hepatic tumours in the rat by feeding dimethylnitrosamine. Br J Cancer. 1956;10(1):114.
45. Mennel HD, Zulch KJ. Formal pathogenesis of experimentally induced brain tumors. Acta Neuropathol. 1972;21(2):140–53.
46. Mukai N, Kobayash S, Oguri M. Ultrastructural findings in medulloepitheliomatous neoplasms induced by human adenovirus 12 in rodents. Acta Neuropathol. 1974;28(4):293–304.
47. Kumanish T, Ikuta F, Yamamoto T. Brain tumors induced by Rous-sarcoma virus, Schmidt-Ruppin strain.3. Morphology of brain tumors induced in adult mice. J Natl Cancer Inst. 1973;50(1):95–109.
48. Jänisch W, Schreiber D. In: Bigner DD, Swenberg JA, editors. Methods of inducing experimental nervous system tumors: Simian sarcoma virus. Experimental tumors of the central nervous system. Kalamazoo, Michigan: Upjohn Company; 1977. p. 36.
49. Rabotti GF, Anderson WR, Sellers RL. Oncogenic activity of Mill Hill (Harris) strain of Rous sarcoma virus for hamsters. Nature. 1965;206(4987):946.
50. Brinster RL, Chen HY, Messing A, Vandyke T, Levine AJ, Palmiter RD. Transgenic mice harboring Sv40 T-antigen genes develop characteristic brain-tumors. Cell. 1984;37(2):367–79.
51. Palmiter RD, Chen HY, Messing A, Brinster RL. Sv40 Enhancer and large-T antigen are instrumental in development of choroid-plexus tumors in transgenic mice. Nature. 1985;316(6027):457–60.
52. Weissenberger J, Steinbach JP, Malin G, Spada S, Rulicke T, Aguzzi A. Development and malignant progression of astrocytomas in GFAP-v-src transgenic mice. Oncogene. 1997;14(17):2005–13.
53. Maddalena AS, Hainfellner JA, Hegi ME, Glatzel M, Aguzzi A. No complementation between TP53 or RB-1 and v-src in astrocytomas of GFAP-v-src transgenic mice. Brain Pathol. 1999;9(4):627–37.
54. Holland EC, Celestino J, Dai C, Schaefer L, Sawaya RE, Fuller GN. Combined activation of Ras and Akt in neural progenitors induces glioblastoma formation in mice. Nat Genet. 2000;25(1):55–7.
55. Uhrbom L, Kastemar M, Johansson FK, Westermark B, Holland EC. Cell type-specific tumor suppression by Ink4a and Arf in Kras-induced mouse gliomagenesis. Cancer Res. 2005;65(6):2065–9.
56. Robinson JP, Vanbrocklin MW, Lastwika KJ, McKinney AJ, Brandner S, Holmen SL. Activated MEK cooperates with Ink4a/Arf loss or Akt activation to induce gliomas in vivo. Oncogene. 2011;30(11):1341–50.
57. Holmen SL, Williams BO. Essential role for Ras signaling in glioblastoma maintenance. Cancer Res. 2005;65(18):8250–5.
58. Robinson JP, VanBrocklin MW, Guilbeault AR, Signorelli DL, Brandner S, Holmen SL. Activated BRAF induces gliomas in mice when combined with Ink4a/Arf loss or Akt activation. Oncogene. 2010;29(3):335–44.
59. Reilly KM, Loisel DA, Bronson RT, McLaughlin ME, Jacks T. Nf1; Trp53 mutant mice develop glioblastoma with evidence of strain-specific effects. Nat Genet. 2000;26(1):109–13.
60. Xiao A, Wu H, Pandolfi PP, Louis DN, Van Dyke T. Astrocyte inactivation of the pRb pathway predisposes mice to malignant astrocytoma development that is accelerated by PTEN mutation. Cancer Cell. 2002;1(2):157–68.

61. Xiao A, Yin C, Yang C, Di Cristofano A, Pandolfi PP, Van Dyke T. Somatic induction of Pten loss in a preclinical astrocytoma model reveals major roles in disease progression and avenues for target discovery and validation. Cancer Res. 2005;65(12):5172–80.
62. Zheng H, Ying H, Yan H, Kimmelman AC, Hiller DJ, Chen AJ, et al. p53 and Pten control neural and glioma stem/progenitor cell renewal and differentiation. Nature. 2008;455(7216):1129–33.
63. Kwon CH, Zhao D, Chen J, Alcantara S, Li Y, Burns DK, et al. Pten haploinsufficiency accelerates formation of high-grade astrocytomas. Cancer Res. 2008;68(9):3286–94.
64. Alcantara Llaguno S, Chen J, Kwon CH, Jackson EL, Li Y, Burns DK. Malignant astrocytomas originate from neural stem/progenitor cells in a somatic tumor suppressor mouse model. Cancer Cell. 2009;15(1):45–56.
65. Chow LM, Endersby R, Zhu X, Rankin S, Qu C, Zhang J, et al. Cooperativity within and among Pten, p53, and Rb pathways induces high-grade astrocytoma in adult brain. Cancer Cell. 2011;19(3):305–16.
66. Boorman GA. Brain-tumors in Man and animals – report of a workshop. Environ Health Perspect. 1986;68:155–73.
67. Fuller C, Helton K, Dalton J, Fouladi M, Kun L, Burger P. Polar spongioblastoma of the spinal cord: a case report. Pediatr Dev Pathol. 2006;9(1):75–80.
68. Brodie BB, Gillette JR, Ladu BN. Enzymatic metabolism of drugs and other foreign compounds. Annu Rev Biochem. 1958;27:427–54.
69. Druckrey H, Preussmann R, Ivankovic S, Schmahl D. Organotropic carcinogenic effects of 65 various N-nitroso- compounds on BD rats. Z Krebsforsch. 1967;69(2):103–201.
70. Zimmerman HM. Brain tumors – their incidence and classification in man and their experimental production. Ann N Y Acad Sci. 1969;159(A2):337.
71. Trentin JJ, Yabe Y, Taylor G. The quest for human cancer viruses. Science. 1962;137(3533):835–41.
72. Huebner RJ, Rowe WP, Lane WT. Oncogenic effects in hamsters of human adenovirus types 12 and 18. Proc Natl Acad Sci U S A. 1962;48(12):2051.
73. Rabotti GF, Grove AS, Sellers RL, Anderson WR. Induction of multiple brain tumours (gliomata and leptomeningeal sarcomata) in dogs by Rous sarcoma virus. Nature. 1966;209(5026):884.
74. Rabotti GF, Sellers RL, Anderson WA. Leptomeningeal sarcomata and gliomata induced in rabbits by Rous sarcoma virus. Nature. 1966;209(5022):524.
75. Jaenisch R, Mintz B. Simian virus 40 DNA sequences in DNA of healthy adult mice derived from preimplantation blastocysts injected with viral DNA. Proc Natl Acad Sci U S A. 1974;71(4):1250–4.
76. Green MR. When the products of oncogenes and anti-oncogenes meet. Cell. 1989;56(1):1–3.
77. Peden KWC, Srinivasan A, Farber JM, Pipas JM. Mutants with changes within or near a hydrophobic region of simian virus-40 large tumor-antigen are defective for binding cellular protein-P53. Virology. 1989;168(1):13–21.
78. Decaprio JA, Ludlow JW, Figge J, Shew JY, Huang CM, Lee WH, et al. Sv40 Large tumor-antigen forms a specific complex with the product of the retinoblastoma susceptibility gene. Cell. 1988;54(2):275–83.
79. Ewen ME, Ludlow JW, Marsilio E, Decaprio JA, Millikan RC, Cheng SH, et al. An N-terminal transformation-governing sequence of Sv40 large T-antigen contributes to the binding of both P110rb and a 2nd cellular protein, P120. Cell. 1989;58(2):257–67.
80. Symonds H, Krall L, Remington L, Saenzrobles M, Lowe S, Jacks T, et al. P53-Dependent apoptosis suppresses tumor-growth and progression in-vivo. Cell. 1994;78(4):703–11.
81. Gotz W, Theuring F, Schachenmayr W, Korf HW. Midline brain-tumors in Msv-Sv 40-transgenic mice originate from the pineal organ. Acta Neuropathol. 1992;83(3):308–14.
82. Danks RA, Orian JM, Gonzales MF, Tan SS, Alexander B, Mikoshiba K, et al. Transformation of astrocytes in transgenic mice expressing Sv40 T-antigen under the transcriptional control of the glial fibrillary acidic protein promoter. Cancer Res. 1995;55(19):4302–10.
83. Theurillat JP, Hainfellner J, Maddalena A, Weissenberger J, Aguzzi A. Early induction of angiogenetic signals in gliomas of GFAP-v-src transgenic mice. Am J Pathol. 1999;154(2):581–90.
84. Holland EC, Varmus HE. Basic fibroblast growth factor induces cell migration and proliferation after glia-specific gene transfer in mice. Proc Natl Acad Sci U S A. 1998;95(3):1218–23.
85. Schindler G, Capper D, Meyer J, Janzarik W, Omran H, Herold-Mende C. Analysis of BRAF V600E mutation in 1,320 nervous system tumors reveals high mutation frequencies in pleomorphic xanthoastrocytoma, ganglioglioma and extra-cerebellar pilocytic astrocytoma. Acta Neuropathol. 2011;121(3):397–405.
86. Moser AR, Shoemaker AR, Connelly CS, Clipson L, Gould KA, Luongo C, et al. Homozygosity for the Min allele of Apc results in disruption of mouse development prior to gastrulation. Dev Dyn. 1995;203(4):422–33.
87. Oshima M, Oshima H, Kitagawa K, Kobayashi M, Itakura C, Taketo M. Loss of Apc heterozygosity and abnormal tissue building in nascent intestinal polyps in mice carrying a truncated Apc gene. Proc Natl Acad Sci U S A. 1995;92(10):4482–6.
88. Fodde R, Edelmann W, Yang K, Vanleeuwen C, Carlson C, Renault B, et al. A targeted chain-termination mutation in the mouse Apc gene results in multiple intestinal tumors. Proc Natl Acad Sci U S A. 1994;91(19):8969–73.
89. Suzuki A, de la Pompa JL, Stambolic V, Elia AJ, Sasaki T, Barrantes ID, et al. High cancer susceptibility and embryonic lethality associated with mutation of the PTEN tumor suppressor gene in mice. Curr Biol. 1998;8(21):1169–78.

90. Gu H, Zou YR, Rajewsky K. Independent control of immunoglobulin switch recombination at individual switch regions evidenced through Cre-IoxP-mediated gene targeting. Cell. 1993;73(6):1155–64.
91. Gu H, Marth JD, Orban PC, Mossmann H, Rajewsky K. Deletion of a DNA-polymerase-beta gene segment in T-cells using cell-type-specific gene targeting. Science. 1994;265(5168):103–6.
92. Kwon CH, Zhu X, Zhang J, Knoop LL, Tharp R, Smeyne RJ, et al. Pten regulates neuronal soma size: a mouse model of Lhermitte-Duclos disease. Nat Genet. 2001;29(4):404–11.
93. Backman SA, Stambolic V, Suzuki A, Haight J, Elia A, Pretorius J, et al. Deletion of Pten in mouse brain causes seizures, ataxia and defects in soma size resembling Lhermitte-Duclos disease. Nat Genet. 2001;29(4):396–403.
94. Groszer M, Erickson R, Scripture-Adams D, Lesche R, Trumpp A, Zack JA. Negative regulation of neural stem/progenitor cell proliferation by the pten tumor suppressor gene in vivo. Science. 2001;294(5549):2186–9.
95. Marino S, Krimpenfort P, Leung C, van der Korput HA, Trapman J, Camenisch I, et al. PTEN is essential for cell migration but not for fate determination and tumourigenesis in the cerebellum. Development. 2002;129(14):3513–22.
96. Anton M, Graham FL. Site-specific recombination mediated by an adenovirus vector expressing the cre recombinase protein: a molecular switch for control of gene expression. J Virol. 1995;69(8): 4600–6.
97. Kaspar BK, Vissel B, Bengoechea T, Crone S, Randolph-Moore L, Muller R, et al. Adeno-associated virus effectively mediates conditional gene modification in the brain. Proc Natl Acad Sci U S A. 2002;99(4):2320–5.
98. Ahmed BY, Chakravarthy S, Eggers R, Hermens WT, Zhang JY, Niclou SP, et al. Efficient delivery of Cre-recombinase to neurons in vivo and stable transduction of neurons using adeno-associated and lentiviral vectors. BMC Neurosci. 2004;5:4.
99. Wang H, Xie H, Zhang H, Das SK, Dey SK. Conditional gene recombination by adenovirus-driven Cre in the mouse uterus. Genesis. 2006;44(2):51–6.
100. Mori T, Tanaka K, Buffo A, Wurst W, Kuhn R, Gotz M. Inducible gene deletion in astroglia and radial glia–a valuable tool for functional and lineage analysis. Glia. 2006;54(1):21–34.

7 The Application of Novel Ionizing Radiation Species for Glioblastoma

Raj Jena

Abstract

Radiation therapy is the mainstay of postoperative treatment for patients with glioblastoma. However, the percentage of patients who achieve long-term cure with radical radiation therapy remains poor. For this reason, there has been considerable interest in improving the outcomes for radiation therapy by evaluating novel ionizing radiation species, specifically highly energetic particles. This chapter reviews the differences in physical and biological properties between x-rays and particle beams in order to highlight potential benefits both for improved tumor cell kill and reduction of normal tissue effects. In order to understand these differences, some key radiation biology concepts will be discussed, before evaluating the evidence both from in vitro and in vivo studies. Most of the techniques for treatment focus on the use of external beam therapy, where an ionizing radiation originating from outside the patient is used to treat an intracranial tumor. In the final section of this chapter, the emerging technique of boron neutron capture therapy (BNCT) will also be discussed.

Keywords

Glioblastoma • Radiotherapy • Particle therapy • Proton therapy • Neutron capture therapy

Dose and Volume Interactions in Radiation Therapy

Clinical trials from the early 1980s demonstrated that radiotherapy stabilizes neurological function and provides modest improvements in overall survival [1]. However, radiation therapy for glioblastoma is a noncurative treatment for the vast majority of patients. The reason for this observation lies in the underlying radiation biology of the disease. The first issue is that the underlying sensitivity of glioblastoma is low. After a 2 Gy fraction of radiotherapy, the fraction of surviving tumor cells in most epithelial cancers in vitro is in the order of 49–55 % [2]. Glioblastoma is a highly heterogeneous tumor,

R. Jena, M.D., MRCP, FRCR
Department of Oncology,
Cambridge University Hospitals NHS Foundation Trust,
Hills Road, Cambridge CB2 0QQ, UK
e-mail: rjena@nhs.net

C. Watts (ed.), *Emerging Concepts in Neuro-Oncology*,
DOI 10.1007/978-0-85729-458-6_7, © Springer-Verlag London 2013

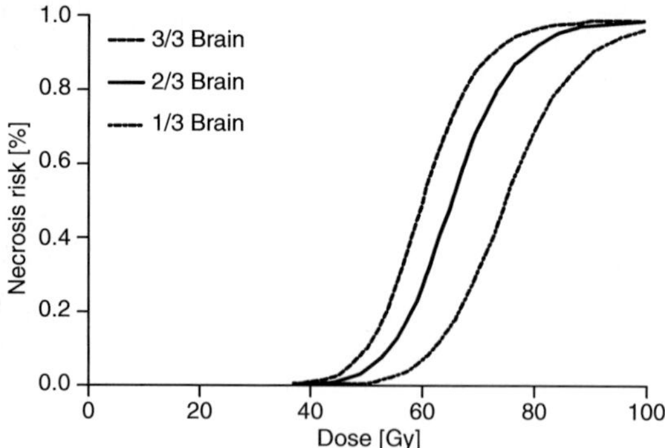

Fig. 7.1 Relationship between radiation dose and the risk of brain necrosis according to the volume of irradiated brain (Adapted from Emami et al. [5]. Used with permission)

and estimates of surviving fraction range from 46 % up to 87 % [3].

In other disease sites, tumor control for radioresistant tumors can be achieved by escalating the total radiation dose. A prime example is the treatment of prostate cancer, where improved control rates in higher-grade tumors have been achieved by escalation of radiation dose [4]. However, glioblastoma is a highly infiltrative tumor, and the target for radiation therapy by necessity must include both the tumor core and the surrounding zone of infiltrated brain tissue. Typically a 2- to 3-cm margin is added to the tumor bed to allow for this infiltration, and much of this volume will contain functioning brain tissue. The risk of injury to brain tissue, specifically radiation-induced brain necrosis, is a function of both radiation dose and irradiated volume (Fig. 7.1) [5]. From this relationship, it is self-evident that in order to achieve safe dose escalation, the volume of brain irradiated to higher dose must be reduced. Particle beams display useful physical and biological properties for this purpose.

X-Rays and Charged Particles for External Beam Therapy

The concept of using charged particle beams for therapeutic purposes is not new. Wilson first established the potential benefit for charged particle beams in 1946, and the first clinical treatments were performed in physics laboratories at the University of California, Berkeley, in 1954. The reason for the interest is apparent from the way in which x-rays and particle beams deposit energy into tissues (Fig. 7.2).

The depth dose curve is a visual representation of the pattern of radiation dose deposition in the body. It is clear that around the point of maximum energy deposition, high-energy x-ray beams have a fairly smooth profile of dose deposition. This means that x-ray beams can be used to provide a homogeneous dose of radiation through a block of tissue. In contrast, a highly energetic particle beam will deposit relatively little dose as it enters tissue. As it interacts with matter, the particle gradually loses momentum, and as it does so, the amount of energy liberated into a volume of tissue increases. Eventually the particle will release its remaining kinetic energy as it comes to a halt at a point known as the Bragg peak. Once the particle has arrested, no further dose will be deposited along its path. If the energy of a particle beam is chosen correctly, it becomes possible for the peak energy deposition to be placed in the center of a tumor target and the dose to surrounding normal tissues to be minimized (Fig. 7.3).

The second benefit for charged particle beams relates to their biological effect. X-ray beams induce damage to DNA through the liberation of secondary electrons, which induce both single-stranded and double-stranded DNA breaks. In the absence of oxygen, the DNA strands may spontaneously reanneal. The time course of this radiochemical interaction is of the order of 10^{-10} s. However, in the presence of oxygen, the DNA breaks may be stabilized by the binding of oxygen to each end of the DNA strand. In contrast, highly energetic particles traversing the nucleus will interact directly with chromatin, producing a larger defect in DNA that is less dependent on the availability of oxygen. The frequency of these larger DNA breaks is dependent on the amount of energy the particle deposits within a given distance of its trajectory. This is known as the linear energy transfer (LET) of the beam. X-ray beams have a uniform linear energy transfer across most of the effective path of the beam. As a result, there

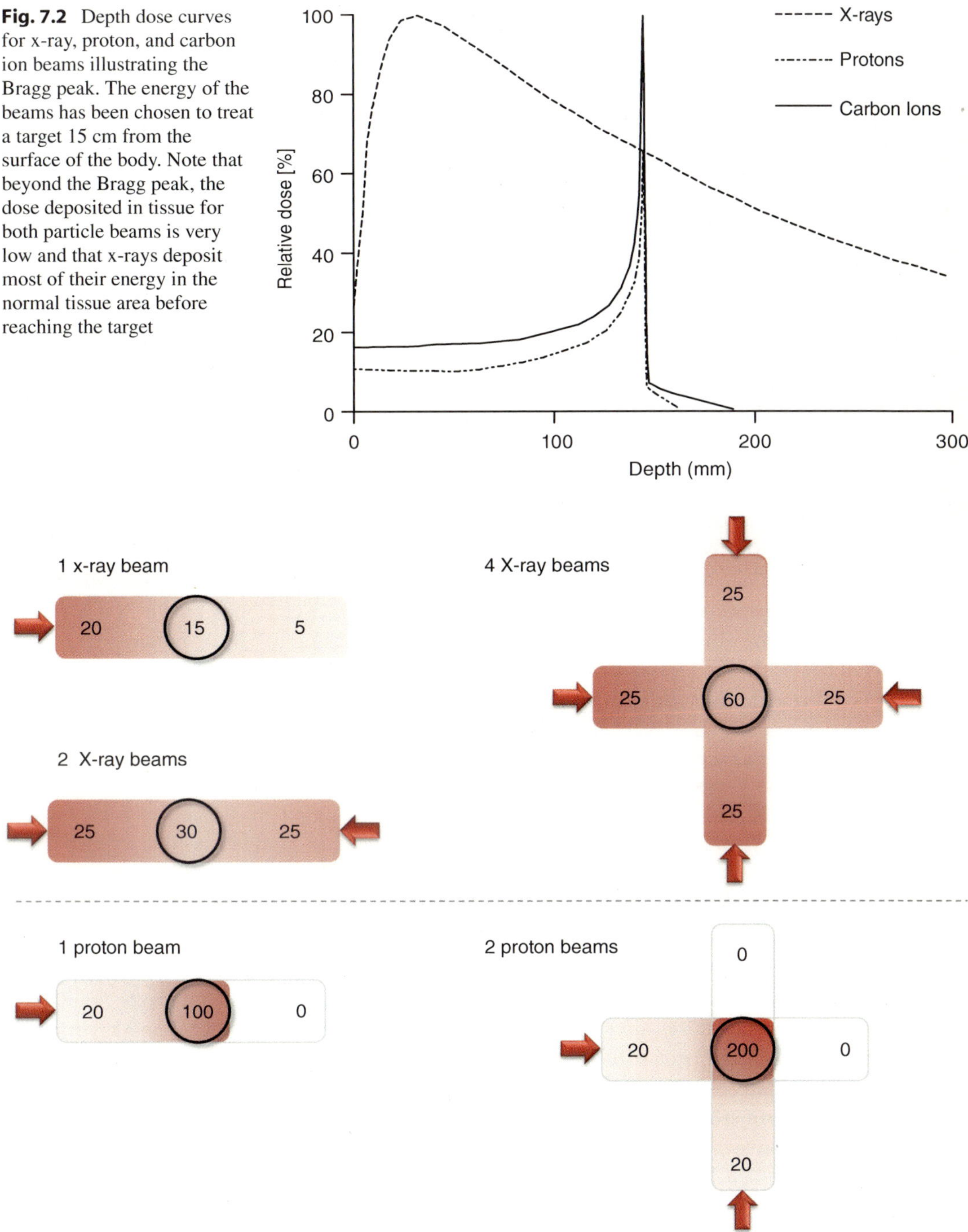

Fig. 7.2 Depth dose curves for x-ray, proton, and carbon ion beams illustrating the Bragg peak. The energy of the beams has been chosen to treat a target 15 cm from the surface of the body. Note that beyond the Bragg peak, the dose deposited in tissue for both particle beams is very low and that x-rays deposit most of their energy in the normal tissue area before reaching the target

Fig. 7.3 Schematic representation of dose deposition in tissues using x-rays and proton beams for irradiation of a deep-seated tumor (*ringed*) surrounded by normal tissue. *Upper panel*: x-rays have a broad dose deposition peak in tissues, illustrated in arbitrary units at three locations in each beam. In order for radiation dose to conform to the target, multiple x-ray beams are required. *Lower panel*: a proton beam deposits lower dose in the entry path as it passes through normal tissue and no dose beyond the Bragg peak. As a result, dose can be conformed to the target using a smaller number of treatment beams. The total dose deposited into normal tissue is much lower than can be achieved using X-rays

is a simple relationship between the physical dose deposited by x-rays and the amount of DNA damage induced in cells. The ratio between physical dose and biological effect is known as the relative biological effectiveness (RBE) and is expressed relative to an x-ray beam. By definition, the RBE of an x-ray beam is 1.

If we now consider a particle beam, the linear energy transfer varies along the path of the beam. The two ion species that have been investigated most thoroughly are protons and carbon ions. Protons have an RBE of around 1.1, making their biological effect similar to x-rays. In contrast, the larger carbon ion has a peak RBE of 3.0–3.3. While it is useful that the peak physical dose and biological effect of a particle beam are found at the Bragg peak, the variation of biological effect for particle beams is also dependent on the radiation dose and the tissue being irradiated. This means that accurate calculation of biological dose for treatments with particle beams is considerably more complex than the equivalent calculation for x-ray beams. In modern particle therapy treatment planning systems, a series of pragmatic simplifications are usually adopted to allow these calculations to be performed [6].

Cell Irradiation and Particle Beams

Cell irradiation experiments are the cornerstone of classical radiation biology. The most common technique that is used is the clonogenic survival assay. In this technique, the surviving fraction of cells exposed to varying doses of radiation is established by irradiating plates of cells at low density and subsequently counting the number of viable colonies per unit area. The surviving fraction is expressed relative to the colony formation of a control plate which is not irradiated. The typical clonogenic survival curve is plotted with the survival fraction on a log scale on the *y*-axis and the radiation dose on a linear scale on the *x*-axis (Fig. 7.4).

Examination of the shape of the survival curves for low-LET radiation such as x-rays reveals a continuous bending curve. At low doses the log surviving fraction is proportional to the radiation dose. This is thought to relate to the increase in double-stranded DNA breaks with increasing dose. As the dose is increased, log surviving fraction becomes proportional to the square of the dose. This is thought to relate to the increased frequency of persistent double-stranded DNA breaks at higher dose. For high-LET radiation, the survival curve takes a near exponential shape with less of the "shoulder" observed for x-rays. The steepness of the survival curve is similar to that observed in the high-dose region of the x-ray survival curve, in keeping with the concept that high-LET radiation generates larger numbers of persistent double-stranded DNA breaks within tissues.

Consistent with observations from x-ray exposures for glioma cell lines, there has been significant observed variation in the sensitivity to proton irradiation from experiments in the modern era. Belli et al. have established cell survival curves for T98G cell lines grown in a normoxic environment using a 7-MeV proton beam which show a marked degree of radiation sensitivity [7]. The data show a steeper dose-response curve than is observed with x-rays, with the surviving fraction at 2 Gy (SF2) reduced to 20 % (Fig. 7.5).

In contrast, the Clatterbridge group have recently performed clonogenic survival curves using a 62-MeV proton beam line using T98G and U373, with observed SF2 values of 73 % for T98G and 0.55 for U373.

Carbon ion beams generate intensely ionizing particle tracks, resulting in high levels of persistent DNA damage. As a result, the RBE of carbon ion beams is between 3 and 5, compared to conventional x-rays. In vitro studies have confirmed an enhanced cell killing effect. In the recent era, Combs et al. have assessed in vitro radiation response to carbon ions using U87 glioma cell lines using the high-LET beam (103 keV/μm) at the Heidelberg Ion-Beam Therapy Center (HIT) [8]. Their results demonstrate a steep dose-response curve, with an SF2 of 12 %. Ando et al. have performed similar irradiations using the 290-MeV carbon ion beam at the National Institute of Radiological Sciences (NIRS) in Japan [9]. Their experiments used a range of LET values by positioning

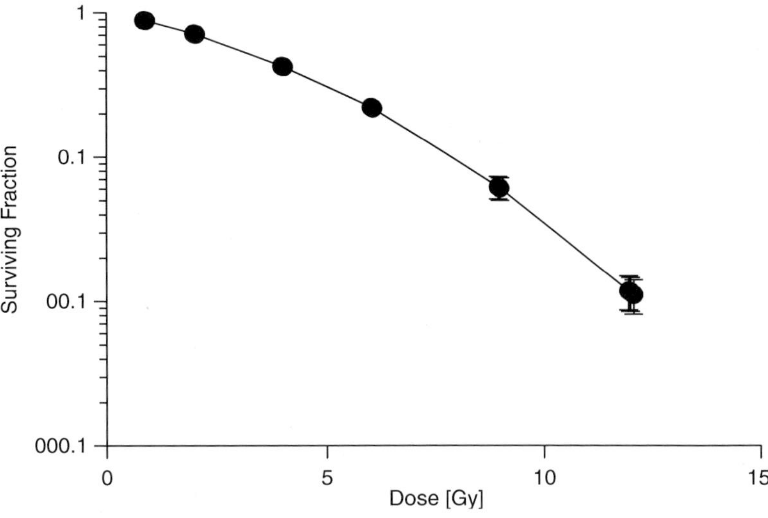

Fig. 7.4 Clonogenic survival curve for T98G human glioma cell line for x-ray irradiation

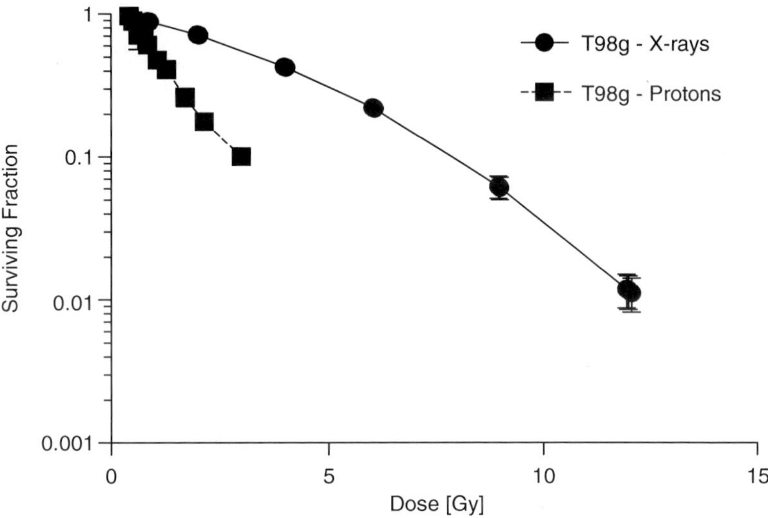

Fig. 7.5 Comparison of clonogenic survival curves for x-rays and proton beams using data from Belli et al. [7]. Note the increased steepness and absence of a shoulder to the survival curve, in keeping with a higher rate of persistent double-stranded DNA damage

cells within different regions of the beam and variation of the beam energy. Using A172 and TK1 human glioblastoma cell lines, the SF2 values observed in the highest LET experiments were also 12 % (Fig. 7.6).

Most mammalian cell lines require a highly regulated environment in order to proliferate in cell culture. Finely balanced requirements must be met for nutrient supply, growth factors, basement membrane for cell adhesion, oxygen, temperature, and pH. However, it has been known since the time of Tomlinson and Gray that the core of most tumors demonstrates low oxygen tensions, low pH, and lower levels of nutrient availability [10]. This microenvironment will drive cells into a state of growth arrest due to the surrounding tissue conditions, and this confers a high degree of radiation resistance. It is extremely difficult to establish stable culture of human cell lines under chronic hypoxic conditions, and as a result, cell handling systems have been designed to allow the effect of short-term hypoxia to be investigated. Many cell irradiation facilities have been adapted to allow cells to be placed in hypoxic environments during the period of radiation exposure. As would be expected, these experiments reveal high levels of tumor cell survival in hypoxic conditions. It is customary to

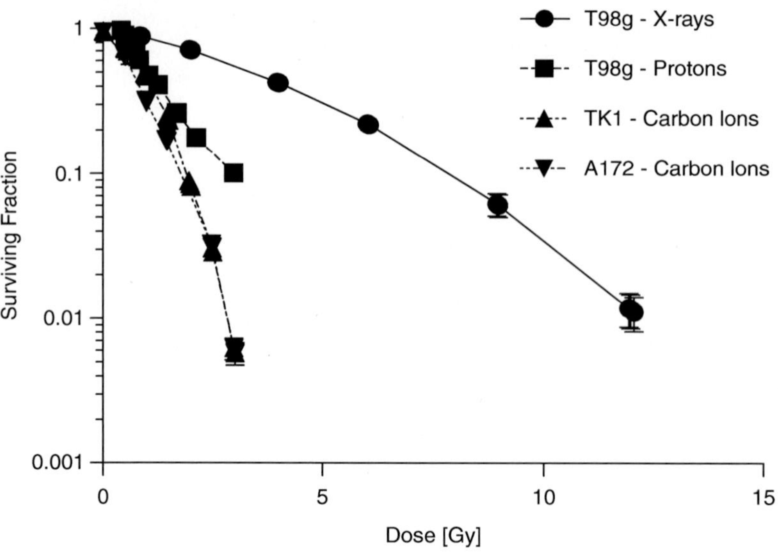

Fig. 7.6 Comparison of clonogenic survival curves for x-rays, protons, and carbon ions using data from Ando et al. Note that the carbon ion irradiations are performed on two distinct glioma cell lines, TK1 and A172. Carbon ion irradiation is associated with the highest level of cell kill

express this difference as the oxygen enhancement ratio (OER), which is the ratio of cell survival in normoxic and hypoxic conditions using x-ray irradiation. Larger, highly energetic particles such as carbon ions and alpha particles are capable of inducing higher levels of cell kill than x-rays, even under hypoxic conditions. This phenomenon has been quantified by Wenzl et al., who have analyzed the effect of hypoxia for high-LET carbon ion beams compared to x-ray irradiation in a range of tumor and normal tissue cell lines [11]. For treatments of 2 Gy, they observe an OER of 1.8–1.9 for carbon ions and 2.3–2.5 for x-rays. To express this in another form, the data show that the variation in cell kill between normoxic and hypoxic conditions is smaller for high-LET carbon ions than for low-LET x-rays.

Barriers to Implementation of Particle Therapy

In order to irradiate monolayers of cells with charged particle beams, the particles need to be accelerated to an energy in the region of 4–20 MeV per nucleon. This modest level of acceleration can be achieved using room-sized devices such as a tandem accelerator or a Van de Graaff generator. In contrast, energies of 260–400 MeV per nucleon are required for a clinical treatment beam, requiring much larger accelerator devices that are typically found in particle physics labs. A typical cyclotron for proton therapy treatment will be in the order of 5–10 m in diameter, while a high-energy synchrotron for a carbon ion team may be up to 40 m in diameter. The requirements for transporting and shielding of the beam require similar high-tolerance engineering techniques. The installation costs of clinical treatment facilities run into the hundreds of millions of dollars, and hence at the time of writing, there are only around 37 centers for proton therapy and 4 centers for carbon ion therapy in the world. Newer techniques such as compact superconducting cyclotrons, electronically accelerated proton beams, and laser-induced particle beams are all being developed in order to try and reduce the build cost of these facilities. Notwithstanding the paucity of treatment facilities, the potential benefits of particle beams appear very promising in the first clinical applications for glioblastoma.

Clinical Studies with Particle Beams

Fitzek et al. conducted a phase II study of post-operative radiotherapy in 23 glioblastoma patients, using photon beam treatment followed by a proton beam boost to an equivalent dose of

90 Gy given in 10×3 Gy fractions [12]. The median survival in this patient cohort was 20 months, compared to 14 months in historical controls. However, increased tumor control came at the cost of increased toxicity, with 30 % of patients developing radiation necrosis within the irradiated volume. Mizumoto et al. performed a hyperfractionated treatment with a tumor boost to 96.6 Gy equivalent in 56 fractions, treating twice a day. In this schedule, treatment was delivered entirely with proton beams, and the median survival was 21.6 months [13]. The observed rate of radiation necrosis for this treatment was 5 %. What might account for the difference in radiation necrosis between the two studies? A difference in treatment volume might be one explanation. Many American centers tend to define the extent of the tumor as the extent of edema visualized on T2-weighted magnetic resonance (MR) imaging and then add a margin for subclinical spread. European and Japanese practice tends to use a 2- to 3-cm margin round the contrast-enhancing tumor or resection cavity. This leads to a significant difference in the size of the irradiated volume. The second possible explanation relates to the use of x-rays for the first phase of treatment in the Fitzek study. Due to the additional entry and exit dose from x-rays, a larger volume of brain would be irradiated than observed with a purely proton ion-based treatment.

Mizoe et al. conducted a study in 48 patients with conventional x-ray irradiation followed by a carbon ion boost of 16.8–24.8 Gy equivalent [14]. Median survival was extended to 17 months, similar to the benefit observed for combined temozolomide and x-ray radiation therapy. Combs et al. have established a series of clinical trials at the HIT facility to evaluate the effectiveness of carbon ion therapy in both primary and recurrent glioblastoma. The CLEOPATRA study seeks to compare the effect of a proton boost and a carbon ion boost following conventional x-ray irradiation to a dose of 50 Gy in resected glioblastoma [15]. The CINDERELLA study compares the efficacy of hypofractionated carbon ion therapy against conventional stereotactic radiotherapy to a dose of 36 Gy in 18 fractions for the treatment of recurrent glioblastoma. The carbon ion dose will be escalated from 10×3 Gy equivalent to 16×3 Gy equivalent according to toxicity data [16].

Combined Chemo-Radiation Therapy with High-LET Radiation

Cell irradiation studies have demonstrated a similar additive effect for temozolomide with high-LET radiation as has been observed with x-ray therapy [17]. Combs et al. have published experimental data for combination of carbon ions with cisplatinum, camptothecin, gemcitabine, and paclitaxel in vitro and observe similar additive rather than synergistic effects [8]. No clinical studies confirming these observations have been performed to date.

Boron Neutron Capture Therapy for Glioblastoma

Boron neutron capture therapy (BNCT) is an experimental technique that may hold promise for the treatment of glioblastoma. The principle of the technique is to give a boron-rich carrier substance to the patient (such as borophenylalanine), which would be targeted to the tumor. A low-energy neutron beam would be targeted at the tumor, to be captured by the boron atoms. The interaction leads to fission of the boron nucleus, resulting in helium and lithium ions. The two ions produce intense ionization within a close proximity to the original boron atom, typically within a distance of 10 μm. This means that the ionizing species are restricted to the tumor focus. The high LET of these induced ions yields a high biological effectiveness, with a peak RBE of 3.8 within the tumor [18]. Early studies from Japan using BNCT both alone and in combination with low-dose x-ray irradiation for glioblastoma yielded median survival times of 15.6 months. A clinical trial of BNCT in glioblastoma was conducted at Helsinki and closed in 2008 due to slow accrual. Similar research initiatives are in place at the Massachusetts Institute of Technology (MIT) and in Birmingham, United

Kingdom. A significant barrier to the widespread implementation of BNCT as a clinical therapy is the need for an appropriate nuclear reactor to generate thermal neutrons.

Conclusion

This chapter reviews some of the experience with the use in novel ionizing radiation species in glioblastoma. While the physical properties and in vitro radiation sensitivity measurements appear promising, results in clinical practice are not yet forthcoming. The reason behind this lies partly in economics [19] and partly in tumor biology. The high installation and running costs of particle therapy treatment facilities have resulted in limited global availability of this treatment. However, even if particle therapy costs could be reduced to the same level as x-ray-based treatment, treatment efficacy is limited by the highly infiltrative nature of these tumors and the need to irradiate large volumes of functioning brain tissue. If particle therapy is to become effective in this disease, it will most likely be used as part of a combined modality therapy, where radiation is used to sterilize the tumor core and other targeted systemic therapies are used to control infiltrative disease.

Further Reading

This chapter has given very brief details of radiation biology concept related to particle therapy. For a more detailed overview, the reader is suggested to review one of the following texts:

Joiner MC, van der Kogel A. Basic clinical radiobiology. London: Hodder Arnold; 2009.

Hall EJ, Giaccia A. Radiobiology for the radiologist. New York: Lippincott Williams & Wilkins; 2011.

References

1. Walker MD, Alexander E, Hunt WE, MacCarty CS, Mahaley MS, Mealey J, Norrell HA, Owens G, et al. Evaluation of BCNU and/or radiotherapy in the treatment of anaplastic gliomas. J Neurosurg. 1978;49(3):333–43. doi:10.3171/jns.1978.49.3.0333. PMID355604.
2. Eschrich S, Zhang H, Zhao H, Boulware D, Lee J, Bloom G, Torres-Roca J. Systems biology modeling of the radiation sensitivity network: a biomarker discovery platform. Int J Radiat Oncol Biol Phys. 2009; 75:497–505.
3. Taghian A, Suit H, Pardo F, Gioioso D, Tomkinson K, DuBois W, Gerweck L. In vitro intrinsic radiation sensitivity of glioblastoma multiforme. Int J Radiat Oncol Biol Phys. 1992;23(1):55–62.
4. Zelefsky MJ, Yamada Y, Fuks Z, Zhang Z, Hunt M, Cahlon O, Park J, Shippy A. Long-term results of conformal radiotherapy for prostate cancer: impact of dose escalation on biochemical tumor control and distant metastases-free survival outcomes. Int J Radiat Oncol Biol Phys. 2008;71(4):1028–33.
5. Emami B, Lyman J, Brown A, Coia L, Goitein M, Munzenrider JE, Shank B, Solin LJ, Wesson M. Tolerance of normal tissue to therapeutic irradiation. Int J Radiat Oncol Biol Phys. 1991;21(1):109–22.
6. Elsässer T, Weyrather WK, Friedrich T, Durante M, Iancu G, Krämer M, Kragl G, Brons S, Winter M, Weber KJ, Scholz M. Quantification of the relative biological effectiveness for ion beam radiotherapy: direct experimental comparison of proton and carbon ion beams and a novel approach for treatment planning. Int J Radiat Oncol Biol Phys. 2010;78(4):1177–83.
7. Belli M, Bettega D, Calzolari P, Cherubini R, Cuttone G, Durante M, Esposito G, Furusawa Y, Gerardi S, Gialanella G, Grossi G, Manti L, Marchesini R, Pugliese M, Scampoli P, Simone G, Sorrentino E, Tabocchini MA, Tallone L. Effectiveness of monoenergetic and spread-out bragg peak carbon-ions for inactivation of various normal and tumour human cell lines. J Radiat Res. 2008;49(6):597–607.
8. Combs SE, Zipp L, Rieken S, Habermehl D, Brons S, Winter M, Haberer T, Debus J, Weber KJ. In vitro evaluation of photon and carbon ion radiotherapy in combination with chemotherapy in glioblastoma cells. Radiat Oncol. 2012;7:9.
9. Tsuchida Y, Tsuboi K, Ohyama H, Ohno T, Nose T, Ando K. Cell death induced by high-linear-energy transfer carbon beams in human glioblastoma cell lines. Brain Tumor Pathol. 1998;15(2):71–6.
10. Tomlinson RH, Gray LH. The histological structure of some human lung cancers and the possible implications for radiotherapy. Br J Cancer. 1955;9:539–49.
11. Wenzl T, Wilkens JJ. Theoretical analysis of the dose dependence of the oxygen enhancement ratio and its relevance for clinical applications. Radiat Oncol. 2011;6:171.
12. Fitzek MM, Thornton AF, Rabinov JD, Lev MH, Pardo FS, Munzenrider JE, Okunieff P, Bussière M, Braun I, Hochberg FH, Hedley-Whyte ET, Liebsch NJ, Harsh 4th GR. Accelerated fractionated proton/photon irradiation to 90 cobalt gray equivalent for glioblastoma multiforme: results of a phase II prospective trial. J Neurosurg. 1999;91(2):251–60.
13. Mizumoto M, Tsuboi K, Igaki H et al., Phase I/II trial of hyperfractionated concomitant boost proton radiotherapy for supratentorial glioblastoma multiforme. Int J Radiat Oncol Biol Phys. 2010 May 1;77(1):98–105.

14. Mizoe JE, Tsujii H, Hasegawa A, Yanagi T, Takagi R, Kamada T, Tsuji H, Takakura K, Organizing Committee of the Central Nervous System Tumor Working Group. Phase I/II clinical trial of carbon ion radiotherapy for malignant gliomas: combined X-ray radiotherapy, chemotherapy, and carbon ion radiotherapy. Int J Radiat Oncol Biol Phys. 2007; 69(2):390–6.
15. Combs SE, Kieser M, Rieken S, Habermehl D, Jäkel O, Haberer T, Nikoghosyan A, Haselmann R, Unterberg A, Wick W, Debus J. Randomized phase II study evaluating a carbon ion boost applied after combined radiochemotherapy with temozolomide versus a proton boost after radiochemotherapy with temozolomide in patients with primary glioblastoma: the CLEOPATRA trial. BMC Cancer. 2010;10:478.
16. Combs SE, Burkholder I, Edler L, Rieken S, Habermehl D, Jäkel O, Haberer T, Haselmann R, Unterberg A, Wick W, Debus J. Randomised phase I/II study to evaluate carbon ion radiotherapy versus fractionated stereotactic radiotherapy in patients with recurrent or progressive gliomas: the CINDERELLA trial. BMC Cancer. 2010;10:533.
17. Barrazzuol L, Jena R, Burnet NG, Jeynes JC, Merchant MJ, Kirkby KJ, Kirkby NF. In vitro evaluation of combined temozolomide and radiotherapy using x-rays and high-linear energy transfer radiation for glioblastoma. Radiat Res. 2012;177(5):651–62.
18. Kawabata S, Miyatake S, Kuroiwa T, Yokoyama K, Doi A, Iida K, Miyata S, Nonoguchi N, Michiue H, Takahashi M, Inomata T, Imahori Y, Kirihata M, Sakurai Y, Maruhashi A, Kumada H, Ono K. Boron neutron capture therapy for newly diagnosed glioblastoma. J Radiat Res. 2009;50(1):51–60.
19. Epstein K. Is spending on proton beam therapy for cancer going too far, too fast? BMJ. 2012;344:e2488.

Advances in Imaging Brain Cancer

8

Stephen J. Price and Adam D. Waldman

Abstract

Imaging has been one of the most important methods of understanding brain tumors, but it is clear that conventional methods are too insensitive to understand tumor biology. New positron emission tomography (PET) and MR imaging biomarkers that can assess biological processes in tumors are being developed. These biomarkers can assess tumor metabolism, proliferation, cellularity, angiogenesis and vascularity, hypoxia, invasion, and gene and molecular marker expression have been developed. These biomarkers will provide the tools to better characterize and monitor brain tumors.

Keywords

Magnetic resonance imaging • Positron emission tomography • Proton spectroscopy • Diffusion-weighted MRI • Diffusion tensor MRI • Perfusion imaging

Introduction

Advances in imaging have revolutionized the way medicine is practiced and have arguably been the most important single development in the management of brain tumors. Gone are the days when the brain was explored on the basis of clinical neurological examination alone [1], frequently resulting in a fruitless search which revealed no tumor [2].

The advent of ventriculography by Dandy in 1918 and subsequently angiography by Moniz in 1927 greatly improved localization of mass lesions, based on their displacement of surrounding structures. Cross-sectional tumor imaging began with the development of computed tomography (CT) by Hounsfield in 1972. CT provides soft tissue

S.J. Price, B.Sc., MBBS, Ph.D., FRCS (Neuro. Surg.) (✉)
Neurosurgery Division and Wolfson Brain Imaging Centre, Department of Clinical Neurosciences, University of Cambridge,
Box 167, Addenbrooke's Hospital, Hill Road, Cambridge CB2 0QQ, UK
e-mail: sjp58@cam.ac.uk

A.D. Waldman, Ph.D., FRCP, FRCR
Department of Imaging, Imperial College, London, UK

C. Watts (ed.), *Emerging Concepts in Neuro-Oncology*,
DOI 10.1007/978-0-85729-458-6_8, © Springer-Verlag London 2013

contrast that allows tumors to be localized and characterized, which had early and direct benefits for surgical and radiotherapy planning [3], and allowed a better understanding of tumor recurrence patterns [4]. The subsequent introduction of magnetic resonance imaging (MRI) in the 1980s provided further improvements in tumor imaging through better characterization of normal and pathological tissues and the ability to generate images in multiple planes. MRI and CT provide excellent anatomical information and remain the mainstays of clinical imaging.

"Conventional" MRI provides high-resolution structural images in multiple planes and yields exquisite anatomical detail, but the signal available from conventional T_2-weighted and contrast-enhanced T_1-weighted sequences lacks biological specificity and is insensitive to subtle but biologically important changes in tumors. There are a number of specific limitations of conventional structural MRI relevant to the evaluation of gliomas.

Characterization

Knowledge of the histological type and grade of brain tumor is essential for stratifying and planning treatment. Even tumors of a given histological type and grade behave highly variably, and understanding tumor genotype is also increasingly important in prognosis and predicting response to different treatment modalities. Features visible on conventional MRI (e.g., contrast enhancement following administration of IV contrast agent) yield some crude indicators of tumor aggressiveness; however, prediction of grade, cellular type, genetic profile, and biological behavior is overall poor [5]. Moreover, gliomas are frequently spatially heterogeneous, and MRI is unable to identify reliably the most biologically active components.

Anatomical Localization

Anatomical localization and identification of tumor boundaries are important for optimizing surgery and radiotherapy. Direct spread of gliomas by invasion of white matter tracts varies between patients and tumor types and is a key reason for the failure of current treatments. Both postmortem and biopsy studies have shown that tumor extends beyond the visible margin on CT [6–9], contrast-enhanced T_1-weighted MRI [9, 10], and T_2-weighted MR [9–12]. The nonspecific nature of the signal and innate insensitivity of MRI to small numbers of individual tumor cells limit determination of the true margin or the invasiveness of these tumors.

Detection of Tumor Progression and Treatment Response

Reliable early detection of tumor response to treatment and tumor progression is critical for establishing the efficacy of novel regimens in therapeutic trials and for optimizing the treatment of individual patients. The limitations of conventional MRI for such assessment are from (1) insensitivity to response early in treatment, resulting in delay of several months before response can be evaluated; (2) lack of specificity of MRI-visible signal to active tumor; and (3) inaccuracies in assessing the size of irregular infiltrating lesions by visual inspection or linear measurements.

Conventional response criteria have been based on linear measurements of contrast-enhancing tumor components. There are a number of difficulties with this approach. First, the enhancing component is not a reliable indicator of the volume of active tumor, as outlined above. Moreover, contrast enhancement is nonspecific and only reflects local disruption of the blood–brain barrier. The latter is illustrated by the increasingly recognized phenomena of pseudoprogression and pseudoresponse. *Pseudoprogression* is associated with combination chemotherapy and radiotherapy regimens, notably using temozolomide, and is characterized by a temporary increase in contrast enhancement and mass effect during the early phase of treatment [13]. (Fig. 8.1) Although the tumor initially appears to be progressing, this effect is seen more commonly in tumors with favorable MGMT methylation status and is associated with better prognosis [14]. Conversely,

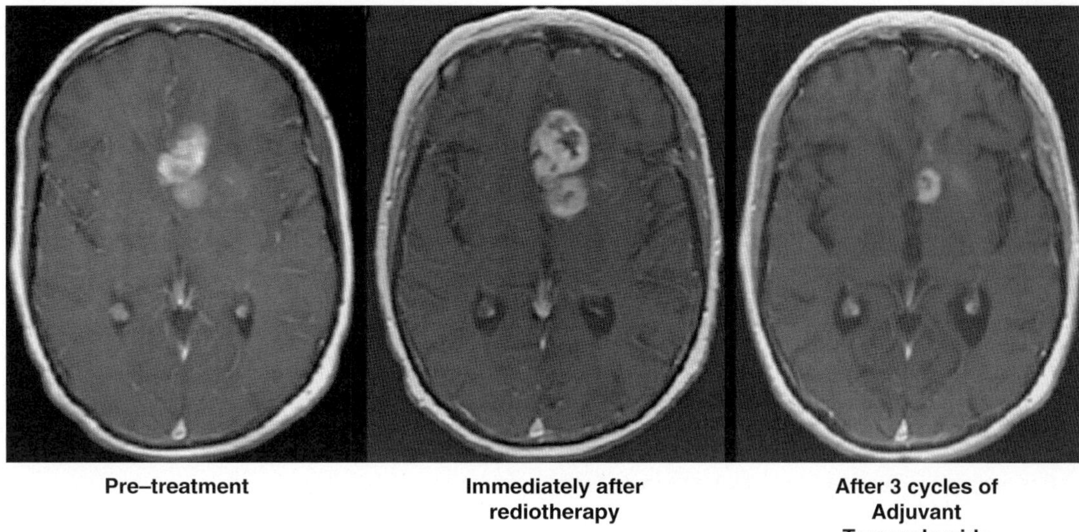

Fig. 8.1 An example of pseudoprogression. This 56-year-old man with a biopsy-confirmed glioblastoma was treated with radiotherapy with concomitant temozolomide chemotherapy. Imaging immediately following radiotherapy showed an apparent increase in tumor size. The patient continued on with the adjuvant phase of temozolomide and repeated imaging after three cycles showed marked reduction in enhancement

pseudoresponse is associated with drugs that reduce blood–brain barrier permeability (such as antiangiogenic agents like bevacizumab) and is characterized by loss of contrast enhancement. This does not predict response, and subsequent progression can occur without contrast enhancement [15].The use of local therapies (i.e., carmustine wafers or brachytherapy) and radiation-induced necrosis can also produce an enhancing mass which resembles tumor.

Finally, irregular tumors are difficult to measure, and this leads to marked interobserver variability in tumor measurements [16].

More recently proposed response criteria such as RANO [17, 18] incorporate assessment of non-enhancing components and clinical features but are still limited by the inability of MRI to demonstrate active tumor reliably.

There is therefore an impetus to develop noninvasive imaging methods that can better characterize gliomas and guide and evaluate treatment.

Physiological Imaging

In vivo data on tumor physiology was initially limited to that provided by radionucleotide-labeled compounds that probe metabolic activity using planar and early tomographic techniques. The advent of positron emission tomography (PET) in the early 1970s allowed the use of imaging probes more chemically similar to the native compound under examination and improved sensitivity and spatial localization of tracers. Advances in understanding of biochemical pathways and genetic process relevant to tumor behavior and refinements in radiochemistry now allow a variety of radiopharmaceuticals to probe more specific pathways in tumor pathology. Since the early 1990s, a number of noninvasive MRI techniques that also probe processes central to tumor pathophysiology have emerged. These physiological and molecular techniques augment the high-resolution structural information provided by conventional MRI and CT.

The purpose of this chapter is to provide an outline of how these new imaging tools provide information on pathological changes in glioma patients and provide noninvasive methods of monitoring these tumors. Detailed discussion of the principles of these imaging methods has not been included as they are better described in other sources [19] and emphasis will be given to clinical application in human gliomas.

Imaging Tumor Metabolism

Malignant cells are more metabolically active than normal cells. Imaging methods that probe energy metabolism include PET (which can show changes in glucose and amino acid metabolism) and MR spectroscopy (which allows levels of some intermediary metabolites to be measured).

Imaging Glucose Metabolism: FDG PET

Otto Warburg first noticed the relationship between aggressive tumor behavior and increased glycolysis [20]. This Warburg effect is now known to be modulated by HIF-1, a substance upregulated in hypoxia that is responsible for increasing expression of numerous genes related to energy metabolism, iron metabolism, and vasoactive proteins, as well as angiogenesis. The hyperglycolysis seen in tumors is due to increases in glucose transport across the blood–brain barrier and cell membranes.

The fluorinated glucose analogue 2-[^{18}F]-fluoro-2-deoxy-D-glucose (FDG) has high sensitivity (although poor specificity) for identifying areas of increased glucose metabolism. In brain tumors, FDG uptake appears to correlate with tumor grade, with high-grade gliomas exhibiting increased uptake and low-grade gliomas exhibiting uptake similar to or lower than normal gray matter. Areas of increased uptake within presumed low-grade gliomas predict the presence of anaplasia [21]. Using a tumor-to-white matter ratio > 1.5 and tumor-to-gray matter ratio > 0.6 could differentiate high- and low-grade tumors with a sensitivity of 94 % and specificity of 77 % [22].

Since FDG uptake can demonstrate areas of increased anaplasia, it has been used to guide treatment. Levivier et al. found that it could identify targets for stereotactic biopsies far better than contrast-enhanced CT [23]. They found that 6/35 targets selected by CT were nondiagnostic, whereas 0/55 targets selected by FDG PET were nondiagnostic. In glioblastomas, radiotherapy volumes predicted by FDG were more than 25 % different from volumes from MRI [24]. 83 % of early recurrence after radiotherapy occurred in regions with increased FDG.

Currently, one of the main applications of FDG PET is the differentiation of recurrent tumor from radiation necrosis, which both show contrast enhancement and may be difficult to differentiate using standard MRI. Recurrent tumors have increased metabolic activity, whereas areas of radiation necrosis are hypometabolic [25]. FDG can differentiate between these two pathologies with a sensitivity of 75 % and specificity of 81 % [26], but can be misleading, as increased activity can also result from accumulation of activated macrophages.

Limitations and Problems with FDG PET

FDG PET has limitations in the assessment of brain tumors:

- FDG uptake is nonspecific and can occur in any region with an increase in metabolic activity. In the normal brain, the cortex, basal ganglia, thalami, cerebellum, and brainstem have increased uptake, while white matter and CSF have low uptake. Similarly, inflammatory processes can also increase FDG uptake.
- FDG uptake will compete with normal glucose; hyperglycemia will decrease the amount of FDG that will be taken up. As a result, it is important that the study is performed at least 4 h after a meal.
- The use of dexamethasone can decrease cerebral glucose metabolism in normal brain [27], but not within the tumor. In fact the tumors themselves can decrease the metabolism of the normal, contralateral cortex [28]. The size of the tumor appears to be a major factor in determining this degree of decreased metabolism.

Imaging Protein Synthesis: Amino Acid PET Tracers

All cancer cells show elevated amino acid uptake due to both an increased demand for amino acids from increased protein synthesis and an increase in the transport of amino acid as a result of malignant transformation [29].

Many amino acid PET studies have used ^{11}C-labeled compounds; these can be considered minimally invasive tracers as the label involves only substitution of a carbon within the native molecule, with no change in its chemical properties. Most work has been done with ^{11}C-methionine (L-[methyl-^{11}C]-methionine) or ^{11}C-tyrosine (L-1-[^{11}C]-tyrosine). A significant drawback, however, is the short half-life of only 20.4 min, limiting use to centers with an on-site cyclotron.

This has stimulated interest in ^{18}F PET tracers with a longer half-life of 109 min and can be produced in a central cyclotron facility and then transported to other centers for use later in the day. Recent studies have used L-3-[^{18}F]fluoro-α(alpha)-methyltyrosine (FMT), *O*-2-[^{18}F]fluoroethyl-L-tyrosine (FET), and L-3-[^{123}I]iodo-α(alpha)-methyltyrosine (IMT).

^{11}C-methionine is the most commonly used amino acid tracer. It can be rapidly produced with high yields without the need for complex purification. It is involved, however, in considerable nonprotein metabolism and produces many nonprotein metabolites making quantification of protein synthesis impossible. There is some evidence that uptake may have a component that is perfusion dependent and involves passive transfer across the blood–brain barrier [30]. Uptake in gliomas correlates closely with microvascular density suggesting that transport into tissues, rather than incorporation into proteins, is the main factor determining its uptake [31]; the half-life of ^{11}C is too short to allow significant incorporation into proteins. Autoradiography has shown that ^{11}C-methionine accumulates mainly in viable tumor cells rather than macrophages and its uptake correlates with tumor proliferation better than FDG uptake [32].

^{11}C-tyrosine is far more difficult to synthesize but is not involved in significant nonprotein metabolism. It therefore is better suited to noninvasive quantification of protein synthesis. Animal studies have suggested that ^{11}C-tyrosine uptake correlates far better with tumor growth rates than FDG [33].

Patterns of amino acid uptake differ with different grades of tumors. Increased uptake in a heterogeneous pattern is seen in high-grade gliomas [34] and similar uptake values to normal brain in low-grade gliomas [35]. Uptake in low-grade oligodendrogliomas, however, appears to be greater than that of low-grade astrocytomas [35]. Increased uptake in patients with WHO grade II and grade III gliomas predicts a shorter survival time [36]. ^{11}C-methionine PET has been reported as showing a 97 % sensitivity for detecting high-grade and 61 % sensitivity for detecting low-grade gliomas [34] and able to differentiate gliomas from nonneoplastic lesions in 79 % of cases [37]. As there is little uptake into inflammatory cells, methionine appears to be particularly sensitive to differentiating radiation necrosis from recurrent tumor [38].

Since methionine uptake appears to correlate with proliferation and is increased in the higher-grade areas, various groups have used it to guide image-guided brain biopsies. Biopsies taken from areas of increased uptake of L-[1-^{11}C]-tyrosine provided better diagnostic yield than conventional MRI in lesions that did not enhance with gadolinium [39]. Biopsies of regions showing increased FET-PET activity identify anaplastic components [40]. Combining MRI and PET for biopsy targeting improves the diagnostic yield in brainstem tumors [41], pediatric brain tumors [42], and tumors with little uptake of FDG [43]. It has also been shown that PET-guided resections could remove the part of the tumor with the largest potential for transforming into a more malignant form, although no improvement in clinical outcome was reported [44].

MR and PET fusion studies have also shown that the volume of increased methionine uptake is greater than the volume of gadolinium enhancement on T_1-weighted MR, and although smaller than the volume of increased T_2-weighted signal, it extends beyond it in most cases [45]. More recent studies suggest that FET-PET can identify the area exhibiting 5-ALA fluorescence [46].

Proton Spectroscopy: Tumor Metabolism

In vivo MR proton spectroscopy (^{1}H MRS) allows the measurement of major metabolites in defined regions of the brain. Data can be acquired from a single voxel or multiple voxels (magnetic resonance spectroscopic imaging, MRSI; chemical shift imaging, CSI); MRSI/CSI allows metabolite abnormalities to be "mapped."

Metabolite levels are determined from the amplitude of resonances ("peaks" in the spectrum) corresponding to "assigned" protons within that metabolite which are known to resonate at specific frequencies; metabolite levels are frequently expressed as ratios, although absolute concentration approximations are also possible. At standard clinical field strengths (1–3 T), metabolites present at approximately mM concentrations are detectable. At short echo time acquisition (<35 ms), the repertoire of visible normal metabolites includes N-acetylaspartate (NAA; found in healthy neurons and their processes and considered to be marker of neuronal integrity), glutamate/glutamine (Glx; neurotransmitters), creatine/phosphocreatine (Cr; reflecting cellular energetic integrity), choline-containing compounds (Cho; a marker of membrane turnover), and *myo*-inositol (Myo; a pentose sugar). Some tumors also contain detectable lactate and mobile lipids, neither of which is detectable in normal adult brain.

The relevance of Cho as a marker of cellular turnover is discussed further below.

Myo-inositol is a pentose sugar which is elevated in grade II and III gliomas. The exact cause and significance of this in terms of glioma metabolism are unclear, and glycine may also contribute to the visible signal *in vivo* [47]. As with other normal brain metabolites, Myo is generally low in GBM.

An association between elevated citrate levels and aggressive phenotype has been reported in a series of pediatric gliomas, although there is clearly metabolic heterogeneity amongst this tumor group and the prognostic utility of this finding has yet to be determined [48].

MRS can be performed on other nuclei which contain unpaired nuclear spins. For example, specific metabolite tracking can be performed by ^{13}C MRS following introduction of ^{13}C-enriched labeled compounds. The feasibility of application in human gliomas has been demonstrated by detection of tumoral lactate signal following infusion of ^{13}C glucose [49], although even with enrichment, sensitivity is limited by the inherently weak signal.

There has been increasing literature on the use of hyperpolarized ^{13}C, which offers up to five orders of magnitude greater sensitivity. Lactate and pyruvate signals following administration of ^{13}C pyruvate in animal glioma models show, for example, early response to radiotherapy [50] and chemotherapy [51].

Imaging Tumor Cell Proliferation

Cellular proliferation is a cardinal feature of cancers. Imaging of this process has focused on assessing proliferation directly by probing DNA synthesis or indirectly by detecting cell turnover.

Imaging DNA Synthesis: FLT PET Studies

DNA synthesis has largely been imaged using PET. Most tracers focus on the thymidine salvage pathway where thymidine, a base only found in DNA, is recycled. Initial work with ^{11}C-thymidine proved challenging on account of technically difficult synthesis, short emission half-life, and rapid metabolism to numerous active metabolites. 3′-deoxy-3′-[^{18}F]-fluorothymidine (FLT) acts as a selective substrate for thymidine kinase 1—the first enzyme in the thymidine salvage pathway. Phosphorylated FLT is trapped in the cell and accumulates. In brain, there is little uptake outside of proliferating regions (Fig. 8.2). In tumors, uptake correlates well with cellular proliferation indices [52, 53]. Studies with treatment suggest there is a reduction in FLT uptake when patients respond to therapy, but this cannot be detected at very early time points [54]. Unfortunately, kinetic analysis of FLT uptake has shown that transfer across the blood–brain barrier is a rate-limiting step and might be the dominant factor affecting uptake [55].

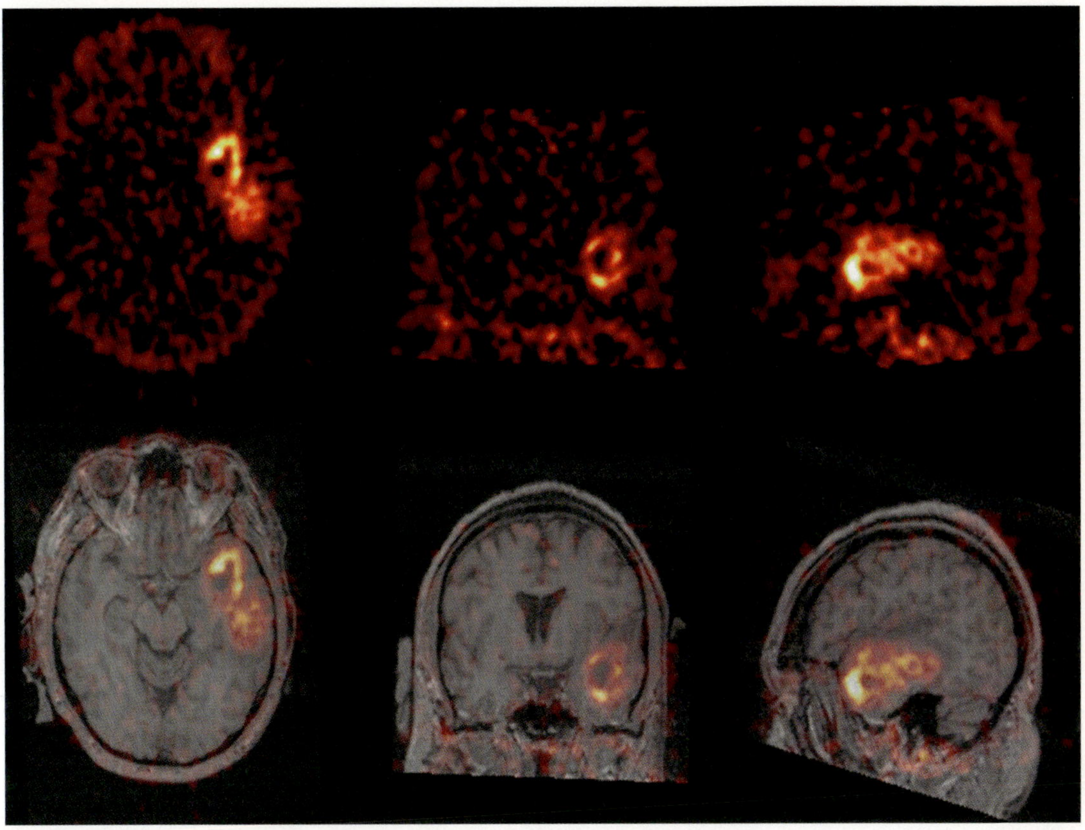

Fig. 8.2 [^{18}F]-fluorothymidine (FLT) PET image of a *left* temporal glioblastoma. There is very little uptake in the normal brain but high uptake in the most active part of the tumor

Imaging Cell Turnover: MR Spectroscopy and Choline PET

Choline metabolism provides a marker of membrane turnover, and a plethora of MRS studies of brain tumors have focused on Cho levels and their ratios to Cr and NAA. The Cho resonance at 3.24 ppm is composed largely of free choline, phosphocholine, and glycerol 3-phosphocholine. These compounds are involved in membrane turnover and are increased with increased cell turnover. In gliomas, this is usually associated with reduction of N-acetylaspartate (NAA), due to tumor infiltrating normal neuronal tissue, and the degree of depletion generally correlates with increasing tumor grade (Fig. 8.3). Although studies suggested that choline alone correlated poorly with the cellular proliferation index [56], the ratio of choline to N-acetylaspartate (NAA) correlates well with proliferation index, irrespective of whether there was contrast enhancement at the site of the voxel [57].

PET tracers of choline metabolism have also been developed. ^{18}F-fluorocholine has little uptake in normal brain and low-grade gliomas but increased uptake in high-grade gliomas [58].

Other MRS studies have shown the presence of detectable mobile lipid moieties in highly aggressive gliomas, notably glioblastoma [56]. These are characterized by broad resonances at 0.9 ppm (methyl lipid) and 1.3 ppm (methylene lipids) and are markers of lipid droplet formation, membrane breakdown, and the development of necrosis [59]. They are best detected in short echo spectra but may sometimes be seen at longer echo times, if levels are high. These may be

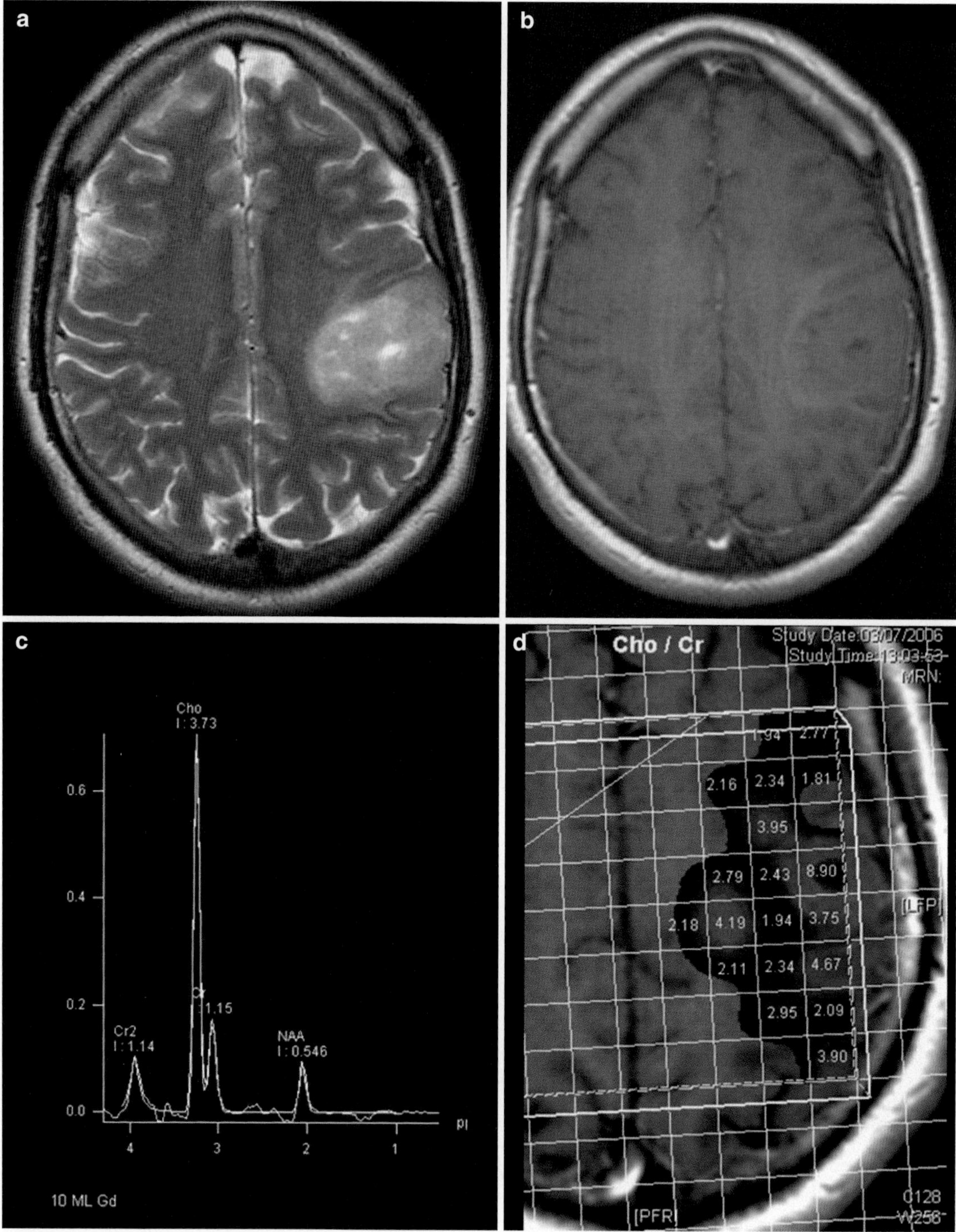

Fig. 8.3 This presumed low-grade glioma seen on T_2-weighted imaging (**a**) did not enhance (**b**). MR spectroscopy identified a region with increased choline peak (**c**). Biopsies of this choline "hot spot" showed this tumor to be an anaplastic astrocytoma

thought of as secondary markers of rapid cellular turnover, which has resulted in necrosis.

Imaging Tumor Cellularity

The increased proliferation and reduced apoptosis that occur in tumors result in increased cellular density—a feature that is seen in all grades of gliomas—when compared with normal brain parenchyma.

Diffusion-Weighted MRI

Diffusion-weighted imaging allows the Brownian motion of water within tissues to be probed. As the freedom of motion of intracellular water is markedly limited by internal cytoarchitecture, the DWI signal is dominated by extracellular water and hence yields ultrastructural information on interstitial volume and properties of the extracellular matrix. The degree of freedom of diffusion is quantified as apparent diffusion coefficient (ADC), a derived parameter which is independent of the relaxation properties which determine signal intensity on conventional T1- and T2-weighted sequences. "Restricted" water diffusion is characterized by low ADC values and elevated or "free" diffusion by high values.

The biophysical principles explaining diffusion-weighted imaging are described in detail elsewhere [60]. Regions with a high ADC have increased diffusion (e.g., in vasogenic regions), whereas regions with restricted diffusion have a low ADC. The increase in vasogenic edema in tumors means that brain tumors have a higher ADC than normal brain.

The Effect of Tumor Cellularity on Diffusion

It is now well established that a major determinant of the diffusion-weighted signal is the volume of the extracellular space [61]. In tumors two conflicting processes affect this: tumor cellular density and vasogenic edema. As cellular density increases, the volume of the extracellular space decreases, thereby reducing the ADC, hence the inverse relationship between cellularity and ADC. More cellular tumors (e.g., lymphomas [62] or primitive neuroectodermal pediatric tumors [63]) have a lower ADC than the less cellular gliomas. Within gliomas, ADC has been shown to be inversely related to tumor cellularity [64].

In contrast, vasogenic edema which is associated with high-grade tumors represents increased interstitial fluid, which increases ADC. Treatment with steroids has been shown to decrease the ADC in both regions of enhancing and non-enhancing tumor [65].

Because of the opposing effects of edema and cellularity, necrotic regions in aggressive lesions, and the overlap in ADC values with those of normal brain, individual ADC measurements have limited role in tumor assessment. Cellular components of higher-grade tumors have lower ADC values than lower-grade tumors, although the ADC value ranges overlap, precluding their use in tumor grading [64, 66].

The use of ADC to monitor response to therapy has been more promising. Studies in animal models of tumors showed an increase in ADC following treatment that correlated with reduction in both tumor volume [67] and cellularity [68]. Similar responses were seen in brain tumor patients [69]. As response to therapy is very heterogeneous, Moffatt *et al.* developed a *functional diffusion map* that compares changes in ADC on a voxel-by-voxel basis. This method was shown to be highly sensitive in determining response to therapy in an animal model [32]. It has also been able to identify changes in glioblastoma patients within 3 weeks of starting chemoradiotherapy that predict survival at 1 year—this assessment is not usually possible for 10–12 weeks using conventional MR [33]. Similarly, it appears to be the most useful measure of assessing response to antiangiogenic therapy[34], as ADC measures are less confounded by the effects of these agents on tumor vasculature.

Macromolecular Composition of the Extracellular Matrix

In tumors, the extracellular matrix differs from normal brain due to the displacement or destruction of the brain architecture and abnormal connective tissue composition, notably hyaluronic acids. This affects the diffusion properties of free water, which is reflected in ADC measurements, and may account for the higher ADC values seen in low-grade glioma compared with normal brain, despite the former having higher cellular density [70].

Other methods provide information on macromolecule composition in tissue. Magnetization transfer (MT) imaging reflects this indirectly via the exchange of magnetization between free protons in water and those bound in proteins. A preliminary quantitative MT study in gliomas has shown differences between tumor, immediate peritumoral tissue, and normal-appearing white matter that reflect differences in regional macromolecular composition [71]. Ultrashort T2-weighted imaging can yield similar information from direct detection of macromolecule-bound water [72] but has not yet been studied systematically in this context.

Imaging Tumor Vascularity and Angiogenesis

The development of an adequate blood supply is a key feature of tumor growth and development and has become a major target for glioma therapy. Regional perfusion can be measured using CT, MR, and PET.

Dynamic CT and MRI methods allow perfusion parameters to be derived from the kinetics of signal intensity change during transit of a bolus of intravenously injected contrast agent through brain tissue.

The most widely used and best-validated clinical perfusion technique is dynamic susceptibility contrast MRI (DSC-MRI), which exploits the relatively long-range T2*-dependent effects of intravascular gadolinium chelates. These cause a decrease in signal due to susceptibility-dependent dephasing and allow measurement of relative cerebral blood volume (rCBV), relative cerebral blood flow (rCBF), and mean transit time.

Dynamic contrast-enhanced MRI (DCE-MRI) uses a T1-weighted sequence to detect the T1-shortening effects of the contrast agent, which causes increased signal. rCBV, MTT, and rCBF can be derived from the first-pass kinetics, and subsequent evolution of signal due to leakage of agent into the interstitium allows permeability of the blood–brain barrier to be evaluated. The transfer coefficient, K-trans, reflects endothelial permeability, vascular surface area, and blood flow. Extravascular volume and vessel size may also be important parameters.

Arterial spin labeling (ASL) is noninvasive and uses MR signal based on influx of magnetically labeled water in blood to quantify absolute levels of cerebral blood flow; current methods allow measurement of rCBF.

Neoangiogenesis is a key feature of aggressive glioma phenotype, and vascular proliferation forms part of histological criteria for high-grade glioma diagnosis. rCBV has been shown to correlate with angiographic and histological markers of tumor vascularity [73] and the expression of vascular endothelial growth factor [74]. High-grade glial tumors tend to have higher relative cerebral blood volume values than low-grade tumors, and perfusion MRI significantly increases the specificity and sensitivity of glioma classification [75]. Sensitivity of 95 % and positive predictive value of 87 % for distinguishing low-grade gliomas from high-grade gliomas have been reported [76].

Neovascularization is also associated with increased vessel leakiness, and K-trans is also independently related to tumor grade, although this correlation is not as strong as for rCBV [77].

Presurgical relative rCBV measurements stratify for progression-free survival in resected low-grade (WHO grade II) gliomas [78] and are a powerful prognostic marker in this context. Rising rCBV from longitudinal measures in individual patients supports the hypothesis that this parameter provides an early noninvasive marker of malignant transformation in untreated low-grade lesions [79] (Fig. 8.4).

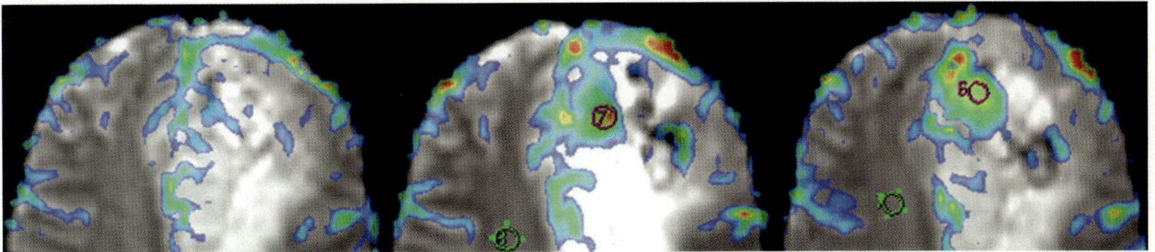

Fig. 8.4 Longitudinal perfusion imaging of a low-grade astrocytoma. There is a progressive increase in the rCBV in these tumors. There were no features on the conventional imaging to suggest transformation

Oligodendrogliomas have significantly higher rCBV values than astrocytomas [80], a feature that concurs with histological features of increased vascular density in oligodendrogliomas; this also correlates with tumor genotype (see below). It is worth noting that this is a potential confounder of elevated relative cerebral blood volume as a marker of grade and prognosis in gliomas, which can only be interpreted reliably in the light of tumor histology or genotype [81].

The absence of tumor neovascularization in lymphomas leads to low relative cerebral blood volume compared with malignant gliomas, and preliminary studies with arterial spin labeling also show effective differentiation between these two tumor types [82].

In terms of evaluation of treatment response to radiotherapy and cytotoxic regimens, rCBV measures appear to be less confounded by pseudoprogression and provide early surrogate markers of outcome [83].

There is evidence that perfusion imaging and permeability imaging provide indicators of the action of antiangiogenic agents on tumor vasculature, with marked reduction in rCBV and permeability within days of initiating treatment [84]. As with conventional contrast-enhanced MRI, the degree to which this provides a surrogate for outcome, however, appears to be limited.

PET Imaging of Angiogenesis and Integrin Expression

Integrins are a family of cell adhesion molecules that play an important role in angiogenesis. The integrins $\alpha(\text{alpha})_v\beta(\text{beta})_3$ and $\alpha(\text{alpha})_v\beta(\text{beta})_5$ are expressed in low levels in normal vasculature and are upregulated in both tumor vasculature and glioma cells. These integrins act as receptors to a number of proteins with an arginine-glycine-aspartate (RGD) tripeptide sequence, for example, laminin, vitronectin, and fibronectin. RGD peptides have been labeled with a variety of radionuclides to image integrin expression. ^{18}F-fluciclitide is an RGD peptide tracer that binds with high affinity to both $\alpha(\text{alpha})_v\beta(\text{beta})_3$ and $\alpha(\text{alpha})_v\beta(\text{beta})_5$ integrins. Animal studies show good uptake into glioma cells that correlates with microvascular density of the tumors and changes with response to antiangiogenic therapy [85]. Studies in patients have shown uptake in the tumor (unpublished data, see Fig. 8.5)

Hypoxia

Malignant tumors exhibit hypoxia as a result of high metabolic activity and relatively poor perfusion. In gliomas the presence of necrosis is a diagnostic feature of glioblastomas. Hypoxia also has important therapeutic implications in radiotherapy as both tumor and normal cells are 2–3 times more sensitive to radiotherapy when well oxygenated than when hypoxic [86]. Oxygen enhances radiation damage and is referred to as the oxygen-fixation hypothesis. The absorption of ionizing radiation by biological tissues leads to the production of free radicals. These act either directly by damaging DNA or indirectly on other molecules (mainly water) causing damage at other critical metabolic sites. The presence of hypoxia has been shown to reduce the radiation damage in a number of cancer sites.

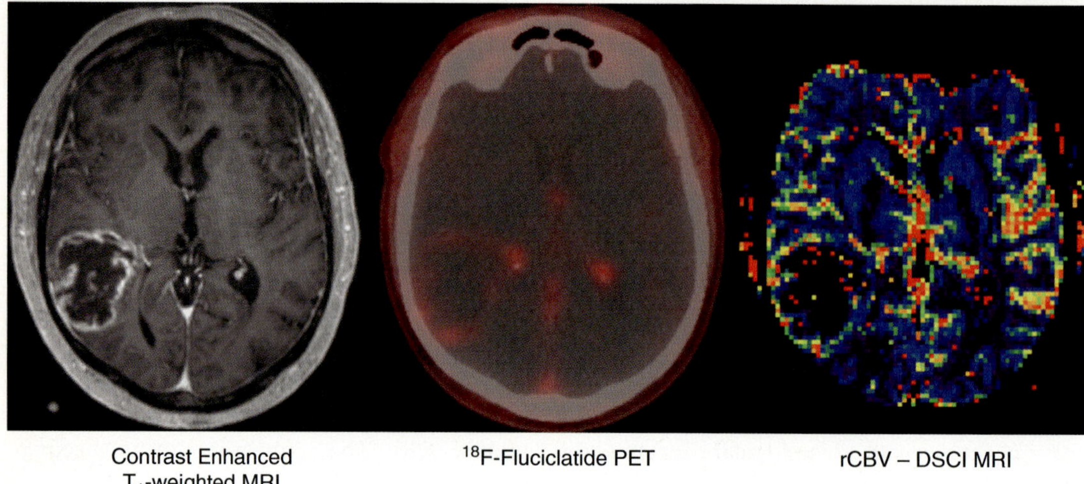

Fig. 8.5 An example of a [18F]-fluciclatide PET image to show expression of α(alpha)$_v\beta$(beta)$_3$ integrin expression in angiogenesis. The enhancing rim of the tumor expresses the highest amount of this integrin, and this correlates with the rim of increased rCBV in this region

Hypoxia can be imaged as follows.

Bold MRI

Blood oxygen level-dependent (BOLD) imaging exploits local changes in the signal generated between diamagnetic oxyhemoglobin and paramagnetic deoxyhemoglobin. As a result, deoxyhemoglobin shortens the T2* of the blood and its surroundings. Increasing oxygenation of the blood reduces this effect and increases the T2* signal. This BOLD effect can be detected by heavily T2*-weighted gradient-recalled echo (GRE) echo planar imaging (EPI) sequences.

Assessment of hypoxia requires "challenging" tissue by assessing the cerebrovascular response to a change in oxygen delivery to the tissue. This can be achieved by:

- Breathing 100 % oxygen and carbogen (95 % oxygen, 5 % CO_2)—this does cause "air hunger" and can be very unpleasant.
- Breath holding to provide a hypercapnia-induced vasodilatation stimulus. Studies in patients with brain tumors showed an increase in normal gray matter signal intensity, not seen in tumors. In high-grade gliomas, there was a decrease in signal. The different cerebrovascular response was likely due to severe hypoxia (unlikely in LGG) or inadequate/absent hypercapnia-induced vasodilatation. In HGG it may represent a hypercapnic steal phenomenon [87].
- Acetazolomide vasodilatation challenge—in animal studies this produces a reduction in BOLD signal in hypoxic regions and an increase in BOLD signal in the non-hypoxic component of the tumor.

18F-misonidazole PET

^{18}F-misonidazole (^{18}F-MISO) PET uses a nitroimidazole derivative that is a fluorinated analogue to the radiosensitizer drug misonidazole. It accumulates in viable hypoxic cells as it is selectively reduced and bound in these cells. Clinical studies have shown that it accumulates in hypoxic areas of brain tumors [88]. Uptake is only seen in high-grade gliomas and not low-grade gliomas and is found in areas that expressed tissue markers of hypoxia [89]. Uptake correlates with survival in patients with glioblastomas [90].

CU-diacetyl-bis(N4-methylthiosemicarbazone) (CU-ATSM)

Cu-diacetyl-bis(N4-methylthiosemicarbazone) (Cu-ATSM) is a relatively new tracer that is labeled with ^{60}Cu (half-life 23.4 mint), ^{61}Cu (half-life 3.32 h), or ^{64}Cu (half-life 12.8 h). Cu-ATSM accumulates in hypoxic tissues and is rapidly washed out of normal tissues, providing good signal contrast. To date, studies in brain tumors have been limited to animal models.

Detecting Lactate Production with MR Spectroscopy

Increased lactate production occurs in disorders of energy metabolism and an increase in nonoxidative glycolysis. Elevated lactate levels may be seen in all grades of glioma but are generally higher in more aggressive lesions [91, 92]. At short echo times, lactate is difficult to evaluate due to overlap with the 1.3-ppm mobile lipid resonances but can be readily measured at longer echo time.

Imaging Tumor Invasion

Local invasion of the surrounding brain is a key feature of gliomas. Imaging this invasion is important for a number of reasons:

- *Better targeting of radiotherapy:* Radiotherapy planning currently uses clinical target volumes that encompass the gross tumor with a 2.5-cm margin of normal brain to account for invasive cells. To avoid toxicity of this normal brain, the dose is reduced to 60 Gy which is insufficient to kill all tumor cells resulting in tumor recurrence usually within the treated volume [93, 94]. Better delineation of invasive regions might allow improved planning of treatment volumes to reduce the risk of radiation necrosis and allow dose escalation [95].
- *Understanding heterogeneity of invasiveness:* Postmortem studies have shown that there is marked heterogeneity of invasion in gliomas. These studies suggest 20–25 % of tumors exhibit marked invasiveness, whereas 20 % extend less than 1 cm from the gross tumor edge. Understanding this will allow early assessment of whether an individual patient has a diffuse tumor needing systemic therapy or localized tumor requiring aggressive local therapy.
- *Monitoring anti-invasive drugs:* A better understanding of the biology of invasion has allowed development of drugs to combat these processes. For example, cilengitide selectively blocks both α(alpha)$_v\beta$(beta)$_3$ and α(alpha)$_v\beta$(beta)$_5$ integrin receptors. This prevents both invasion and angiogenesis. Phase I/IIa trials using this drug in combination with standard chemoradiotherapy have suggested activity [96] and have led to the Phase III CENTRIC study. Monitoring changes in invasion will be difficult and need new imaging methods.

As mentioned previously, invasion is present in areas that appear normal on conventional MRI. Imaging strategies have either looked for white matter disruption with diffusion tensor MRI or looked for evidence of tumor in normal-appearing white matter.

Imaging White Matter Disruption with Diffusion Tensor MRI

Diffusion tensor imaging can be considered as an extension of diffusion-weighted imaging; ADC is calculated from the average of diffusion in different directions reflects (isotropic diffusion), analysis of multidirectional acquisition allows "diffusion directionality" to be quantified as fractional anisotropy. In the brain, water preferentially moves along white matter tracts. This directional or anisotropic diffusion can be described by a tensor. Standard analysis provides a summary parameter describing the tensor. Fractional anisotropy is the most commonly used parameter as it maps anisotropy with the best detail and signal-to-noise ratio [97]. Initial studies showed that DTI could identify a larger white matter abnormality than compared to conventional imaging—an abnormality not seen in edema-producing but noninvasive

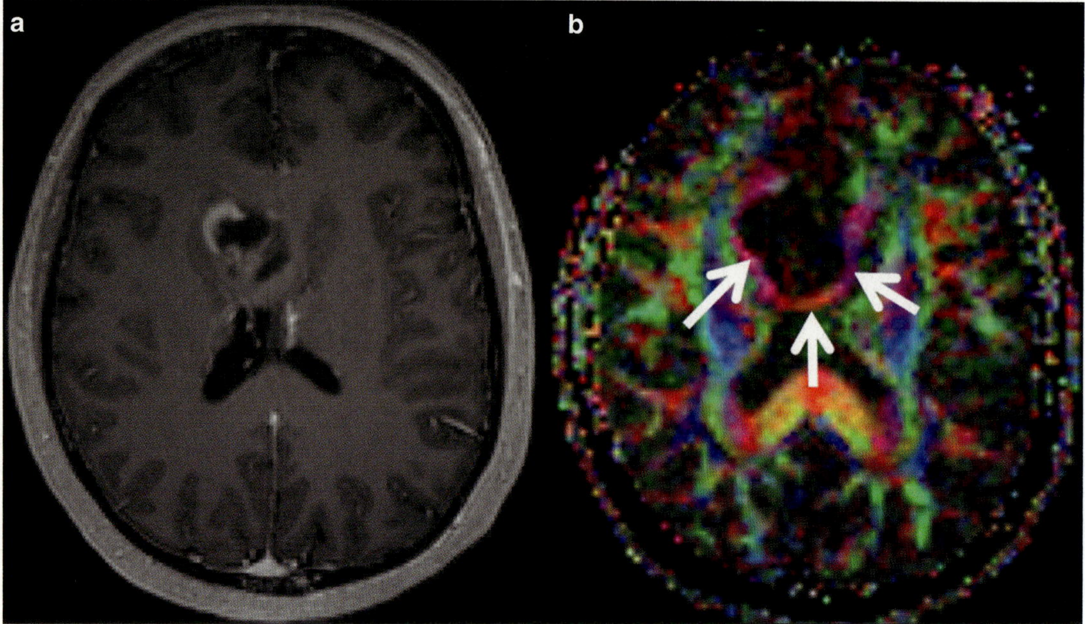

Fig. 8.6 (**a**) diffusion tensor color image (**b**) of a glioblastoma extending into the corpus callosum. It shows there is still intact, noninfiltrated corpus callosum behind the tumor (*arrowed*)

tumors (e.g., meningiomas and metastases) [98]. Further characterization of the invasive region using directionally encoded color maps [99] and tensor decomposition into isotropic (*p*) and anisotropic (*q*) components [100] have allowed differentiation of tumor invasion with tumor destruction of a white matter tract (Fig. 8.6). Using the latter method, it is possible to identify DTI abnormalities in normal-appearing brain that predict the site and pattern of tumor recurrence [101]. This method has also confirmed the presence of invading tumor cells using image-guided brain biopsies [102].Computational models also allow dominant "directional components" in adjacent voxels to be connected to form 3-dimensional maps, which display the dominant white matter fiber tracts (tractography).

Studies using MR spectroscopy have also shown evidence of white matter disruption around tumors. N-acetylaspartate (NAA) is a marker of intact neurons and is reduced within the center of tumors. NAA remains lower than normal brain suggesting loss of neurons in these regions [103–105]. The reduction in NAA correlated well with DTI measures of invasion [103, 105]. Similarly MR spectroscopy can identify disruption of the glial matrix due to tumor invasion by reductions in the glutamate/glutamine (Glx) peak adjacent to tumors [106]. Normal glial cells take up glutamate that is converted to glutamine as part of the glutamate-glutamine shuttle between glial cells and neurons. Tumor invasion causes disruption of the glial matrix, which in turn results in a reduction in intracellular glutamate.

Other Imaging Methods to Detect Invasive Glioma

As we have already discussed, high-grade gliomas differ from normal brain in a number of ways that can be identified by imaging. Other imaging methods have shown areas with the imaging characteristics of glioma outside the boundaries of the tumor defined by conventional imaging.

Proton Spectroscopy Studies of Tumor Proliferation

Studies with MR proton spectroscopy have demonstrated that there is a region outside of the tumor boundaries that exhibits increased choline and reduced NAA [104, 107]. Using the Cho/NAA ratio, Pirzkall et al. showed that there was a metabolic abnormality outside of the contrast-enhancing tumor boundary in most cases and an abnormality outside of the T_2-weighted boundary in over 60 % of patients [104]. These findings are not found in noninvasive tumors (e.g., meningiomas) and only seen in invasive gliomas [107]. Biopsy studies have confirmed that normalized choline is a marker for tumor invasion, and although it is better than conventional imaging in determining the tumor margin, it cannot differentiate between normal brain and mild infiltration [91]. Changes in Cho/NAA and Cho/Cr appear to correlate well with expression of metalloproteinase-2 [108].

PET Studies of Abnormalities in Metabolism

Most of the PET studies trying to assess tumor margins have used amino acid PET tracers as they are more sensitive than FDG PET. Studies have shown that methionine uptake occurs outside the area of contrast enhancement on MRI but within the T_2-weighted abnormality [45, 109, 110]. Biopsy studies have confirmed that this peripheral uptake is due to tumor infiltration [111]. This area of abnormality correlates well with MR spectroscopy [112] and DTI [58, 113].

Increased Perfusion

A number of studies have shown that there is an increase in rCBV in the region adjacent to the glioma margin defined on conventional imaging [114–118] that is not seen in noninvasive tumors [115, 118]. These abnormalities predict where enhancing tumor will recur [116], and biopsy studies have shown that it extends up to 2 cm into the region of tumor infiltration [117].

Labeled Marker Cells

Mesenchymal stem cells have been shown to demonstrate marked tropism toward actively dividing glioma cells. Preliminary work in animal models suggests that iron oxide or radiolabeling of such stem cells could provide a highly sensitive method of detecting cellular invasion and focused therapeutic delivery [119].

Imaging Gene Expression and Expression of Molecular Markers

As our understanding of prognostic and predictive markers increases, there has been great interest in seeing if imaging can identify expression of these markers.

Loss of 1p19q in Oligodendroglial Tumors

Studies have used conventional imaging and have suggested that oligodendroglial tumors with intact 1p19q have homogeneous T_1-weighted and T_2-weighted appearances and have a sharp border between tumor and normal-appearing brain [120]. Tumors with 1p19q loss also have a higher rCBV [121–123] and lower maximal ADC [124] compared to those with intact 1p19q. No difference in spectroscopic characteristics has been demonstrated between the two genotypes [125]. PET studies have suggested loss of 1p19q is seen more commonly in patients with increased FDG uptake [126].

MGMT Methylation Status

Recent studies have tried to differentiate tumors with MGMT methylation from unmethylated tumors. On conventional imaging, unmethylated tumors were significantly associated with ring enhancement [127], and methylated tumors exhibited a more ill-defined margin [128]. Texture analysis of the T_2-weighted images could differentiate the two methylation states, but blinded classification could correctly predict methylation status in 71 % of patients. Methylated tumors also had increased ADC ratios and reduced FA—although overlap in values would make it difficult to classify individual tumors on this basis [128]. Perfusion characteristics were similar in both groups [128].

IDH-1 Mutations

A consequence of IDH-1 mutations is the production of excess 2-hydroxyglutarate within tumors. Recent studies using ex vivo and clinical proton spectroscopy have shown that it is possible to identify 2-hydroxyglutarate noninvasively within tumors [129, 130]. Further work using this to predict IDH-1 status is required.

EGFR Receptor Overexpression

Studies with conventional imaging suggest that EGFR-overexpressing tumors demonstrate an increase in the ratio of the volume of T_1-weighted to T_2-weighted abnormality and reduction in border sharpness on T_2-weighted imaging compared to other tumors and have suggested these findings may reflect increased angiogenesis, edema, and/or invasion in EGFR-overexpressing tumors [131]. New PET probes that label EGFR inhibitors are being developed [132, 133] that may be able to assess EGFR status more accurately.

Conclusions

Quantitative physiological MRI and molecular imaging with PET provide noninvasive biomarkers relevant to key aspects of glioma biology. Despite their potential "added value" for tumor characterization, treatment guidance, and therapeutic evaluation, these techniques are only just beginning to emerge into clinical research and wider clinical practice. The development of new specific imaging probes and the integration of validated parameters into prospective studies will further our understanding of glioma pathophysiology and aid therapeutic development. Quantitative imaging is likely to play an increasing role in both clinical trials and the management of individual patients.

References

1. Bennett AH, Godlee RJ. Excision of a tumour from the brain. Lancet. 1884;124(3199):1090–1.
2. Sachs E. The problem of the glioblastomas. J Neurosurg. 1950;7:185–9.
3. Todd IDH. Choice of volume in the X-ray treatment of supratentorial gliomas. Br J Radiol. 1963;36:645–9.
4. Hochberg FH, Pruitt A. Assumptions in the radiotherapy of glioblastoma. Neurology. 1980;30:907–11.
5. Upadhyay N, Waldman AD. Conventional MRI evaluation of gliomas. Br J Radiol. 2011;84:S107–11.
6. Lilja A, Bergstrom K, Spannare B, Olsson Y. Reliability of computed tomography in assessing histopathological features of malignant supratentorial gliomas. J Comput Assist Tomogr. 1981;5:625–36.
7. Selker RG, Mendelow H, Walker M, Sheptak PE, Phillips JG. Pathological correlation of CT ring in recurrent, previously treated gliomas. Surg Neurol. 1982;17:251–4.
8. Burger PC, Dubois PJ, Schold Jr SC, Smith Jr KR, Odom GL, Crafts DC, Giangaspero F. Computerized tomographic and pathologic studies of the untreated, quiescent, and recurrent glioblastoma multiforme. J Neurosurg. 1983;58:159–69.
9. Kelly PJ, Daumas-Duport C, Kispert DB, Kall BA, Scheithauer BW, Illig JJ. Imaging-based stereotaxic serial biopsies in untreated intracranial glial neoplasms. J Neurosurg. 1987;66:865–74.
10. Lunsford LD, Martinez AJ, Latchaw RE. Magnetic resonance imaging does not define tumor boundaries. Acta Radiol Suppl. 1986;369:154–6.
11. Johnson PC, Hunt SJ, Drayer BP. Human cerebral gliomas: correlation of postmortem MR imaging and neuropathologic findings. Radiology. 1989;170:211–7.
12. Watanabe M, Tanaka R, Takeda N. Magnetic resonance imaging and histopathology of cerebral gliomas. Neuroradiology. 1992;34:463–9.
13. Brandsma D, Stalpers L, Taal W, Sminia P, van den Bent MJ. Clinical features, mechanisms, and management of pseudoprogression in malignant gliomas. Lancet Oncol. 2008;9:453–61.
14. Brandes AA, Franceschi E, Tosoni A, Blatt V, Pession A, Tallini G, Bertorelle R, Bartolini S, Calbucci F, Andreoli A, Frezza G, Leonardi M, Spagnolli F, Ermani M. MGMT promoter methylation status Can predict the incidence and outcome of pseudoprogression after concomitant radiochemotherapy in newly diagnosed glioblastoma patients. J Clin Oncol. 2008;26:2192–7.
15. Norden AD, Young GS, Setayesh K, Muzikansky A, Klufas R, Ross GL, Ciampa AS, Ebbeling LG, Levy B, Drappatz J, Kesari S, Wen PY. Bevacizumab for recurrent malignant gliomas: efficacy, toxicity, and patterns of recurrence. Neurology. 2008;70:779–87.
16. Vos MJ, Uitdehaag BM, Barkhof F, Heimans JJ, Baayen HC, Boogerd W, Castelijns JA, Elkhuizen PH, Postma TJ. Interobserver variability in the radiological assessment of response to chemotherapy in glioma. Neurology. 2003;60:826–30.
17. Wen PY, Macdonald DR, Reardon DA, Cloughesy TF, Sorensen AG, Galanis E, DeGroot J, Wick W, Gilbert MR, Lassman AB, Tsien C, Mikkelsen T, Wong ET, Chamberlain MC, Stupp R, Lamborn KR, Vogelbaum MA, van den Bent MJ, Chang SM. Updated response

assessment criteria for high-grade gliomas: response assessment in neuro-oncology working group. J Clin Oncol. 2010;28:1963–72.
18. van den Bent MJ, Wefel JS, Schiff D, Taphoorn MJ, Jaeckle K, Junck L, Armstrong T, Choucair A, Waldman AD, Gorlia T, Chamberlain M, Baumert BG, Vogelbaum MA, Macdonald DR, Reardon DA, Wen PY, Chang SM, Jacobs AH. Response assessment in neuro-oncology (a report of the RANO group): assessment of outcome in trials of diffuse low-grade gliomas. Lancet Oncol. 2011;12:583–93.
19. Louis DN, Ohgaki H, Wiestler OD, Cavenee WK. WHO classification of tumours of the central nervous system. Lyon: IARC; 2007.
20. Warburg O. On the origin of cancer cells. Science. 1956;123:309–14.
21. Goldman S, Levivier M, Pirotte B, Brucher JM, Wikler D, Damhaut P, Stanus E, Brotchi J, Hildebrand J. Regional glucose metabolism and histopathology of gliomas. A study based on positron emission tomography-guided stereotactic biopsy. Cancer. 1996;78:1098–106.
22. Delbeke D, Meyerowitz C, Lapidus RL, Maciunas RJ, Jennings MT, Moots PL, Kessler RM. Optimal cutoff levels of F-18 fluorodeoxyglucose uptake in the differentiation of low-grade from high-grade brain tumors with PET. Radiology. 1995;195:47–52.
23. Levivier M, Goldman S, Pirotte B, Brucher JM, Baleriaux D, Luxen A, Hildebrand J, Brotchi J. Diagnostic yield of stereotactic brain biopsy guided by positron emission tomography with [18 F]fluorodeoxyglucose. J Neurosurg. 1995; 82:445–52.
24. Tralins KS, Douglas JG, Stelzer KJ, Mankoff DA, Silbergeld DL, Rostomily RC, Hummel S, Scharnhorst J, Krohn KA, Spence AM. Volumetric analysis of 18 F-FDG PET in glioblastoma multiforme: prognostic information and possible role in definition of target volumes in radiation dose escalation. J Nucl Med. 2002;43:1667–73.
25. Di Chiro G, Oldfield E, Wright DC, De Michele D, Katz DA, Patronas NJ, Doppman JL, Larson SM, Ito M, Kufta CV. Cerebral necrosis after radiotherapy and/or intraarterial chemotherapy for brain tumors: PET and neuropathologic studies. AJR Am J Roentgenol. 1988;150:189–97.
26. Chao ST, Suh JH, Raja S, Lee SY, Barnett G. The sensitivity and specificity of FDG PET in distinguishing recurrent brain tumor from radionecrosis in patients treated with stereotactic radiosurgery. Int J Cancer. 2001;96:191–7.
27. Fulham MJ, Brunetti A, Aloj L, Raman R, Dwyer AJ, Di Chiro G. Decreased cerebral glucose metabolism in patients with brain tumors: an effect of corticosteroids. J Neurosurg. 1995;83:657–64.
28. Roelcke U, von Ammon K, Hausmann O, Kaech DL, Vanloffeld W, Landolt H, Rem JA, Gratzl O, Radu EW, Leenders KL. Operated low grade astrocytomas: a long term PET study on the effect of radiotherapy. J Neurol Neurosurg Psychiatry. 1999;66:644–7.
29. Isselbacher KJ. Sugar and amino acid transport by cells in culture—differences between normal and malignant cells. N Engl J Med. 1972;286:929–33.
30. Roelcke U, Radu E, Ametamey S, Pellikka R, Steinbrich W, Leenders KL. Association of rubidium and C-methionine uptake in brain tumors measured by positron emission tomography. J Neurooncol. 1996;27:163–71.
31. Kracht LW, Friese M, Herholz K, Schroeder R, Bauer B, Jacobs A, Heiss WD. Methyl-[11C]- l-methionine uptake as measured by positron emission tomography correlates to microvessel density in patients with glioma. Eur J Nucl Med Mol Imaging. 2003;30: 868–73.
32. Minn H, Clavo AC, Grenman R, Wahl RL. In vitro comparison of cell proliferation kinetics and uptake of tritiated fluorodeoxyglucose and L-methionine in squamous-cell carcinoma of the head and neck. J Nucl Med. 1995;36:252–8.
33. Daemen BJ, Elsinga PH, Paans AM, Lemstra W, Konings AW, Vaalburg W. Suitability of rodent tumor models for experimental PET with L-[1–11C]tyrosine and 2-[18 F]fluoro-2-deoxy-D-glucose. Int J Rad Appl Instrum B. 1991;18:503–11.
34. Ogawa T, Shishido F, Kanno I, Inugami A, Fujita H, Murakami M, Shimosegawa E, Ito H, Hatazawa J, Okudera T. Cerebral glioma: evaluation with methionine PET. Radiology. 1993;186:45–53.
35. Derlon JM, Petit-Taboue MC, Chapon F, Beaudouin V, Noel MH, Creveuil C, Courtheoux P, Houtteville JP. The in vivo metabolic pattern of low-grade brain gliomas: a positron emission tomographic study using 18F-fluorodeoxyglucose and 11C-L-methylmethionine. Neurosurgery. 1997;40:276–87.
36. De Witte O, Goldberg I, Wikler D, Rorive S, Damhaut P, Monclus M, Salmon I, Brotchi J, Goldman S. Positron emission tomography with injection of methionine as a prognostic factor in glioma. J Neurosurg. 2001;95:746–50.
37. Herholz K, Holzer T, Bauer B, Schroder R, Voges J, Ernestus RI, Mendoza G, Weber-Luxenburger G, Lottgen J, Thiel A, Wienhard K, Heiss WD. 11C-methionine PET for differential diagnosis of low-grade gliomas. Neurology. 1998;50:1316–22.
38. Thiel A, Pietrzyk U, Sturm V, Herholz K, Hovels M, Schroder R. Enhanced accuracy in differential diagnosis of radiation necrosis by positron emission tomography-magnetic resonance imaging coregistration: technical case report. Neurosurgery. 2000;46:232–4.
39. Go KG, Keuter EJ, Kamman RL, Pruim J, Metzemaekers JD, Staal MJ, Paans AM, Vaalburg W. Contribution of magnetic resonance spectroscopic imaging and L-[1–11C]tyrosine positron emission tomography to localization of cerebral gliomas for biopsy. Neurosurgery. 1994;34:994–1002.
40. Floeth FW, Pauleit D, Sabel M, Stoffels G, Reifenberger G, Riemenschneider MJ, Jansen P, Coenen HH, Steiger HJ, Langen KJ. Prognostic value of O-(2–18F-fluoroethyl)-L-tyrosine PET and MRI in low-grade glioma. J Nucl Med. 2007;48:519–27.

41. Massager N, David P, Goldman S, Pirotte B, Wikler D, Salmon I, Nagy N, Brotchi J, Levivier M. Combined magnetic resonance imaging- and positron emission tomography-guided stereotactic biopsy in brainstem mass lesions: diagnostic yield in a series of 30 patients. J Neurosurg. 2000;93:951–7.
42. Pirotte B, Goldman S, Salzberg S, Wikler D, David P, Vandesteene A, Van Bogaert P, Salmon I, Brotchi J, Levivier M. Combined positron emission tomography and magnetic resonance imaging for the planning of stereotactic brain biopsies in children: experience in 9 cases. Pediatr Neurosurg. 2003;38:146–55.
43. Pirotte B, Goldman S, Massager N, David P, Wikler D, Lipszyc M, Salmon I, Brotchi J, Levivier M. Combined use of 18F-fluorodeoxyglucose and 11C-methionine in 45 positron emission tomography-guided stereotactic brain biopsies. J Neurosurg. 2004;101:476–83.
44. Pirotte B, Goldman S, Dewitte O, Massager N, Wikler D, Lefranc F, Taib NOB, Rorive S, David P, Brotchi J, Levivier M. Integrated positron emission tomography and magnetic resonance imaging-guided resection of brain tumors: a report of 103 consecutive procedures. J Neurosurg. 2006;104:238–53.
45. Miwa K, Shinoda J, Yano H, Okumura A, Iwama T, Nakashima T, Sakai N. Discrepancy between lesion distributions on methionine PET and MR images in patients with glioblastoma multiforme: insight from a PET and MR fusion image study. J Neurol Neurosurg Psychiatry. 2004;75:1457–62.
46. Floeth FW, Sabel M, Ewelt C, Stummer W, Felsberg J, Reifenberger G, Steiger HJ, Stoffels G, Coenen HH, Langen KJ. Comparison of (18)F-FET PET and 5-ALA fluorescence in cerebral gliomas. Eur J Nucl Med Mol Imaging. 2011;38:731–41.
47. Candiota AP, Majos C, Julia-Sape M, Cabanas M, Acebes JJ, Moreno-Torres A, Griffiths JR, Arus C. Non-invasive grading of astrocytic tumours from the relative contents of myo-inositol and glycine measured by in vivo MRS. JBR-BTR. 2011;94:319–29.
48. Bluml S, Panigrahy A, Laskov M, Dhall G, Krieger MD, Nelson MD, Finlay JL, Gilles FH. Elevated citrate in pediatric astrocytomas with malignant progression. Neuro Oncol. 2011;13:1107–17.
49. Wijnen JP, Van der Graaf M, Scheenen TW, Klomp DW, de Galan BE, Idema AJ, Heerschap A. In vivo 13C magnetic resonance spectroscopy of a human brain tumor after application of 13C-1-enriched glucose. Magn Reson Imaging. 2010;28:690–7.
50. Day SE, Kettunen MI, Cherukuri MK, Mitchell JB, Lizak MJ, Morris HD, Matsumoto S, Koretsky AP, Brindle KM. Detecting response of rat C6 glioma tumors to radiotherapy using hyperpolarized [1- 13C] pyruvate and 13C magnetic resonance spectroscopic imaging. Magn Reson Med. 2011;65:557–63.
51. Park I, Bok R, Ozawa T, Phillips JJ, James CD, Vigneron DB, Ronen SM, Nelson SJ. Detection of early response to temozolomide treatment in brain tumors using hyperpolarized 13C MR metabolic imaging. J Magn Reson Imaging. 2011;33:1284–90.
52. Chen W, Cloughesy T, Kamdar N, Satyamurthy N, Bergsneider M, Liau L, Mischel P, Czernin J, Phelps ME, Silverman DHS. Imaging proliferation in brain tumors with 18F-FLT PET: comparison with 18F-FDG. J Nucl Med. 2005;46:945–52.
53. Price SJ, Fryer TD, Cleij MC, Dean AF, Joseph J, Salvador R, Wang DD, Hutchinson PJ, Clark JC, Burnet NG, Pickard JD, Aigbirhio FI, Gillard JH. Imaging regional variation of cellular proliferation in gliomas using 3'-deoxy-3'-[18F]fluorothymidine positron-emission tomography: an image-guided biopsy study. Clin Radiol. 2009;64:52–63.
54. Schiepers C, Dahlbom M, Chen W, Cloughesy T, Czernin J, Phelps ME, Huang SC. Kinetics of 3,Ä›-deoxy-3,Ä›-18F-fluorothymidine during treatment monitoring of recurrent high-grade glioma. J Nucl Med. 2010;51:720–7.
55. Jacobs AH, Thomas A, Kracht LW, Li H, Dittmar C, Garlip G, Galldiks N, Klein JC, Sobesky J, Hilker R, Vollmar S, Herholz K, Wienhard K, Heiss WD. 18F-fluoro-L-thymidine and 11C-methylmethionine as markers of increased transport and proliferation in brain tumors. J Nucl Med. 2005;46:1948–58.
56. Nafe R, Herminghaus S, Raab P, Wagner S, Pilatus U, Schneider B, Schlote W, Zanella F, Lanfermann H. Preoperative proton-MR spectroscopy of gliomas—correlation with quantitative nuclear morphology in surgical specimen. J Neurooncol. 2003;63:233–45.
57. McKnight TR, Lamborn KR, Love TD, Berger MS, Chang S, Dillon WP, Bollen A, Nelson SJ. Correlation of magnetic resonance spectroscopic and growth characteristics within Grades II and III gliomas. J Neurosurg. 2007;106:660–6.
58. Lam WW, Ng DC, Wong WY, Ong SC, Yu SW, See SJ. Promising role of [18F] fluorocholine PET/CT vs [18F] fluorodeoxyglucose PET/CT in primary brain tumors-early experience. Clin Neurol Neurosurg. 2011;113:156–61.
59. Remy C, Fouilhe N, Barba I, Sam-Lai E, Lahrech H, Cucurella MG, Izquierdo M, Moreno A, Ziegler A, Massarelli R, Decorps M, Arus C. Evidence that mobile lipids detected in rat brain glioma by 1H nuclear magnetic resonance correspond to lipid droplets. Cancer Res. 1997;57:407–14.
60. Price SJ, Tozer DJ, Gillard JH. Methodology of diffusion-weighted, diffusion tensor and magnetisation transfer imaging. Br J Radiol. 2011;84:S121–6.
61. Latour LL, Svoboda K, Mitra PP, Sotak CH. Time-dependent diffusion of water in a biological model system. Proc Natl Acad Sci U S A. 1994;91:1229–33.
62. Guo AC, Cummings TJ, Dash RC, Provenzale JM. Lymphomas and high-grade astrocytomas: comparison of water diffusibility and histologic characteristics. Radiology. 2002;224:177–83.
63. Gauvain KM, McKinstry RC, Mukherjee P, Perry A, Neil JJ, Kaufman BA, Hayashi RJ. Evaluating pediatric brain tumor cellularity with diffusion-tensor imaging. AJR Am J Roentgenol. 2001;177:449–54.
64. Sugahara T, Korogi Y, Kochi M, Ikushima I, Shigematu Y, Hirai T, Okuda T, Liang L, Ge Y, Komohara Y,

Ushio Y, Takahashi M. Usefulness of diffusion-weighted MRI with echo-planar technique in the evaluation of cellularity in gliomas. J Magn Reson Imaging. 1999;9:53–60.
65. Bastin ME, Carpenter TK, Armitage PA, Sinha S, Wardlaw JM, Whittle IR. Effects of dexamethasone on cerebral perfusion and water diffusion in patients with high-grade glioma. AJNR Am J Neuroradiol. 2006;27:402–8.
66. Lam WW, Poon WS, Metreweli C. Diffusion MR imaging in glioma: does It have Any role in the pre-operation determination of grading of glioma? Clin Radiol. 2002;57:219–25.
67. Chenevert TL, McKeever PE, Ross BD. Monitoring early response of experimental brain tumors to therapy using diffusion magnetic resonance imaging. Clin Cancer Res. 1997;3:1457–66.
68. Mardor Y, Roth Y, Lidar Z, Jonas T, Pfeffer R, Maier SE, Faibel M, Nass D, Hadani M, Orenstein A, Cohen JS, Ram Z. Monitoring response to convection-enhanced taxol delivery in brain tumor patients using diffusion-weighted magnetic resonance imaging. Cancer Res. 2001;61:4971–3.
69. Chenevert TL, Stegman LD, Taylor JM, Robertson PL, Greenberg HS, Rehemtulla A, Ross BD. Diffusion magnetic resonance imaging: an early surrogate marker of therapeutic efficacy in brain tumors. J Natl Cancer Inst. 2000;92:2029–36.
70. Sadeghi N, Camby I, Goldman S, Gabius HJ, Baleriaux D, Salmon I, Decaesteckere C, Kiss R, Metens T. Effect of hydrophilic components of the extracellular matrix on quantifiable diffusion-weighted imaging of human gliomas: preliminary results of correlating apparent diffusion coefficient values and hyaluronan expression level. AJR Am J Roentgenol. 2003;181:235–41.
71. Tozer DJ, Rees JH, Benton CE, Waldman AD, Jager HR, Tofts PS. Quantitative magnetisation transfer imaging in glioma: preliminary results. NMR Biomed. 2011;24:492–8.
72. Waldman A, Rees JH, Brock CS, Robson MD, Gatehouse PD, Bydder GM. MRI of the brain with ultra-shortecho-time pulse sequences. Neuroradiology. 2003;45:887–92.
73. Sugahara T, Korogi Y, Kochi M, Ikushima I, Hirai T, Okuda T, Shigematsu Y, Liang L, Ge Y, Ushio Y, Takahashi M. Correlation of MR imaging-determined cerebral blood volume maps with histologic and angiographic determination of vascularity of gliomas. AJR Am J Roentgenol. 1998;171:1479–86.
74. Maia AC, Malheiros SM, da Rocha AJ, da Silva CJ, Gabbai AA, Ferraz FA, Stavale JN. MR cerebral blood volume maps correlated with vascular endothelial growth factor expression and tumor grade in nonenhancing gliomas. AJNR Am J Neuroradiol. 2005;26:777–83.
75. Law M, Yang S, Babb JS, Knopp EA, Golfinos JG, Zagzag D, Johnson G. Comparison of cerebral blood volume and vascular permeability from dynamic susceptibility contrast-enhanced perfusion MR imaging with glioma grade. AJNR Am J Neuroradiol. 2004;25:746–55.
76. Law M, Yang S, Wang H, Babb JS, Johnson G, Cha S, Knopp EA, Zagzag D. Glioma grading: sensitivity, specificity, and predictive values of perfusion MR imaging and proton MR spectroscopic imaging compared with conventional MR imaging. AJNR Am J Neuroradiol. 2003;24:1989–98.
77. Patankar TF, Haroon HA, Mills SJ, Baleriaux D, Buckley DL, Parker GJM, Jackson A. Is volume transfer coefficient (Ktrans) related to histologic grade in human gliomas? AJNR Am J Neuroradiol. 2005;26:2455–65.
78. Law M, Oh S, Babb JS, Wang E, Inglese M, Zagzag D, Knopp EA, Johnson G. Low-grade gliomas: dynamic susceptibility-weighted contrast-enhanced perfusion MR imaging—prediction of patient clinical response. Radiology. 2006;238:658–67.
79. Danchaivijitr N, Waldman AD, Tozer DJ, Benton CE, Brasil Caseiras G, Tofts PS, Rees JH, Jager HR. Low-grade gliomas: do changes in rCBV measurements at longitudinal perfusion-weighted MR imaging predict malignant transformation? Radiology. 2008;247:170–8.
80. Cha S, Tihan T, Crawford F, Fischbein NJ, Chang S, Bollen A, Nelson SJ, Prados M, Berger MS, Dillon WP. Differentiation of low-grade oligodendrogliomas from Low-grade astrocytomas by using quantitative blood-volume measurements derived from dynamic susceptibility contrast-enhanced MR imaging. AJNR Am J Neuroradiol. 2005;26:266–73.
81. Lev MH, Ozsunar Y, Henson JW, Rasheed AA, Barest GD, Harsh GR, Fitzek MM, Chiocca EA, Rabinov JD, Csavoy AN, Rosen BR, Hochberg FH, Schaefer PW, Gonzalez RG. Glial tumor grading and outcome prediction using dynamic spin-echo MR susceptibility mapping compared with conventional contrast-enhanced MR: confounding effect of elevated rCBV of oligodendrogliomas. AJNR Am J Neuroradiol. 2004;25:214–21.
82. Weber MA, Zoubaa S, Schlieter M, Juttler E, Huttner HB, Geletneky K, Ittrich C, Lichy MP, Kroll A, Debus J, Giesel FL, Hartmann M, Essig M. Diagnostic performance of spectroscopic and perfusion MRI for distinction of brain tumors. Neurology. 2006;66:1899–906.
83. Mangla R, Singh G, Ziegelitz D, Milano MT, Korones DN, Zhong J, Ekholm SE. Changes in relative cerebral blood volume 1 month after radiation-temozolomide therapy Can help predict overall survival in patients with glioblastoma1. Radiology. 2010;256:575–84.
84. Sorensen AG, Batchelor TT, Zhang WT, Chen PJ, Yeo P, Wang M, Jennings D, Wen PY, Lahdenranta J, Ancukiewicz M, di Tomaso E, Duda DG, Jain RK. A "vascular normalization index" as potential mechanistic biomarker to predict survival after a single dose of cediranib in recurrent glioblastoma patients. Cancer Res. 2009;69:5296–300.
85. Battle MR, Goggi JL, Allen L, Barnett J, Morrison MS. Monitoring tumor response to antiangiogenic

sunitinib therapy with 18F-fluciclatide, an 18F-labeled αVβ3-integrin and αVβ5-integrin imaging agent. J Nucl Med. 2011;52:424–30.
86. Gray LH, Conger AD, Ebert M, Hornsey S, Scott OCA. The concentration of oxygen dissolved in tissues at the time of irradiation as a factor in radiotherapy. Br J Radiol. 1953;26:638–48.
87. Hsu YY, Chang CN, Jung SM, Lim KE, Huang JC, Fang SY, Liu HL. Blood oxygenation level-dependent MRI of cerebral gliomas during breath holding. J Magn Reson Imaging. 2004;19:160–7.
88. Valk PE, Mathis CA, Prados MD, Gilbert JC, Budinger TF. Hypoxia in human gliomas: demonstration by PET with fluorine-18-fluoromisonidazole. J Nucl Med. 1992;33:2133–7.
89. Cher LM, Murone C, Lawrentschuk N, Ramdave S, Papenfuss A, Hannah A, O'Keefe GJ, Sachinidis JI, Berlangieri SU, Fabinyi G, Scott AM. Correlation of hypoxic cell fraction and angiogenesis with glucose metabolicrateingliomasusing18F-fluoromisonidazole, 18F-FDG PET, and immunohistochemical studies. J Nucl Med. 2006;47:410–8.
90. Spence AM, Muzi M, Swanson KR, O'Sullivan F, Rockhill JK, Rajendran JG, Adamsen TCH, Link JM, Swanson PE, Yagle KJ, Rostomily RC, Silbergeld DL, Krohn KA. Regional hypoxia in glioblastoma multiforme quantified with [18F]fluoromisonidazole positron emission tomography before radiotherapy: correlation with time to progression and survival. Clin Cancer Res. 2008;14:2623–30.
91. Croteau D, Scarpace L, Hearshen D, Gutierrez J, Fisher JL, Rock JP, Mikkelsen T. Correlation between magnetic resonance spectroscopy imaging and image- guided biopsies: semiquantitative and qualitative histopathological analyses of patients with untreated glioma. Neurosurgery. 2001;49:823–9.
92. Murphy M, Loosemore A, Clifton AG, Howe FA, Tate AR, Cudlip SA, Wilkins PR, Griffiths JR, Bell BA. The contribution of proton magnetic resonance spectroscopy (1HMRS) to clinical brain tumour diagnosis. Br J Neurosurg. 2002;16:329–34.
93. Lee SW, Fraass BA, Marsh LH, Herbort K, Gebarski SS, Martel MK, Radany EH, Lichter AS, Sandler HM. Patterns of failure following high-dose 3-D conformal radiotherapy for high-grade astrocytomas: a quantitative dosimetric study. Int J Radiat Oncol Biol Phys. 1999;43:79–88.
94. Oppitz U, Maessen D, Zunterer H, Richter S, Flentje M. 3D-recurrence-patterns of glioblastomas after CT-planned postoperative irradiation. Radiother Oncol. 1999;53:53–7.
95. Jena R, Price SJ, Baker C, Jefferies SJ, Pickard JD, Gillard JH, Burnet NG. Diffusion tensor imaging: possible implications for radiotherapy treatment planning of patients with high-grade glioma. Clin Oncol. 2005;17:581–90.
96. Stupp R, Hegi ME, Neyns B, Goldbrunner R, Schlegel U, Clement PMJ, Grabenbauer GG, Ochsenbein AF, Simon M, Dietrich PY, Pietsch T, Hicking C, Tonn JC, Diserens AC, Pica A, Hermisson M, Krueger S, Picard M, Weller M. Phase I/IIa study of cilengitide and temozolomide with concomitant radiotherapy followed by cilengitide and temozolomide maintenance therapy in patients with newly diagnosed glioblastoma. J Clin Oncol. 2010;28:2712–8.
97. Papadakis NG, Xing D, Houston GC, Smith JM, Smith MI, James MF, Parsons AA, Huang CL, Hall LD, Carpenter TA. A study of rotationally invariant and symmetric indices of diffusion anisotropy. Magn Reson Imaging. 1999;17:881–92.
98. Price S, Burnet N, Donovan T, Green H, Pena A, Antoun N, Pickard J, Carpenter T, Gillard J. Diffusion tensor imaging of brain tumours at 3T: a potential tool for assessing white matter tract invasion? Clin Radiol. 2003;58:455–62.
99. Witwer BP, Moftakhar R, Hasan KM, Deshmukh P, Haughton V, Field A, Arfanakis K, Noyes J, Moritz CH, Meyerand ME, Rowley HA, Alexander AL, Badie B. Diffusion-tensor imaging of white matter tracts in patients with cerebral neoplasm. J Neurosurg. 2002;97:568–75.
100. Price S, Pena A, Burnet N, Jena R, Green H, Carpenter T, Pickard J, Gillard J. Tissue signature characterisation of diffusion tensor abnormalities in cerebral gliomas. Eur Radiol. 2004;14:1909–17.
101. Price S, Jena R, Burnet N, Carpenter T, Pickard J, Gillard J. Predicting patterns of glioma recurrence using diffusion tensor imaging. Eur Radiol. 2007;17:1675–84.
102. Price SJ, Jena R, Burnet NG, Hutchinson PJ, Dean AF, Pena A, Pickard JD, Carpenter TA, Gillard JH. Improved delineation of glioma margins and regions of infiltration with the use of diffusion tensor imaging: an image-guided biopsy study. AJNR Am J Neuroradiol. 2006;27:1969–74.
103. Goebell E, Fiehler J, Ding XQ, Paustenbach S, Nietz S, Heese O, Kucinski T, Hagel C, Westphal M, Zeumer H. Disarrangement of fiber tracts and decline of neuronal density correlate in glioma patients—a combined diffusion tensor imaging and 1H-MR spectroscopy study. AJNR Am J Neuroradiol. 2006;27:1426–31.
104. Pirzkall A, Li X, Oh J, Chang S, Berger MS, Larson DA, Verhey LJ, Dillon WP, Nelson SJ. 3D MRSI for resected high-grade gliomas before RT: tumor extent according to metabolic activity in relation to MRI. Int J Radiat Oncol Biol Phys. 2004;59:126–37.
105. Stadlbauer A, Nimsky C, Gruber S, Moser E, Hammen T, Engelhorn T, Buchfelder M, Ganslandt O. Changes in fiber integrity, diffusivity, and metabolism of the pyramidal tract adjacent to gliomas: a quantitative diffusion tensor fiber tracking and MR spectroscopic imaging study. AJNR Am J Neuroradiol. 2007;28:462–9.

106. Kimura T, Ohkubo M, Igarashi H, Kwee IL, Nakada T. Increase in glutamate as a sensitive indicator of extracellular matrix integrity in peritumoral edema: a 3.0-tesla proton magnetic resonance spectroscopy study. J Neurosurg. 2007;106:609–13.
107. Di Costanzo A, Scarabino T, Trojsi F, Popolizio T, Catapano D, Giannatempo GM, Bonavita S, Portaluri M, Tosetti M, d'Angelo VA, Salvolini U, Tedeschi G. Proton MR spectroscopy of cerebral gliomas at 3 T: spatial heterogeneity, and tumour grade and extent. Eur Radiol. 2008;18:1727–35.
108. Zhang K, Li C, Liu Y, Li L, Ma X, Meng X, Feng D. Evaluation of invasiveness of astrocytoma using 1H-magnetic resonance spectroscopy: correlation with expression of matrix metalloproteinase-2. Neuroradiology. 2007;49:913–9.
109. Mosskin M, Bergstrom M, Collins VP, Ehrin E, Eriksson L, von Holst H, Noren G. Positron emission tomography with 11C-methionine of intracranial tumours compared with histology of multiple biopsies. Acta Radiol. 1986;369:157–60.
110. Mosskin M, Ericson K, Hindmarsh T, von Holst H, Collins VP, Bergstrom M, Eriksson L, Johnstrom P. Positron emission tomography compared with magnetic resonance imaging and computed tomography in supratentorial gliomas using multiple stereotactic biopsies as reference. Acta Radiol. 1989;30:225–32.
111. Kracht LW, Miletic H, Busch S, Jacobs AH, Voges J, Hoevels M, Klein JC, Herholz K, Heiss WD. Delineation of brain tumor extent with [11C] L-methionine positron emission tomography: local comparison with stereotactic histopathology. Clin Cancer Res. 2004;10:7163–70.
112. Stadlbauer A, Prante O, Nimsky C, Salomonowitz E, Buchfelder M, Kuwert T, Linke R, Ganslandt O. Metabolic imaging of cerebral gliomas: spatial correlation of changes in O-(2–18F-fluoroethyl)-L-tyrosine PET and proton magnetic resonance spectroscopic imaging. J Nucl Med. 2008;49:721–9.
113. Stadlbauer A, Pölking E, Prante O, Nimsky C, Buchfelder M, Kuwert T, Linke R, Doelken M, Ganslandt O. Detection of tumour invasion into the pyramidal tract in glioma patients with sensorimotor deficits by correlation of 18 F-fluoroethyl-L-tyrosine PET and magnetic resonance diffusion tensor imaging. Acta Neurochir (Wien). 2009;151:1061–9.
114. Henry RG, Vigneron DB, Fischbein NJ, Grant PE, Day MR, Noworolski SM, Star-Lack JM, Wald LL, Dillon WP, Chang SM, Nelson SJ. Comparison of relative cerebral blood volume and proton spectroscopy in patients with treated gliomas. AJNR Am J Neuroradiol. 2000;21:357–66.
115. Blasel S, Franz K, Mittelbronn M, Morawe G, Jurcoane A, Pellikan S, Zanella F, Hattingen E. The striate sign: peritumoural perfusion pattern of infiltrative primary and recurrent gliomas. Neurosurg Rev. 2010;33:193–203. discussion 203–194.
116. Blasel S, Franz K, Ackermann H, Weidauer S, Zanella F, Hattingen E. Stripe-like increase of rCBV beyond the visible border of glioblastomas: site of tumor infiltration growing after neurosurgery. J Neurooncol. 2011;103:575–84.
117. Price SJ, Green HA, Dean AF, Joseph J, Hutchinson PJ, Gillard JH. Correlation of MR relative cerebral blood volume measurements with cellular density and proliferation in high-grade gliomas: an image-guided biopsy study. AJNR Am J Neuroradiol. 2011;32:501–6.
118. Lehmann P, Vallee JN, Saliou G, Monet P, Bruniau A, Fichten A, De MG. Dynamic contrast-enhanced T2*-weighted MR imaging: a peritumoral brain oedema study. J Neuroradiol. 2009;36:88–92.
119. Bennewitz MF, Tang KS, Markakis EA, Shapiro EM: Specific Chemotaxis of Magnetically Labeled Mesenchymal Stem Cells: Implications for MRI of Glioma. Molecular imaging and biology : MIB : the official publication of the Academy of Molecular Imaging, 2012
120. Jenkinson MD, du Plessis DG, Smith TS, Joyce KA, Warnke PC, Walker C. Histological growth patterns and genotype in oligodendroglial tumours: correlation with MRI features. Brain. 2006;129:1884–91.
121. Jenkinson MD, Smith TS, Joyce KA, Fildes D, Broome J, du Plessis DG, Haylock B, Husband DJ, Warnke PC, Walker C. Cerebral blood volume, genotype and chemosensitivity in oligodendroglial tumours. Neuroradiology. 2006;48:703–13.
122. Law M, Brodsky JE, Babb J, Rosenblum M, Miller DC, Zagzag D, Gruber ML, Johnson G. High cerebral blood volume in human gliomas predicts deletion of chromosome 1p: Preliminary results of molecular studies in gliomas with elevated perfusion. J Magn Reson Imaging. 2007;25:1113–9.
123. Kapoor G, Gocke T, Chawla S, Whitmore R, Nabavizadeh A, Krejza J, Lopinto J, Plaum J, Maloney-Wilensky E, Poptani H, Melhem E, Judy K, O'Rourke D. Magnetic resonance perfusion-weighted imaging defines angiogenic subtypes of oligodendroglioma according to 1p19q and EGFR status. J Neurooncol. 2009;92:373–86.
124. Jenkinson MD, Smith TS, Brodbelt AR, Joyce KA, Warnke PC, Walker C. Apparent diffusion coefficients in oligodendroglial tumors characterized by genotype. J Magn Reson Imaging. 2007;26: 1405–12.
125. Jenkinson MD, Smith TS, Joyce K, Fildes D, du Plessis DG, Warnke PC, Walker C. MRS of oligodendroglial tumors: correlation with histopathology and genetic subtypes. Neurology. 2005;64:2085–9.
126. Stockhammer F, Thomale UW, Plotkin M, Hartmann C, Von DA. Association between fluorine-18-labeled fluorodeoxyglucose uptake and 1p and 19q loss of heterozygosity in World Health Organization Grade II gliomas. J Neurosurg. 2007;106:633–7.
127. Drabycz S, Roldan G, de Robles P, Adler D, McIntyre JB, Magliocco AM, Cairncross JG, Mitchell JR. An analysis of image texture, tumor location, and MGMT promoter methylation in glioblastoma using

magnetic resonance imaging. Neuroimage. 2010;49: 1398–405.

128. Moon WJ, Choi J, Roh H, Lim S, Koh YC. Imaging parameters of high grade gliomas in relation to the MGMT promoter methylation status: the CT, diffusion tensor imaging, and perfusion MR imaging. Neuroradiology. 2012;54(6):555–63.
129. Pope WB, Prins RM, Albert Thomas M, Nagarajan R, Yen KE, Bittinger MA, Salamon N, Chou AP, Yong WH, Soto H, Wilson N, Driggers E, Jang HG, Su SM, Schenkein DP, Lai A, Cloughesy TF, Kornblum HI, Wu H, Fantin VR, Liau LM. Noninvasive detection of 2-hydroxyglutarate and other metabolites in IDH1 mutant glioma patients using magnetic resonance spectroscopy. J Neurooncol. 2012;107:197–205.
130. Jalbert L, Elkhaled A, Phillips J, Yoshihara H, Srinivasan R, Bourne G, Chang S, Cha S, Nelson S. Presence of 2-Hydroxyglutarate in IDH-1 mutated low-grade glioma using ex vivo proton HR-MAS. Proc Int Soc Mag Reson Med. 2011;19:183.
131. Aghi M, Gaviani P, Henson JW, Batchelor TT, Louis DN, Barker II FG. Magnetic resonance imaging characteristics predict epidermal growth factor receptor amplification status in glioblastoma. Clin Cancer Res. 2005;11:8600–5.
132. Abourbeh G, Dissoki S, Jacobson O, Litchi A, Ben Daniel R, Laki D, Levitzki A, Mishani E. Evaluation of radiolabeled ML04, a putative irreversible inhibitor of epidermal growth factor receptor, as a bioprobe for PET imaging of EGFR-overexpressing tumors. Nucl Med Biol. 2007;34:55–70.
133. Mishani E, Abourbeh G, Rozen Y, Jacobson O, Laky D, Ben David I, Levitzki A, Shaul M. Novel carbon-11 labeled 4-dimethylamino-but-2-enoic acid [4-(phenylamino)-quinazoline-6-yl]-amides: potential PET bioprobes for molecular imaging of EGFR-positive tumors. Nucl Med Biol. 2004;31:469–76.

Part III

Emerging Concepts in Clinical Practice

Surgical Management of Glial Cancers

9

Walter Stummer

Abstract

The value of surgery for low- and high-grade gliomas alike has long been debated due to the lack of randomized controlled data. Recent studies are providing more and more insight into different benefits derived from surgery and in particular the value extensive cytoreduction in both low- and high-grade gliomas, even in the elderly with malignant gliomas. However, there is ample evidence indicating that prevention of neurological deficits is paramount to radicality.

A plethora of intraoperative methods are available to maximize radicality while minimizing the risks for neurological deficits. These methods should be used judiciously and as a complement to immaculate surgical technique and an intimate familiarity with neuroanatomy.

Nevertheless, not many general recommendations can be made regarding resectability of individual tumors, since the decision of how far to reduce tumor burden depends on many factors, including surgical experience, availability of surgical adjuncts, and patient preference. If extensive cytoreduction is the aim of surgery, technical adjuncts should be used as much as possible, even in high-grade gliomas.

Keywords

Low-grade glioma • High-grade glioma • Surgery • Intraoperative mapping • Intraoperative imaging

Rationale for Surgery

In general oncology, tumor cytoreductive surgery is considered to enable a rapid log kill of tumor cells [1] and also can be used for diagnosis and palliation of symptoms [2]. Cytoreductive surgery for the treatment of gliomas has been performed for decades, in its beginnings with a

W. Stummer
Department of Neurosurgery, University of Münster, Albert-Schweitzer Campus 1A, Munster 48161, Germany
e-mail: Walter.stummer@ukmuenster.de

C. Watts (ed.), *Emerging Concepts in Neuro-Oncology*,
DOI 10.1007/978-0-85729-458-6_9, © Springer-Verlag London 2013

very radical approach, including lobectomies and even hemispherectomies (as reviewed by Quigely et al. [3]).

However, experts in the brain tumor field have questioned the validity of using surgery as a first-line therapy to improve survival [4, 5], citing also that craniotomy still can be associated with serious complications [6]. In fact, some studies suggest that if there is a survival benefit, it is only modest and it may be limited to a select group of patients [3, 7–9].

On the other hand, while it was always acknowledged that surgery cannot cure gliomas in the vast majority of cases because of the a priori diffuse nature of the disease, there was always the perception that it provides some survival advantage. This perception has a long tradition, based, in part, on early data, for instance, a study compiling 2,600 of patients with glioblastoma collected in the 1950s. This study contains a comparison of two patient cohorts who either received lobectomy as a form of very radical surgery or biopsy together with radiation. More extensive resection corresponded to prolonged survival, particularly when patients also received radiotherapy [10].

At present, the beneficial perception of resection is largely based on a collection of clinical data from retrospective and observational studies that assessed survival or disease progression by the extent of resection in patients with or without early postoperative imaging [11–20].

It has been debated whether the cohorts analyzed in such studies are balanced by known prognostic factors such as age and Karnofsky performance status (KPS), among others [21]. In fact, studies in which the distribution of such factors was assessed demonstrated that younger patients or patients with high KPS scores received more extensive resections [22, 23]. Such unbalanced data could potentially influence the postulated effect of resection and would thus confound any conclusions regarding the prognostic role of resection as estimated by multivariate analysis. An adequately powered, well-controlled randomized study on the extent of resection could theoretically overcome these problems for both high- and low-grade gliomas, yet it is difficult to envision a study design which would not be confounded by crossover, the difficulty for controlling the degree of resection, and patients' understandable desire to not participate in a surgical trial in which partial and not best possible resection be a goal in the control group. Nevertheless, evidence is accumulating which supports a strong role of surgery for patients with low- and high-grade gliomas alike.

In principal, patients may not profit from extensive surgical cytoreduction alone, which is one aim of surgery. There are several other goals that are achievable by craniotomy and tumor resection:

- Enable representative histology and avoid undergrading
- Relieve mass effect
- Facilitate adjuvant therapies
- Positively influence seizure activity
- Apply local therapies
- Prolong survival through cytoreduction and prevent malignant deterioration

Enable Representative Histology and Avoid Undergrading

Imaging alone does not suffice for reliably determining histology of cerebral lesions, even though a glioma may appear likely. The ALA study [23] was a surgical study on malignant gliomas where randomization depended on MR imaging and not biopsy prior to open surgery. In this study, surgeons were asked to enter patients if malignant glioma was likely on preoperative MRI. Ten percent of histologies in this series of 270 patients were not malignant gliomas. These erroneous histologies would have resulted in wrong treatment decisions had tissue samples not been obtained. Examples for erroneous histologies were (expectedly) abscesses and metastasis, but there were also pilocytic astrocytomas, low-grade gliomas (oligodendroglioma, oligoastrocytoma), a meningioma with regressive changes, necrotizing vasculitis, and lymphoma. While stereotactic biopsies might serve to reliably and safely establish histology in such cases, collecting a representative tissue sample during surgery is a welcome additional benefit.

In the context of low-grade gliomas, the possible differential diagnosis is similarly diverse, including inflammation, ischemia, AIDS-related lesions, demyelinating lesions, or cortical dysplasia [24, 25]. Some of the confounding diagnosis can be resolved by multiparametric imaging, including MRI, MRI spectroscopy, and PET, but assumptions from imaging may still be erroneous so that at least stereotactic biopsies are warranted.

Gliomas are notoriously heterogeneous. Whereas one part of the glioma may histologically present as a classical low grade, other parts may well have deteriorated to high-grade gliomas without standard MRI imaging making this likely. The likelihood that a presumed low-grade glioma contains anaplastic foci increases with age [26], but younger age does not preclude anaplastic characteristics if a low-grade glioma is presumed on imaging [26]. Such anaplastic foci may not be particularly conspicuous on standard MRI imaging. Recently, amino acid PET has emerged as a tool that may be of help in identifying occult high-grade gliomas in seemingly low-grade gliomas, even if these areas are not contrast enhancing [27, 28].

With extended resections, tissue samples can be obtained from all areas of the tumor, providing more tissue and thus rendering histopathological and molecular characterization of tumor more precise and, importantly, minimizing the risk of undergrading gliomas. It has been demonstrated that more extensive resections more frequently result in the diagnosis of high-grade glioma [29, 30] and are superior to stereotactic biopsy alone. To this end, Jackson et al. [31] compiled 81 patients referred to a major neuro-oncological center after stereotactic biopsy that were then treated by resection. Thirty-eight percent of these patients ended up having a histology, of which 26 % would affect treatment. Similarly, Muragaki et al. [32] observed 30 % undergrading in patients with anaplastic astrocytomas with initial biopsies that then went on to have resections at a neuro-oncological referral center.

Relieve Mass Effect

It is a common experience for neurosurgeons that surgery in patients with neurological symptoms will be of value by increasing KPS and function [33]. This observation has been substantiated in larger studies. About 30 % of 400 patients demonstrated improvement of symptoms after surgery for parenchymal tumors in a large series by Sawaya et al. [34]. Similarly, the large randomized ALA study showed improvement of symptoms as defined by the NIH stroke score in approximately 30 % of cases [23]. When stratified by degree of resection, patients with complete resections maintained functional independence (as defined by a KPS >=70) significantly longer, with significantly deferred deterioration of neurological function, as assessed by the NIH stroke score [35].

One group found a prolongation in survival when comparing surgery to stereotactic biopsy only when patients had preexisting midline shift indicating mass effect but not when midline shift was lacking [5], suggesting that resection is only of value in the presence of a substantial mass lesion. In this study, no postoperative imaging was performed to understand the influence of residual tumor on these results, nor were intraoperative methods such as neuronavigation or other methods for localizing residual tumor used. Essentially therefore, this study is an indicator of the value of debulking rather than an indicator of the missing value of extensive cytoreduction in malignant glioma surgery.

Two additional studies indicate surgical debulking to be of benefit. Vuorinen et al. performed a small surgical study in 30 elderly patients randomizing between biopsy and craniotomy [9]. Only two patients in the craniotomy arm in that study were stated to have had "total" resections, without earlier postoperative imaging having been performed. Three patients in the biopsy arm did not go on to have radiotherapy due to their bad clinical condition. Patients with craniotomies deteriorated later and had prolonged survival. Together, these results support a role for decompressive surgery also in the elderly.

Finally, Fadul et al. [36] analyzed the number of serious complications such as hemorrhages and herniations after biopsy or open surgery. They found these serious complications to occur after biopsy or partial resection, but the fewest to occur after "total" resections, with the degree of radicality being assessed by surgeons.

Facilitating Adjuvant Therapies

It is unlikely that cytoreduction by resection of glioma tissue, if followed by adjuvant therapies, carries a benefit exclusively through reduction of tumor cell burden. Rather, interactions between surgery and adjuvant therapies may be assumed, especially for malignant gliomas, for instance, by removal of hypoxic tumor tissue in the gross tumor mass. Hypoxia influences the behavior of tumor cells by activating genes involved in the adaptation to hypoxic stress (e.g., angiogenesis), representing an important indicator of cancer prognosis [37–40]. Hypoxia is associated with aggressive growth, metastasis, and poor response to treatment [41, 42]. Apart from hypoxia, other factors may play a role. Central GBM cells appear to proliferate more readily than they invade [43] which may affect short-term prognosis. The marginal region of the tumor has more angiogenesis [44], and tumor cells in this region may be less proliferative, more migratory, and more resistant to apoptosis, which likely contributes to treatment resistance [43].

In addition, experimental studies have observed that interstitial fluid pressure (IFP) may affect the penetration and residence time of chemotherapy agents. Increased IFP in glioma tumors with increased permeability may produce a pressure gradient that distributes chemotherapy agents into necrotic areas or into surrounding brain tissue where it is absorbed and cleared, thus limiting the exposure time to the cytotoxic agents in the marginal region [45]. Following resection, there is an initial increase in IFP in the immediate postoperative period because of cerebral edema. Once edema is resolved, IFP decreases and the direction of fluid flow reverses toward the resection cavity. With intracavitary chemotherapy, the increase in IFP in the immediate postoperative period because of edema, particularly with BCNU wafers [46], may improve the limited penetration of antitumor agents into the infiltrating zone [47, 48]. Systemic chemotherapy, on the other hand, is generally delivered a few weeks after resection when edema has been resolved and there is a decrease in IFP, particularly in the marginal zone, and permeability is limited from cytokine production of tumor cells within that zone. Thus, decreasing IFP may prolong the drug residence time in the marginal zone [45].

These experimental observations are supported by clinical studies, in particular the results of three large prospectively randomized phase 3 studies in newly diagnosed malignant glioma. In all three studies, surgical resection was employed, followed by various adjuvant therapies (as compiled in [49]). The primary objectives of these 3 studies were to assess the support of resection by intraoperative fluorescence guidance using 5-aminolevulinic acid (Gliolan®, medac GmbH) followed by radiotherapy [23], the efficacy and safety of concomitant radiochemotherapy with temozolomide (Temodar®, Schering Corp., Kenilworth) [50], and intracavitary chemotherapy with carmustine (BCNU) wafers (Gliadel® Wafer, MGI Pharma) [51]. The results of these three studies supported the approval of these products for the treatment of patients with malignant glioma. Within all 3 trials, data regarding the extent of resection, as well as other prognostic factors, were collected prospectively. Exploratory analyses of all three trials with stratification by degree of resection demonstrated greater survival in patients with "complete" resection, either as assessed by early postoperative imaging or by surgeon's assessment of degree of resection, than in patients with "incomplete" resections. In addition, there was an indication that more extensive surgery was enhancing the effect of the adjuvant interventions.

ALA Study

In the 5-ALA study [23], patients with resectable, newly diagnosed malignant glioma were randomized to receive 5-ALA fluorescence-guided resection and radiotherapy ($n = 139$) or conventional non-enhanced microsurgery and radiotherapy ($n = 131$). A significant survival difference was observed in the whole study cohort between patients who received complete resection and radiotherapy with partial resection and radiotherapy: 17.9 months (95 % CI, 14.3–19.4) vs. 12.9 months (95 % CI, 10.6–14.0), respectively ($P < 0.0001$).

BCNU Wafer Study

In a randomized, double-blind, placebo-controlled phase 3 trial by Westphal et al. [51], 240 patients with newly diagnosed malignant glioma were randomized to receive resection with BCNU wafer or placebo wafer implantation followed by radiotherapy. Implantation of BCNU wafers resulted in significantly longer survival compared with placebo in the ITT population (13.9 vs. 11.6 months; HR=0.71, 95 % CI, 0.52–0.96; $P=0.03$).

Complete resection was defined as removal of ≥90 % of tumor tissue as measured on postoperative radiographs compared with preoperative scans. In the complete resection subgroup, median survival was significantly greater for BCNU and radiotherapy wafer than for placebo and radiotherapy (14.8 vs. 12.6 months; $P=0.01$). In the partial resection subgroup, on the other hand, median survival was 12.1 and 11.2 months, respectively ($P=0.39$), indicating greater efficacy of wafers in ≥90 % resections.

For patients with GBM, the results were very similar, with a median survival of 14.5 months for BCNU and radiotherapy wafer versus 12.4 months for placebo and radiotherapy ($P=0.02$) in the complete resection subgroup and 11.7 and 10.6 months, respectively ($P=0.98$), in the partial resection subgroup.

Concomitant Radiochemotherapy Followed by Adjuvant Temozolomide

The EORTC-NCIC trial (EORTC 26981–22981; NCIC CE.3) study was a randomized, multicenter trial in 573 patients with newly diagnosed malignant glioma [50]. Patients were randomized to receive radiotherapy alone or radiotherapy with continuous daily temozolomide followed by 6 cycles of temozolomide. In the primary report, Stupp et al. observed a significant survival benefit with temozolomide. Median survival was 14.6 months in the radiochemotherapy group versus 12.1 months in the radiotherapy only group (HR=0.63; 95 % CI, 0.52–0.75; $P<0.001$), and the 2-year survival rate was 26.5 % (95 % CI, 21.2–31.7) vs. 10.4 % (95 % CI, 6.8–14.1), respectively. Long-term follow-up demonstrated that the survival benefit with temozolomide was maintained at 3 years (16.7 % vs. 4.3 %, respectively; $P<0.0001$) [52].

During this study, the extent of resection was reported per surgeon assessment on clinical and/or radiological grounds. The clinical impact of resection was examined further in a post hoc analysis of survival by the extent of resection. Absolute median gain in survival time with radiochemotherapy versus radiotherapy alone was greatest in patients with complete resections (+4.1 months) compared with those with incomplete resections (+1.8 months) or biopsies (+1.5 months), although this increase did not meet statistical significance in tests for interaction (test for heterogeneity, $P=0.24$). The Kaplan-Meier survival data corresponded to a risk reduction for radiochemotherapy versus radiotherapy alone of 43 % in patients who received complete resection ($P=0.0001$), 35 % for partial resection ($P=0.0001$), and a nonsignificant decrease of 31 % for biopsy ($P=0.088$). Adjuvant temozolomide appeared to have the greatest survival impact in patients who received complete resection, suggesting that the effectiveness of temozolomide may be related to the extent of resection [52].

Thus, these studies provide evidence on the relationship between resection and the efficacy of adjuvant therapies. More extensive resection was associated with improved outcome, and in the temozolomide and BCNU wafer studies, the survival benefit of the study treatment appeared to be most favorable in patients who received extensive resection.

Positive Influence on Epilepsy

Seizures are a frequent symptom of gliomas, occurring very commonly in low-grade gliomas, often as presenting symptom. In high-grade gliomas, seizures are less common but are more difficult to control than in low-grade gliomas [53]. It has been demonstrated that patients that were treated by craniotomy with gross total resections in low-grade tumors are seizure-free or have meaningful improvement of seizure severity and frequency [54]. A beneficial influence of surgery in seizures for patients with high-grade gliomas was similarly observed [55]. Importantly, even partial and subtotal tumor resections are helpful in selected cases, i.e., for gliomas involving the insula (as reviewed by Kurzwelly) [56].

Applying Local Therapies

Craniotomies with tumor resection allow access to the zone of infiltration in gliomas and give the opportunity for applying drugs locally, circumventing the blood–brain barrier. Two archetypes for this type of treatment have been tested in phase 3 trials, BCNU wafers and gene therapy. BCNU wafers (Gliadel®) [51] have demonstrated basic efficacy, as discussed above. Gene therapy has been tested in several clinical trials in glioblastomas. In these trials, either retrovirus vector-producer cells or adenovirus vectors mediating transduction of the herpes simplex virus thymidine kinase gene for activating the prodrug ganciclovir were injected intraoperatively after glioblastoma resection into the walls of the resection cavity. Subsequently, patients were treated with ganciclovir. After promising early results [57], no influence on survival was noted in two large randomized trials, the first being performed during the 1990s [58]. The second trial, the Cerepro® trial, was recently concluded but did not reach the predefined end point (unpublished data). Benefit was noted in some subgroups in augmental analyses.

Prolonging Survival Through Cytoreduction

Low-Grade Gliomas

Most of the controversy regarding surgery for gliomas has surrounded the issue of maximal cytoreduction in the face of surgical risks, given that typical patients with low-grade gliomas are of young age and without deficits, having presented with single seizures, and that data proving the benefit of cytoreductive surgery are limited.

On the other hand, low-grade gliomas tend to grow or to degenerate into high-grade gliomas, with up to 75 % of patients dying within 5–10 years after diagnosis [59, 60] despite long-term survival of many patients with this disease. Furthermore, *all* low-grade gliomas grow and their growth rate has been determined to be approximately 4 mm/year (95 % CI: 3.8–4.4 mm) [61].

Apart from surgery, there is little in terms of alternatives for the treatment of these progressive lesions. Radiotherapy, when given after diagnosis, results in a prolongation of progression-free survival, but not of survival, as demonstrated in the EORTC 22845 trial [62], with a possible price: Douw et al. [63] followed 65 patients with low-grade gliomas, 32 of which had received early radiotherapy and the rest without. Twenty-seven percent of patients without early radiotherapy suffered significant cognitive deficits compared to 53 % with radiotherapy. Although previously considered to be of little value, chemotherapy with temozolomide has recently been found to elicit tumor response in a number of patients [64–66] and has been observed to have a positive effect on seizure frequency [67, 68]. However, chemotherapy cannot be considered standard for the treatment of low-grade glioma patients.

At present, the only standard that is generally agreed upon in the context of low-grade gliomas is to obtain tissue diagnosis. No randomized controlled trials are available regarding the extent of resection and outcome. Outcome in low-grade gliomas depends on a number of covariates, such as patient age and histology [59, 69–71], which confound the interpretation of outcome data. Furthermore, there are interactions between prognostic factors and resectability and concurrently between prognostic factors and survival [21]. For instance, tumor involvement of the corticospinal tract, large tumor volume, and oligodendroglioma subtype have been found to independently predict resectability [72] while shorter survival times were observed when tumors overlapped eloquent brain regions [73, 74]. Figure 9.1 shows examples of "low-grade gliomas" with differing degrees of resectability depending on size and location.

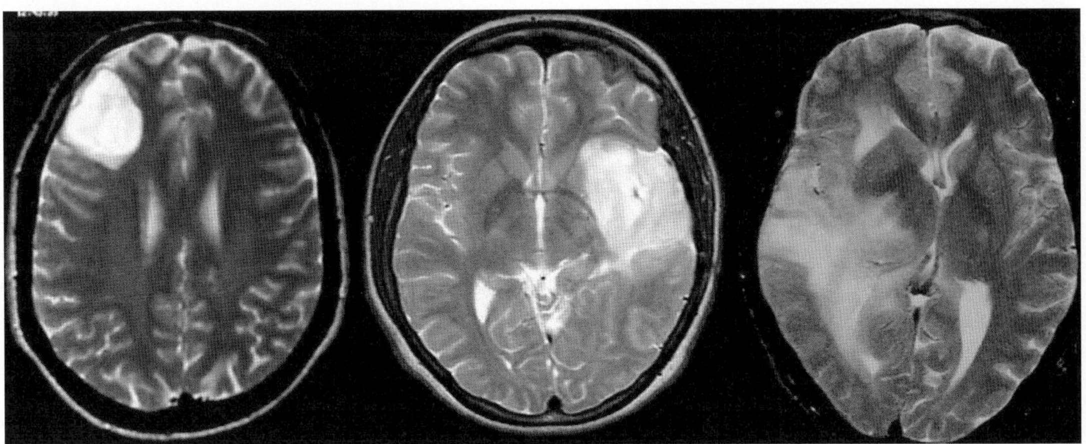

Fig. 9.1 Three different gliomas, each with low-grade gliomas histology, but with completely different degrees of "resectability." The glioma on the *left* hand image is easily resected by most neurosurgeons; the *center* image shows an insular glioma, which might be considered resectable in highly specialized centers using intraoperative mapping under awake conditions. The glioma on the *right* is not resectable. It is likely that such gliomas per se have a different biology. Such variations in presentation, resectability, and biology confound the interpretation of survival data generated in retrospective studies

Of the available studies on the value or resection, most have design-related limitations including a small study size, a small number of events (i.e., deaths for survival studies), or missing postoperative imaging. Only few studies actually analyze the influence of prognostic factors on resection apart from their influence on survival. Two recent studies overcome many, but not all, of these limitations. The first series with 170 patients operated on for low-grade gliomas [60] in whom resection was stratified by gross total, near total, and subtotal resections as defined by fluid-attenuated inversion recovery (FLAIR) imaging. Significant differences in progression-free survival, overall survival, and malignant degeneration-free survival were found for patients with gross total resections compared to patients with incomplete resections. However, a careful analysis of factors predicting the degree of resection demonstrated preoperative tumor size to have been significantly smaller in patients with gross total resections (diameter 3.7), compared to near total (4.6) or subtotal resections (4.8 cm), again rendering interpretation of survival data difficult, because preoperative tumor size has independently been linked to survival. In another recent study [74] with an even larger number of patients, overall survival and progression-free survival were independently influenced by *postoperative* tumor volume while there was no significant relationship between pre- and postoperative tumor volumes. This observation suggests an influence of surgery rather than preexisting tumor size to be important. A confounder in this study, again, was that outcome (overall survival, progression-free survival) was also independently predicted by *preoperative* tumor volumes, rendering interpretation difficult, since it has been suggested that the biology of smaller lesions may differ from the biology of larger lesions [12, 75].

Identifying Low-Grade Glioma Patients at Risk

Although there is a growing evidence suggesting that more extensive resections are favorable, other management strategies are still acceptable, including simple biopsy and watchful waiting, especially in light of possible permanent neurological deficits of up to 6 % in recent series [60]. Prerequisite for a conservative strategy, however, is that tumors are not on the verge of malignant transformation.

In deciding when to resect and when to observe, it should be kept in mind that low-grade gliomas are biologically diversified. Not all will remain quiescent and predictable over extended periods of time and will require earlier action because of impending malignant degeneration. Amino acid PET may help in defining those patients with low-grade gliomas with a high risk for transformation [28, 76]. Floeth et al. [76] demonstrated the risk for malignant degeneration and death to be highest in patients with amino acid uptake and diffuse lesions. If lesions were circumscript, prognosis was better and best in patients with circumscript lesion on T2 MR imaging without amino acid uptake.

Other factors have also been identified on which to base decisions regarding "aggressive" therapies such as surgery or radiotherapy, as suggested by Pignatti et al., based on two large EORTC studies on low-grade gliomas [77]. In this analysis, age >=40, astrocytoma versus oligodendroglioma or oligoastrocytoma, tumor size >=6 cm, crossing of the midline, and neurological deficits prior to surgery were all independent risk factors for shorter survival. Depending on the number of risk factors, patient survival ranged between less than 1 year and 10 years. Pallud et al. [78] demonstrated worse prognosis for patients with low-grade gliomas which grew >8 mm per year as opposed to patients with <8-mm tumor growth per year. More recently, MRI-derived markers of metabolism, such as MR spectroscopy and chemical shift imaging or perfusion imaging with assessment of cerebral blood volume, have become available as additional indicators of imminent malignant degeneration and neo-angiogenesis in previously low-grade gliomas [79–81].

Cytoreductive Surgery in High-Grade Gliomas

In the absence of randomized controlled clinical trials, the controversy concerning the value of cytoreductive surgery in high-grade gliomas has been much the same as in low-grade gliomas.

There is a substantial collection of prospective and retrospective data supporting resection over biopsy in malignant glioma (summarized by Laws et al. [15]) and one small prospectively randomized trial addressing this question [9]. The latter examined the effectiveness of resection in malignant glioma in a small series of 30 patients over the age of 65 years, randomizing these patients to have either biopsy or resection. There was a significant improvement in median survival for resection versus biopsy (171 days vs. 85 days, respectively; $P=0.035$). Unfortunately, this study was not adequately powered to draw definitive conclusions and excluded younger patients who have been shown to benefit from resection more than older patients [15].

Historically, data on the *extent of resection* were not captured in the majority of studies for about 40 years, as most centers did not have the infrastructure to accurately assess the extent of resection by early postoperative imaging. More recently, studies with such imaging have become available to offer a more reliable basis to determine the relationship between the extent of resection and outcome, despite these studies being retrospective and observational [16, 22, 82, 83]. Because of their uncontrolled design, however, these studies have been criticized for potentially confounded results. Biased distribution of patients may have caused differences between study groups for known prognostic factors, such as age, tumor location, and KPS [4, 16, 84, 85]. For example, patients that are treated by craniotomies and resections are often younger than those that have tumor biopsies [15]. The degree of resection has been demonstrated to depend on KPS and age [22, 86]. Biased distribution of prognostic factors may have concurrently influenced survival and the extent of resection, which cannot be corrected by multivariate analysis [21]. With minor exceptions [22], few investigators analyze the impact of these prognostic factors on the extent of resection, even in a very new series by McGirt et al. [20]. This study clearly demonstrates the *prognostic* value of degree of resection of contrast-enhancing tumor and survival of patients with gross total resections to live longest, but fails to analyze age, KPS, and resectability as possible confounders.

More reliable information has come from the ALA trial [23]. In this trial, 270 patients were randomized to have surgery either using conventional white-light microscopy ($n = 131$) or fluorescence-guided resections ($n = 139$).

The complete resection rates (i.e., of contrast-enhancing tumor) were 65 % and 36 % in the 5-ALA and conventional surgery groups, respectively. In logistic-regression models, the use of 5-ALA had the most important effect on the probability of complete resection (odds ratio = 3.41; 95 % CI, 2.03–5.71; $P < 0.0001$), followed by age and tumor location, while performance status did not reach statistical significance. The 29 % difference in the frequency of complete resections translated into a significantly improved 6-month PFS rate (41.0 % vs. 21.0 %, respectively; $P = 0.0003$). Analysis of overall survival also favored surgery with 5-ALA over conventional surgery (15.2 vs. 13.5 months; hazard ratio [HR] = 0.82; 95 % CI, 0.62–1.07; $P = 0.1$). Although this difference was not statistically significant, it should be realized that the study was not powered to show differences in overall survival. A significant survival difference was observed in the whole study cohort between patients who received complete resection and partial resection: 17.9 months (95 % CI, 14.3–19.4) vs. 12.9 months (95 % CI, 10.6–14.0), respectively ($P < 0.0001$).

To further investigate the impact of resection, a post hoc analysis of these data restratified the per-protocol cohort of patients (those with grade IV tumors) by the extent of resection, i.e., complete versus partial [86]. The resulting groups were balanced regarding a number of possible prognostic factors (e.g., neurological function, preoperative tumor characteristics, additional treatments received), with the exception of age and tumor location (eloquent vs. non-eloquent), the latter of which was assessed by surgeons. Patients with complete resections survived 16.7 compared to 11.8 months for patients with incomplete resections ($P < 0.0001$). The survival advantage with complete resection was maintained when patients were substratified by age <60 years vs. >60 years, and the differences in age distributions in the substrata were no longer detectable. Median age was 52 and 54 years for complete and partial resection, respectively, in the subgroup <60 years of age, and 65 and 66 years, respectively, in the subgroup >60 years of age. The survival advantage also was maintained after substratification of patients by tumor location. Thus, both age and tumor location could be eliminated as confounding factors.

In a subsequent analysis that evaluated survival by the Radiation Therapy Oncology Group recursive partitioning analysis (RPA), the survival advantage associated with complete versus partial resection was maintained in RPA class IV and V subgroups [87]. Complete resection was also favored for the RPA class III subgroup, but this did not reach statistical significance, likely because of the small sample size.

It was of interest to note that survival benefits depended on all of contrast-enhancing tumor being removed; even small volumes of residual tumor of >0–0.7 cm^3 resulted in a significant worsening of prognosis [86]. This finding was similar to the previous experience of [16]. Both observations underscore the importance for removing *all* of the contrast-enhancing tumor, which is hypoxic and particularly aggressive [37, 43], as best possible surgical treatment. On the other hand, in a larger retrospective case series by Sanai et al. [88], a significant survival advantage was seen with as little as 78 % extent of resection, and a stepwise improvement in survival was evident even in the 95–100 % range. It appears that less than complete resections will also be of value to the patient.

Neurological Deficits and Survival

While complete resections of contrast-enhancing tumor should be the goal in malignant glioma surgery, it is paramount that this aim is achieved without neurological deficits. In this regard, McGirt et al. demonstrated patients with major neurological deficits (motor, language) after surgery for glioblastomas to have shorter survival than those patients without deficits [89]. In that study, both surgically acquired motor deficits (median survival: 9.0 months) and surgically acquired language deficits (9.6 months) were independently associated with decreased median survival compared with patients without new neurological deficits after surgery (12.8 months; $P<0.05$). Our own experience has been quite similar [90]. In a multicentric phase 2 safety study using fluorescence-guided resections in 206 patients, median survival was 14.8 months in patients without neurological adverse events recorded 48 h after surgery as compared to 12.4 months if patients had *any* neurological adverse event.

The reasons for these observations are unclear. However, it can be hypothesized that patients with neurological deficits are less heavily treated in the adjuvant phase of therapy, especially after recurrence of their tumors.

Cytoreductive Surgery for High-Grade Glioma in the Elderly

Malignant gliomas are a disease of the elderly. The average age of the typical glioblastoma patient is 63 years [91]. At present, due to rising life expectancies and the overall age of the population in developed countries, the incidence of glioblastoma is likely to increase.

Older people, however, are traditionally subject to therapy bias. Therapy bias is linked to the perception that older patients might suffer more side effects and fewer benefits of accepted therapies and is further strengthened by the rarity of large prospectively randomized trials focused on this subpopulation. Thus, older patients are usually underrepresented in clinical trials [92–94].

Specifically, in glioblastoma patients older than 60 years, prognosis is generally considered poor, but these patients are also likely to receive less therapy than younger patients [95, 96]. From a neurosurgical perspective, older people are frailer and are less likely to recuperate if surgery results in neurological deficit. Thus, older people with suspected malignant gliomas are more likely to be treated by biopsy than by craniotomy [15].

Large neuro-oncological studies have failed to give an answer to the important question of how aggressively the elderly patients with malignant gliomas should be treated. The median age in these trials was well below the expected age for this population. For instance, the median age in the EORTC 26981 trial [50] was 56 years, in the BCNU wafer trial 53 years [51], and in the Glioma Outcomes Project [15] 54 years. Further data regarding the efficacy of therapies in this neglected neuro-oncological subpopulation are therefore urgently required.

Elderly studies suggest that the prognosis of the elderly glioblastoma patient is particularly limited. The Glioma Outcomes Project reported a median survival of 36.1 weeks in 271 patients older than 60 years treated by craniotomy and radiotherapy [15]. Other older studies are difficult to compare, but in general demonstrate a limited prognosis of less than 6 months [9, 97–99] for surgery and radiotherapy.

Other studies, especially of recent origin, suggest a more favorable prognosis for older patients, provided that these patients receive therapies comparable to younger patients. A median survival of 14.9 months was reported for a small series of older patients when surgery and radiotherapy were followed by adjuvant temozolomide [100]. Our own results [90] demonstrated 130 patients beyond 60 years (median 68 years) to profit from aggressive surgery using 5-ALA, especially in conjunction with adjuvant radiochemotherapy. In the former group, median survival was 16.3 months vs. 11.2 months for older patients with radiotherapy only. In comparison, survival within the EORTC 26981 study [50] was 10.9 months in 83 patients older than 60 years, i.e., the same age group, treated by concomitant radiochemotherapy with adjuvant temozolomide (control: 11.3 months) and not increased in older patients. Unfortunately, specifications regarding the type and extent of surgical treatment were missing in that report, and it must be assumed that some of these elderly patients were not treated by resection. If so, it is unknown what types of technical adjuncts were used and in what portion of patients' gross total resections were achieved.

Maximizing Safety of Glioma Resection

The goal of any surgery for gliomas, given the available evidence, is maximal, safe cytoreduction. From a surgical point of view, safe but extensive resections of gliomas are difficult to achieve, because these tumors are often within or in close proximity to eloquent areas [101], and margins between tumor and brain are difficult to identify, because of the diffuse infiltrative nature of these lesions. The impression of the surgeon regarding the degree of resection, as deduced from his microscopic image, is often inaccurate and tends to overestimate the extent of surgery [22]. Even given the increasing knowledge regarding anatomical representation of function, predictions of functional brain from anatomy alone are insufficient due to distortions of brain topography by tumor. Furthermore, in case of language functions, cortical representation may be highly variable [102, 103], and important white matter tracts are virtually impossible to detect visually during surgery.

Thus, simple neurosurgery based on conventional surgical microscopes and anatomical knowledge is not sufficient for safe cytoreductive therapy of gliomas. For these reasons, surgical adjuncts have been explored, with the aim of better identifying tumor and for identifying functionally important brain before inflicting damage.

Intraoperative neuronavigation is now a commonly available tool for intraoperative orientation [104]. Its main weakness however is brain shift [105], resulting in inaccuracies as soon as the dura is opened and, more so, when extensive resections have been performed. Information from functional MRI and tractography can be integrated into neuronavigation systems; however, the user must bear in mind that functional MRI might have a reproducibility problems, especially with naming tasks [106]. Tractography on preoperative imaging is equally subject to brain shift [107, 108]. Thus, confident intraoperative use of neuronavigation for localizing critical structures requires intraoperative imaging for maintaining accuracy, for instance, by 3D ultrasound or intraoperative MRI [109, 110].

Intraoperative ultrasound is a practical tool for assessing the extent of low-grade glioma [111, 112] but has its distinct limitations in surgery for high-grade gliomas due to artifacts by edema and blood, which develop during resection [113].

Finally, the last years has seen an increase in the use of intraoperative MRI [114–116]. Its value for optimizing resection is generally accepted, but its expense is a major barrier for many.

A cheaper and simpler method for intraoperative is by the use of 5-ALA (Gliolan®) for fluorescence-guided resection. This method has been approved in 2007 for malignant gliomas [23] based on a randomized controlled trial. This trial unequivocally demonstrated that the number of complete resections of contrast-enhancing tumor could be doubled using fluorescence-guided resections, which translated into a prolongation of progression-free survival. With a high positive predictive value, fluorescence highlights contrast-enhancing malignant glioma tissue [82] and has potential to highlight anaplastic foci within otherwise low-grade lesions [117, 118]. Diffuse low-grade gliomas do not appear to accumulate fluorescence; however, confocal microscopy has revealed individual cells in low-grade gliomas to contain protoporphyrin IX, the fluorescing metabolite of 5-ALA [119].

None of the other methods have been tested in the context of a randomized trial, with the exception of neuronavigation. In a small randomized trial [120], Willems and coworkers found no advantage in using neuronavigation for maximizing resection of contrast-enhancing lesions. One trial with ultra-low-field MRI is ongoing [121], and a preliminary report in 27 glioma patients has been given, indicating smaller volumes of residual glioma tissue in the group allocated to intraoperative MRI.

Regarding identification and preservation of brain function, it is well accepted that intraoperative cortical and subcortical mapping and monitoring techniques, including awake craniotomies, are helpful and must be considered gold standard for preserving function in glioma surgery [122–128]. In low-grade gliomas, such methods are of additional importance, since function has been observed within the confines of glioma tissue [129].

However, while the value of intraoperative mapping techniques appears obvious, these methods have also not been tested in the context of a randomized study to determine the true impact on neurological safety, radicality, and outcome. It is unlikely that such a study will ever be undertaken, because the use of intraoperative monitoring cannot be controlled. One retrospective comparison between patients operated on with and without intraoperative functional mapping [130], however, did indicate improved survival and a higher degree of radicality if patients were operated on with intraoperative mapping.

Taken together, there is a plethora of intraoperative tools to help the surgeon to safely maximize resection, and each has its merits and limitations. It is crucial to remember that mere availability of a tool is not sufficient. Rather, the surgeon has to use the tool to its fullest potential in order to make a difference. In addition even the best tools will not replace intricate knowledge of neuroanatomy and immaculate microsurgical technique as the basis of successful surgery.

Which Gliomas Are "Resectable"?

In light of the many advances during the last decade regarding intraoperative tumor identification and mapping, general recommendations about which tumors are amenable to complete resections are not easily made. The concept of resectability depends on more than just the location and size of the tumor. Which tumors are deemed resectable is a complex discussion which frequently involves disagreement even among specialized neurosurgeons. To illustrate the different attitudes concerning "resectability," Fig. 9.1 gives an example of an insular glioma, which would be considered largely unresectable in many centers, but mostly resectable in specialized centers [131, 132]. One important element of resectability is the involvement of "eloquent" brain. Investigators have attempted to categorize "eloquent" and "non-eloquent," also defining "near-eloquent" brain [34]. "Non-eloquent" brain in that report encompassed frontal and temporal polar lesions, right parieto-occipital lesions, and cerebellar hemispheres; "near-eloquent" brain, areas near to the motor or sensory cortex calcarine fissure, speech center, dentate nucleus, and brain stem or in the corpus callosum. "Eloquent" brain was defined to encompass motor and sensory cortex, visual centers, speech centers, internal capsule, basal ganglia, hypothalamus/thalamus, brain stem, and dentate nucleus.

However, such classifications depend in part on the perception of the surgeon regarding "eloquent" and "expendable" brain, for instance, regarding the visual pathways, which are involved in many cases of glioblastoma. Are they expendable or not in the face of a malignant glioma? This question cannot be answered in a general sense. The basic consequences regarding surgery by subdividing the brain into "eloquent" and "non-eloquent" are also unclear. The use of intraoperative mapping allows extended and safe resections even in regions of the brain traditionally considered "eloquent" and involving language and motor functions [122, 133]. The perception of "resectability" of individual tumors thus clearly depends on the availability of such methodology and experience in individual centers. Also, brain acute and long-term plasticity is being discussed more and more in conjunction with possible deficits by resection and makes the equation even more complex [134].

Apart from the availability and extensive use of mapping techniques and location tools such as intraoperative imaging, other factors will also influence the individual surgeon's impression of tumors being "resectable," for instance, a surgeon's training, his personal conviction on the value of resection, his willingness to take risks by extending resections, and, last but most important, patient preference. A decision for surgery and decisions on the aspired extent of surgery should always be based on an individual assessment of risks and gains for a particular patient, taking the many factors into account that determine prognosis and quality of life. In young patients with low-grade gliomas that are diagnosed incidentally or after a single seizure, particular care should be taken to educate patients carefully, to allow truly informed decisions on the part of the patient.

The fundamental basic tenet in any form of glioma surgery, however, is safety. To this end, surgeons should use as many technical adjuncts as reasonably possible, also in the field malignant gliomas.

References

1. McCarter MD, Fong Y. Role for surgical cytoreduction in multimodality treatments for cancer. Ann Surg Oncol. 2001;8:38–43.
2. Sills Jr AK, Duntsch C, Weimar J. Therapeutic strategies for local recurrent malignant glioma. Curr Treat Options Oncol. 2004;5:491–7.
3. Quigley MR, Maroon JC. The relationship between survival and the extent of the resection in patients with supratentorial malignant gliomas. Neurosurgery. 1991;29:385–8.
4. Curran Jr WJ, Scott CB, Horton J. Recursive partitioning analysis of prognostic factors in three Radiation Therapy Oncology Group malignant glioma trials. J Natl Cancer Inst. 1993;85:704–10.
5. Kreth FW, Warnke PC, Scheremet R, et al. Surgical resection and radiation therapy versus biopsy and radiation therapy in the treatment of glioblastoma multiforme. J Neurosurg. 1993;78:762–6.
6. Whittle IR. Surgery for gliomas. Curr Opin Neurol. 2002;15:663–9.
7. Ashby LS, Ryken TC. Management of malignant glioma: steady progress with multimodal approaches. Neurosurg Focus. 2006;20:E3.
8. Kowalczuk A, Macdonald RL, Amidei C, et al. Quantitative imaging study of extent of surgical resection and prognosis of malignant astrocytomas. Neurosurgery. 1997;41:1028–36.
9. Vuorinen V, Hinkka S, Farkkila M, et al. Debulking or biopsy of malignant glioma in elderly people—a randomised study. Acta Neurochir (Wien). 2003;145: 5–10.
10. Tönnis W, Walter W. Das Glioblastoma multiforme. (Bericht über 2611 Fälle). Das Glioblastoma Multiforme: Pathologie Klinik, Diagnostik und Therapie. Acta Neurochirurgica Suppl VI. Wien: Springer; 1959. p. 40–-62.
11. Ciric I, Vick NA, Mikhael MA, et al. Aggressive surgery for malignant supratentorial gliomas. Clin Neurosurg. 1990;36:375–83.
12. Berger MS, Deliganis AV, Dobbins J, et al. The effect of extent of resection on recurrence in patients with low grade cerebral hemisphere gliomas. Cancer. 1994;74:1784–91.
13. Devaux BC, O'Fallon JR, Kelly PJ. Resection, biopsy, and survival in malignant glial neoplasms. A retrospective study of clinical parameters, therapy, and outcome. J Neurosurg. 1993;78:767–75.
14. Winger MJ, Macdonald DR, Cairncross JG. Supratentorial anaplastic gliomas in adults. The prognostic importance of extent of resection and prior low-grade glioma. J Neurosurg. 1989;71: 487–93.
15. Laws ER, Parney IF, Huang W, Glioma Outcomes Investigators. Survival following surgery and prognostic factors for recently diagnosed malignant glioma: data from the Glioma Outcomes Project. J Neurosurg. 2003;99:467–73.
16. Lacroix M, Abi-Said D, Fourney DR, et al. A multivariate analysis of 416 patients with glioblastoma multiforme: prognosis, extent of resection, and survival. J Neurosurg. 2001;95:190–8.
17. Keles GE, Chang EF, Lamborn KR, et al. Volumetric extent of resection and residual contrast enhancement on initial surgery as predictors of outcome in adult patients with hemispheric anaplastic astrocytoma. J Neurosurg. 2006;105:34–40.
18. Keles GE, Anderson B, Berger MS. The effect of extent of resection on time to tumor progression and survival in patients with glioblastoma multiforme of the cerebral hemisphere. Surg Neurol. 1999; 52:371–9.
19. Simpson JR, Horton J, Scott C, et al. Influence of location and extent of surgical resection on survival of patients with glioblastoma multiforme: results of three consecutive Radiation Therapy Oncology Group (RTOG) clinical trials. Int J Radiat Oncol Biol Phys. 1993;26:239–44.
20. McGirt MJ, Chaichana KL, Gathinji M, et al. Independent association of extent of resection with survival in patients with malignant brain astrocytoma. J Neurosurg. 2009;110:156–62.
21. Hess KR. Extent of resection as a prognostic variable in the treatment of gliomas. J Neurooncol. 1999;42: 227–31.
22. Albert FK, Forsting M, Sartor K, et al. Early postoperative magnetic resonance imaging after resection of malignant glioma: objective evaluation of residual tumor and its influence on regrowth and prognosis. Neurosurgery. 1994;34:45–60.
23. Stummer W, Pichlmeier U, Meinel T, ALA-Glioma Study Group. Fluorescence-guided surgery with 5-aminolevulinic acid for resection of malignant glioma: a randomised controlled multicentre phase III trial. Lancet Oncol. 2006;7:392–401.
24. Law M, Hamburger M, Johnson G, et al. Differentiating surgical from non-surgical lesions using perfusion MR imaging and proton MR spectroscopic imaging. Technol Cancer Res Treat. 2004;3:557–65.
25. Omuro AM, Leite CC, Mokhtari K, Delattre JY. Pitfalls in the diagnosis of brain tumours. Lancet Neurol. 2006;5:937–48.
26. Barker 2nd FG, Chang SM, Huhn SL, et al. Age and the risk of anaplasia in magnetic resonance-nonenhancing supratentorial cerebral tumors. Cancer. 1997;80:936–41.
27. Smits A, Baumert BG. The clinical value of PET with amino acid tracers for gliomas WHO grade II. Int J Mol Imaging. 2011;2011:372509.
28. Ewelt C, Floeth FW, Felsberg J, et al. Finding the anaplastic focus in diffuse gliomas: the value of Gd-DTPA enhanced MRI, FET-PET, and intraoperative, ALA-derived tissue fluorescence. Clin Neurol Neurosurg. 2011;113:541–7.
29. Glantz MJ, Burger PC, Herndon 2nd JE, et al. Influence of the type of surgery on the histologic diagnosis in patients with anaplastic gliomas. Neurology. 1991;41:1741–4.

30. Woodworth G, McGirt MJ, Samdani A, Garonzik I, Olivi A, Weingart JD. Accuracy of frameless and frame-based image-guided stereotactic brain biopsy in the diagnosis of glioma: comparison of biopsy and open resection specimen. Neurol Res. 2005;27: 358–62.
31. Jackson RJ, Fuller GN, Abi-Said D, Lang FF, Gokaslan ZL, Shi WM, Wildrick DM, Sawaya R. Limitations of stereotactic biopsy in the initial management of gliomas. Neuro Oncol. 2001;3:193–200.
32. Muragaki Y, Chernov M, Maruyama T, et al. Low-grade glioma on stereotactic biopsy: how often is the diagnosis accurate? Minim Invasive Neurosurg. 2008;51:275–9.
33. Ciric I, Ammirati M, Vick N, Mikhael M. Supratentorial gliomas: surgical considerations and immediate postoperative results. Gross total resection versus partial resection. Neurosurgery. 1987;21:21–6.
34. Sawaya R, Hammoud M, Schoppa D, et al. Neurosurgical outcomes in a modern series of 400 craniotomies for treatment of parenchymal tumors. Neurosurgery. 1998;42:1044–55.
35. Stummer W, Tonn JC, Mehdorn HM, ALA-Glioma Study Group. Counterbalancing risks and gains from extended resections in malignant glioma surgery: a supplemental analysis from the randomized 5-aminolevulinic acid glioma resection study. Clinical article. J Neurosurg. 2011;114:613–23.
36. Fadul C, Wood J, Thaler H, et al. Morbidity and mortality of craniotomy for excision of supratentorial gliomas. Neurology. 1988;38:1374–9.
37. Collingridge DR, Piepmeier JM, Rockwell S, Knisely JP. Polarographic measurements of oxygen tension in human glioma and surrounding peritumoural brain tissue. Radiother Oncol. 1999;53:127–31.
38. Evans SM, Judy KD, Dunphy I, Jenkins WT, et al. Hypoxia is important in the biology and aggression of human glial brain tumors. Clin Cancer Res. 2004;10:8177–84.
39. Knisely JP, Rockwell S. Importance of hypoxia in the biology and treatment of brain tumors. Neuroimaging Clin N Am. 2002;12:525–36.
40. Zagzag D, Zhong H, Scalzitti JM, Laughner E, Simons JW, Semenza GL. Expression of hypoxia-inducible factor 1 alpha in brain tumors: association with angiogenesis, invasion, and progression. Cancer. 2000;88: 2606–18.
41. Koritzinsky M, Seigneuric R, Magagnin MG, et al. The hypoxic proteome is influenced by gene-specific changes in mRNA translation. Radiother Oncol. 2005;76:177–86.
42. Lal A, Peters H, St Croix B, Haroon ZA, Dewhirst MW, Strausberg RL, Kaanders JH, van der Kogel AJ, Riggins GJ. Transcriptional response to hypoxia in human tumors. J Natl Cancer Inst. 2001;93: 1337–43.
43. Giese A, Bjerkvig R, Berens ME, Westphal M. Cost of migration: invasion of malignant gliomas and implications for treatment. J Clin Oncol. 2003;21: 1624–36.
44. Holash J, Maisonpierre PC, Compton D, et al. Vessel cooption, regression, and growth in tumors mediated by angiopoietins and VEGF. Science. 1999;284: 1994–8.
45. Navalitloha Y, Schwartz ES, Groothuis EN. Therapeutic implications of tumor interstitial fluid pressure in subcutaneous RG-2 tumors. Neuro Oncol. 2006;8:227–33.
46. Gottfried ON, Deogaonkar M, Way DL. Postoperative cerebral edema after intracavitary implantation of Gliadel(R) wafers for treatment of malignant gliomas [abstract]. J Invest Med. 2000;48(1):79A.
47. Fleming AB, Saltzman WM. Pharmacokinetics of the carmustine implant. Clin Pharmacokinet. 2002;41: 403–19.
48. Fung LK, Ewend MG, Sills A, et al. Pharmacokinetics of interstitial delivery of carmustine, 4-hydroperoxy-cyclophosphamide, and paclitaxel from a biodegradable polymer implant in the monkey brain. Cancer Res. 1998;58:672–84.
49. Stummer W, van den Bent MJ, Westphal M. Cytoreductive surgery of glioblastoma as the key to successful adjuvant therapies: new arguments in an old discussion. Acta Neurochir (Wien). 2011; 153:1211–8.
50. Stupp R, Mason WP, van den Bent MJ, European Organisation for Research and Treatment of Cancer Brain Tumor and Radiotherapy Groups; National Cancer Institute of Canada Clinical Trials Group. Radiotherapy plus concomitant and adjuvant temozolomide for glioblastoma. N Engl J Med. 2005;352:987–96.
51. Westphal M, Ram Z, Riddle V, Hilt D, Bortey E, Executive Committee of the Gliadel Study Group. Gliadel wafer in initial surgery for malignant glioma: long-term follow-up of a multicenter controlled trial. Acta Neurochir (Wien). 2006;148:269–75.
52. Stupp R, Hegi ME, Mason WP, European Organisation for Research and Treatment of Cancer Brain Tumour and Radiation Oncology Groups; National Cancer Institute of Canada Clinical Trials Group. Effects of radiotherapy with concomitant and adjuvant temozolomide versus radiotherapy alone on survival in glioblastoma in a randomised phase III study: 5-year analysis of the EORTC-NCIC trial. Lancet Oncol. 2009;10:459–66.
53. Rosati A, Tomassini A, Pollo B, et al. Epilepsy in cerebral glioma: timing of appearance and histological correlations. J Neurooncol. 2009;93:395–400.
54. Chang EF, Potts MB, Keles GE, et al. Seizure characteristics and control following resection in 332 patients with low-grade gliomas. J Neurosurg. 2008;108: 227–35.
55. Chaichana KL, Parker SL, Olivi A, Quiñones-Hinojosa A. Long-term seizure outcomes in adult patients undergoing primary resection of malignant brain astrocytomas. Clinical article. J Neurosurg. 2009;111:282–92.
56. Kurzwelly D, Herrlinger U, Simon M. Seizures in patients with low-grade gliomas—incidence,

pathogenesis, surgical management, and pharmacotherapy. Adv Tech Stand Neurosurg. 2010;35: 81–111.

57. Immonen A, Vapalahti M, Tyynelä K, et al. AdvHSV-tk gene therapy with intravenous ganciclovir improves survival in human malignant glioma: a randomised, controlled study. Mol Ther. 2004;10:967–72.
58. Rainov NG. A phase III clinical evaluation of herpes simplex virus type 1 thymidine kinase and ganciclovir gene therapy as an adjuvant to surgical resection and radiation in adults with previously untreated glioblastoma multiforme. Hum Gene Ther. 2000;11: 2389–401.
59. Keles GE, Lamborn KR, Berger MS. Low-grade hemispheric gliomas in adults: a critical review of extent of resection as a factor influencing outcome. J Neurosurg. 2001;95:735–45.
60. McGirt MJ, Chaichana KL, Attenello FJ, et al. Extent of surgical resection is independently associated with survival in patients with hemispheric infiltrating low-grade gliomas. Neurosurgery. 2008;63:700–7.
61. Mandonnet E, Delattre JY, Tanguy ML, et al. Continuous growth of mean tumor diameter in a subset of grade II gliomas. Ann Neurol. 2003;53: 524–8.
62. van den Bent MJ, Afra D, de Witte O, EORTC Radiotherapy and Brain Tumor Groups and the UK Medical Research Council. Long-term efficacy of early versus delayed radiotherapy for low-grade astrocytoma and oligodendroglioma in adults: the EORTC 22845 randomised trial. Lancet. 2005;366: 985–90.
63. Douw L, Klein M, Fagel SS, van den Heuvel J, et al. Cognitive and radiological effects of radiotherapy in patients with low-grade glioma: long-term follow-up. Lancet Neurol. 2009;8:810–8.
64. Pouratian N, Gasco J, Sherman JH, et al. Toxicity and efficacy of protracted low dose temozolomide for the treatment of low grade gliomas. J Neurooncol. 2007;82:281–8.
65. Kesari S, Schiff D, Drappatz J, LaFrankie D, et al. Phase II study of protracted daily temozolomide for low-grade gliomas in adults. Clin Cancer Res. 2009;15:330–7.
66. Kaloshi G, Benouaich-Amiel A, Diakite F, et al. Temozolomide for low-grade gliomas: predictive impact of 1p/19q loss on response and outcome. Neurology. 2007;68(21):1831–6.
67. Hu A, Xu Z, Kim RY, Nguyen A, et al. Seizure control: a secondary benefit of chemotherapeutic temozolomide in brain cancer patients. Epilepsy Res. 2011;95:270–2.
68. Sherman JH, Moldovan K, Yeoh HK, et al. Impact of temozolomide chemotherapy on seizure frequency in patients with low-grade gliomas. J Neurosurg. 2011;114:1617–21.
69. Claus EB, Horlacher A, Hsu L, et al. Survival rates in patients with low-grade glioma after intraoperative magnetic resonance image guidance. Cancer. 2005; 103:1227–33.
70. Janny P, Cure H, Mohr M, Heldt N, et al. Low grade supratentorial astrocytomas: management and prognostic factors. Cancer. 1994;73:1937–45.
71. Johannesen TB, Langmark F, Lote K. Progress in long-term survival in adult patients with supratentorial low-grade gliomas: a population-based study of 993 patients in whom tumors were diagnosed between 1970 and 1993. J Neurosurg. 2003;99:854–62.
72. Talos IF, Zou KH, Ohno-Machado L, et al. Supratentorial low grade glioma resectability: statistical predictive analysis based on anatomic MR features and tumor characteristics. Radiology. 2006;239: 506–13.
73. Chang EF, Clark A, Smith JS, et al. Functional mapping-guided resection of low-grade gliomas in eloquent areas of the brain: improvement of long-term survival. J Neurosurg. 2011;114:566–73.
74. Smith JS, Chang EF, Lamborn KR, et al. Role of extent of resection in the long-term outcome of low-grade hemispheric gliomas. J Clin Oncol. 2008;26: 1338–45.
75. Mariani L, Siegenthaler P, Guzman R, et al. The impact of tumour volume and surgery on the outcome of adults with supratentorial WHO grade II astrocytomas and oligoastrocytomas. Acta Neurochir (Wien). 2004;146:441–8.
76. Floeth FW, Pauleit D, Sabel M, et al. Prognostic value of O-(2–18 F-fluoroethyl)-L-tyrosine PET and MRI in low-grade glioma. J Nucl Med. 2007;48:519–27.
77. Pignatti F, van den Bent M, Curran D, European Organization for Research and Treatment of Cancer Brain Tumor Cooperative Group; European Organization for Research and Treatment of Cancer Radiotherapy Cooperative Group. Prognostic factors for survival in adult patients with cerebral low-grade glioma. J Clin Oncol. 2002;20:2076–84.
78. Pallud J, Mandonnet E, Duffau H, et al. Prognostic value of initial magnetic resonance imaging growth rates for World Health Organization grade II gliomas. Ann Neurol. 2006;60:380–3.
79. Law M, Yang S, Wang H, et al. Glioma grading: sensitivity, specificity, and predictive values of perfusion MR imaging and proton MR spectroscopic imaging compared with conventional MR imaging. AJNR Am J Neuroradiol. 2003;24:1989–98.
80. Guillevin R, Menuel C, Abud L. Proton MR spectroscopy in predicting the increase of perfusion MR imaging for WHO grade II gliomas. J Magn Reson Imaging. 2011. doi:10.1002/jmri.22862.
81. Guillevin R, Menuel C, Duffau H, et al. Proton magnetic resonance spectroscopy predicts proliferative activity in diffuse low-grade gliomas. J Neurooncol. 2008;87:181–7.
82. Stummer W, Novotny A, Stepp H, et al. Fluorescence-guided resection of glioblastoma multiforme by using 5-aminolevulinic acid-induced porphyrins: a prospective study in 52 consecutive patients. J Neurosurg. 2000;93:1003–13.
83. Vecht CJ, Avezaat CJ, van Putten WL. The influence of the extent of surgery on the neurological function

and survival in malignant glioma. A retrospective analysis in 243 patients. J Neurol Neurosurg Psychiatry. 1990;53:466–71.
84. Prognostic factors for high-grade malignant glioma: development of a prognostic index. A Report of the Medical Research Council Brain Tumour Working Party. J Neurooncol. 1990;9:47–55.
85. Steltzer KJ, Sauvé KI, Spence AM, et al. Corpus callosum involvement as a prognostic factor for patients with high-grade astrocytoma. Int J Radiat Oncol Biol Phys. 1997;38:27–30.
86. Stummer W, Reulen HJ, Meinel T, ALA-Glioma Study Group. Extent of resection and survival in glioblastoma multiforme: identification of and adjustment for bias. Neurosurgery. 2008;62:564–76.
87. Pichlmeier U, Bink A, Schackert G, Stummer W, ALA Glioma Study Group. Resection and survival in glioblastoma multiforme: an RTOG recursive partitioning analysis of ALA study patients. Neuro Oncol. 2008;10:1025–34.
88. Sanai N, Polley MY, McDermott MW, et al. An extent of resection threshold for newly diagnosed glioblastomas. J Neurosurg. 2011;115:3–8.
89. McGirt MJ, Mukherjee D, Chaichana KL. Association of surgically acquired motor and language deficits on overall survival after resection of glioblastoma multiforme. Neurosurgery. 2009;65:463–9. discussion.
90. Stummer W, Nestler U, Stockhammer F, et al. Favorable outcome in the elderly cohort treated by concomitant temozolomide radiochemotherapy in a multicentric phase II safety study of 5-ALA. J Neurooncol. 2011;103:361–70.
91. Surawicz TS, McCarthy BJ, Kupelian V. Descriptive epidemiology of primary brain and CNS tumors. Results from the central brain tumor registry of the United States, 1990–1994. Neuro Oncol. 1999;1: 14–25.
92. Hutchins LF, Unger JM, Crowley JJ, et al. Underrepresentation of patients 65 years of age or older in cancer-treatment trials. N Engl J Med. 1999;341:2061–7.
93. Gross CP, Herrin J, Wong N, et al. Enrolling older persons in cancer trials: the effect of socio-demographic, protocol, and recruitment center characteristics. J Clin Oncol. 2005;23:4755–63.
94. Yee KW, Pater JL, Pho L, et al. Enrollment of older patients in cancer treatment trials in Canada: why age is a barrier? J Clin Oncol. 2003;21:1618–23.
95. Kita D, Ciernik IF, Vaccarella S, et al. Age as predictive factor in glioblastomas: population-based study. Neuroepidemiology. 2009;33:17–22.
96. Iwamoto FM, Reiner AS, Panageas KS, et al. Patterns of care in elderly glioblastoma patients. Ann Neurol. 2008;64:628–34.
97. Kelly PJ, Hunt C. The limited value of cytoreductive therapy in elderly patients with malignant gliomas. Neurosurgery. 1994;34:62–6.
98. Ampil F, Fowler M, Kookmin K. Intracranial astrocytoma in elderly patients. J Neurooncol. 1992;12: 125–30.
99. Kushnir I, Tzuk-Shina T. Efficacy of treatment for glioblastoma multiforme in elderly patients (65+): a retrospective analysis. Isr Med Assoc J. 2011; 13:290–4.
100. Brandes AA, Vastola F, Basso U, et al. A prospective study on glioblastoma in the elderly. Cancer. 2003;97:657–62.
101. Berger MS, Rostomily RC. Low grade gliomas: functional mapping resection strategies, extent of resection, and outcome. J Neurooncol. 1997;34: 85–101.
102. Herholz K, Thiel A, Wienhard K, et al. Individual functional anatomy of verb generation. Neuroimage. 1996;3:185–94.
103. Ojemann G, Ojemann J, Lettich E, Berger M. Cortical language localization in left, dominant hemisphere: an electrical stimulation mapping investigation in 117 patients. J Neurosurg. 1989;71: 316–26.
104. Rasmussen Jr IA, Lindseth F, Rygh OM, et al. Functional neuronavigation combined with intraoperative 3D ultrasound: initial experiences during surgical resections close to eloquent brain areas and future directions in automatic brain shift compensation of preoperative data. Acta Neurochir (Wien). 2007;149:365–78.
105. Reinges MH, Nguyen HH, Krings T, et al. Course of brain shift during microsurgical resection of supratentorial cerebral lesions: limits of conventional neuronavigation. Acta Neurochir (Wien). 2004;146:369–77.
106. Rau S, Fesl G, Bruhns P, et al. Reproducibility of activations in Broca area with two language tasks: a functional MR imaging study. AJNR Am J Neuroradiol. 2007;28:1346–53.
107. Romano A, D'Andrea G, Calabria LF. A. Pre- and intraoperative tractographic evaluation of corticospinal tract shift. Neurosurgery. 2011;69: 696–704.
108. Berman JI, Berger MS, Chung SW, et al. Accuracy of diffusion tensor magnetic resonance imaging tractography assessed using intraoperative subcortical stimulation mapping and magnetic source imaging. J Neurosurg. 2007;107:488–94.
109. Letteboer MM, Willems PW, Viergever MA, Niessen WJ. Brain shift estimation in image-guided neurosurgery using 3-D ultrasound. IEEE Trans Biomed Eng. 2005;52:268–76.
110. Nimsky C, Ganslandt O, Cerny P, et al. Quantification of, visualization of, and compensating for brain shift using intraoperative magnetic resonance imaging. Neurosurgery. 2000;47:1070–80.
111. Katisko JP, Koivukangas JP. Optically neuronavigated ultrasonography in an intraoperative magnetic resonance imaging environment. Neurosurgery. 2007;60:373–80.
112. Lindner D, Trantakis C, Renner C, et al. Application of intraoperative 3D ultrasound during navigated tumor resection. Minim Invasive Neurosurg. 2006; 49:197–202.

113. Solheim O, Selbekk T, Jakola AS, Unsgård G. Ultrasound-guided operations in unselected high-grade gliomas—overall results, impact of image quality and patient selection. Acta Neurochir (Wien). 2010;152:1873–86.
114. Kuhnt D, Becker A, Ganslandt O. Correlation of the extent of tumor volume resection and patient survival in surgery of glioblastoma multiforme with high-field intraoperative MRI guidance. Neuro Oncol. 2011;13(12):1339–48.
115. Kuhnt D, Ganslandt O, Schlaffer SM. Quantification of glioma removal by intraoperative high-field magnetic resonance imaging: an update. Neurosurgery. 2011;69:852–62.
116. Muragaki Y, Iseki H, Maruyama T, et al. Usefulness of intraoperative magnetic resonance imaging for glioma surgery. Acta Neurochir Suppl. 2006;98: 67–75.
117. Stummer W, Stocker S, Wagner S, et al. Intraoperative detection of malignant gliomas by 5-aminolevulinic acid-induced porphyrin fluorescence. Neurosurgery. 1998;42:518–25.
118. Widhalm G, Wolfsberger S, Minchev G, et al. 5-Aminolevulinic acid is a promising marker for detection of anaplastic foci in diffusely infiltrating gliomas with nonsignificant contrast enhancement. Cancer. 2010;116:1545–52.
119. Sanai N, Snyder LA, Honea NJ, et al. Intraoperative confocal microscopy in the visualization of 5-aminolevulinic acid fluorescence in low-grade gliomas. J Neurosurg. 2011;115:740–8.
120. Willems PW, Taphoorn MJ, Burger H, et al. Effectiveness of neuronavigation in resecting solitary intracerebral contrast-enhancing tumors: a randomized controlled trial. J Neurosurg. 2006;104: 360–8.
121. Senft C, Bink A, Heckelmann M, et al. Glioma extent of resection and ultra-low-field iMRI: interim analysis of a prospective randomized trial. Acta Neurochir Suppl. 2011;109:49–53.
122. Sanai N, Mirzadeh Z, Berger MS. Functional outcome after language mapping for glioma resection. N Engl J Med. 2008;358:18–27.
123. Guggisberg AG, Honma SM, Findlay AM, et al. Mapping functional connectivity in patients with brain lesions. Ann Neurol. 2008;63:193–203.
124. Schiffbauer H, Berger MS, Ferrari P, et al. Preoperative magnetic source imaging for brain tumor surgery: a quantitative comparison with intraoperative sensory and motor mapping. J Neurosurg. 2002;97:1333–42.
125. Walker JA, Quiñones-Hinojosa A, Berger MS. Intraoperative speech mapping in 17 bilingual patients undergoing resection of a mass lesion. Neurosurgery. 2004;54:113–8.
126. Maldonado IL, Moritz-Gasser S, de Champfleur NM, et al. Surgery for gliomas involving the left inferior parietal lobule: new insights into the functional anatomy provided by stimulation mapping in awake patients. J Neurosurg. 2011;115:770–9.
127. Maldonado IL, Moritz-Gasser S, Duffau H. Does the left superior longitudinal fascicle subserve language semantics? A brain electrostimulation study. Brain Struct Funct. 2011;216:263–74.
128. De Benedictis A, Moritz-Gasser S, Duffau H. Awake mapping optimizes the extent of resection for low-grade gliomas in eloquent areas. Neurosurgery. 2010;66:1074–84.
129. Skirboll SS, Ojemann GA, Berger MS, et al. Functional cortex and subcortical white matter located within gliomas. Neurosurgery. 1996;38: 678–84.
130. Duffau H, Lopes M, Arthuis F, et al. Contribution of intraoperative electrical stimulations in surgery of low grade gliomas: a comparative study between two series without (1985–96) and with(1996–2003) functional mapping in the same institution. J Neurol Neurosurg Psychiatry. 2005;76:845–51.
131. Duffau H. A personal consecutive series of surgically treated 51 cases of insular WHO Grade II glioma: advances and limitations. J Neurosurg. 2009;110:696–708.
132. Simon M, Neuloh G, von Lehe M, Meyer B, Schramm J. Insular gliomas: the case for surgical management. J Neurosurg. 2009;110:685–95.
133. Kim SS, McCutcheon IE, Suki D, et al. Awake craniotomy for brain tumors near eloquent cortex: correlation of intraoperative cortical mapping with neurological outcomes in 309 consecutive patients. Neurosurgery. 2009;64:836–45.
134. Ius T, Angelini E, de Thiebaut Schotten M, Mandonnet E, Duffau H. Evidence for potentials and limitations of brain plasticity using an atlas of functional resectability of WHO grade II gliomas: towards a "minimal common brain". Neuroimage. 2011;56:992–1000.

Radiation Oncology in Brain Cancer 10

Susan C. Short

Abstract

Radiotherapy remains a central part of the treatment of most primary and secondary brain cancers and contributes significantly to long-term outcomes. Recent developments in radiotherapy technology are providing the opportunity to deliver more accurate, flexible, and individualized treatment in many contexts. This is possible through use of techniques including intensity-modulated radiotherapy, image guidance, and single- and multiple-dose radiosurgery. Current challenges include how best to evaluate the optimal application of these technologies in different patient groups. An additional issue is that an increasing proportion of brain cancer patients treated with radiotherapy is surviving long enough to be at risk of long-term toxicities which are still inadequately understood and difficult to treat.

Keywords

Radiotherapy • Intensity modulation • Radiosurgery • Toxicity

Principles of Radiotherapy

Radiation is an effective cytotoxic treatment due to cellular damage, principally DNA damage, caused when high-energy radiation interacts with cells. Cancer cells are less tolerant of damage than normal cells and are therefore selectively killed by this treatment. The mechanisms by which cells respond to and repair radiation-induced DNA damage have been studied in detail, and the differences between how cancer cells and relevant normal tissue cells respond are starting to be understood. The principal toxic lesion following irradiation of nucleated cells is a double-strand break (DSB) in DNA, and, using immunofluorescent identification of individual breaks localized by sites of histone H2AX phosphorylation, it has been shown that DSBs are induced in a dose-dependent manner after radiation. An example of this assay in glioma cells is shown in Fig. 10.1. It is also clear that unrepaired DSBs at relatively long times after exposure to

S.C. Short, MBBS, MRCP, FRCR, Ph.D.
Department of Oncology,
Leeds Institute of Molecular Medicine,
Beckett St, Leeds LS9 7TF, UK
e-mail: s.c.short@leeds.ac.uk

C. Watts (ed.), *Emerging Concepts in Neuro-Oncology*,
DOI 10.1007/978-0-85729-458-6_10, © Springer-Verlag London 2013

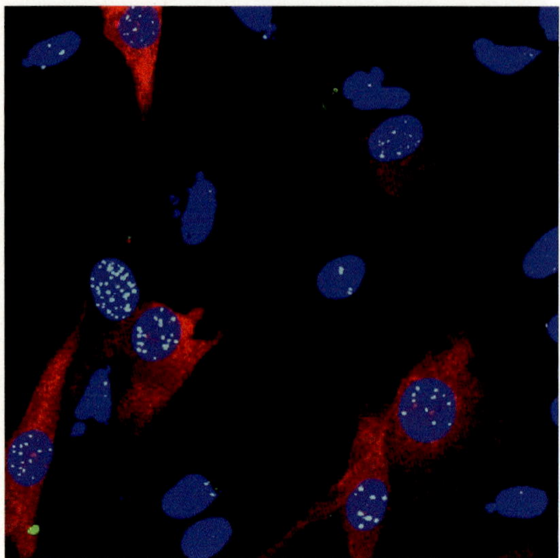

Fig. 10.1 Cell nuclei stained with DAPI (*Blue stain*) and with a fluorescent antibody directed against the phosphorylated protein H2AX, which locates individual double-strand breaks on DNA

radiation correlate with cell death [1, 2]. Glioma cells and normal human astrocytes show a typical dose response using this assay, but there may be subtle differences in how these different cell types repair this damage [3]. The relevance of understanding this biology lies in the potential to increase the cytotoxic effects of radiation specifically in cancer cells by targeting repair pathways unique to them or pathways which they exhibit an over-reliance on compared to normal tissue cells. The concept of synthetic lethality relies on this approach, in which a repair inhibitor is targeted at the functioning repair processes that are active in tumor cells in which some repair capability is lost. Approaches to tumor cell-specific radiosensitization include targeting the processing of single-strand breaks to promote DSB formation at mitosis by poly (ADP-ribose) polymerase (PARP) inhibition, thereby exploiting the high mitotic rate of glioma cells. Tumor cells may also be specifically reliant on certain repair proteins including Rad51 and on specific cellular survival mechanisms including autophagy and antiapoptotic signaling [4–6].

It should be noted that, although combination treatment with radiotherapy and temozolomide chemotherapy has been shown to significantly improve outcome in patients with GBM [7, 8], there are scant data suggesting that this is due to a radiosensitizing effect of this drug. Overall, published data from in vitro studies suggest that the interaction between the two agents is additive. It is also noteworthy that scheduling is relevant since the cytotoxic effects of temozolomide are not expressed for 2–3 cell cycle times after administration, and sub-additivity may be seen when radiation is given before this [9].

Although the relevant details of the molecular biology of DNA repair are still being investigated in order to identify new targets for treatment, it is well established in clinical studies that, in order to preserve an optimal therapeutic ratio, large total radiation doses are better tolerated when delivered as a series of small doses given 5 days per week. This allows normal tissue, which is less able to tolerate large single doses, to repair and regenerate following treatment. Hence, a standard radiotherapy regime is given as daily doses (fractions) of around 2 Gy, and a radical dose (around 60 Gy) is achieved during a 6-week treatment course. In circumstances where normal tissue is excluded from the treatment field, larger doses per day or single large doses as delivered in radiosurgery may be tolerated.

Radiotherapy Side Effects

Despite optimization of imaging for target definition and highly complex radiotherapy delivery methods, there are very few circumstances in which non-tumor tissue can be totally excluded from exposure to radiation during treatment. For brain cancers, the relevant normal tissues are skin and hair on the scalp, non-involved eloquent brain, optic and auditory apparatus, pituitary gland, and cerebral vessels, all of which show dose-dependent expression of radiation toxicity which is apparent in a time course which depends on the biology of the tissue involved. Radiotherapy side effects are conventionally grouped by the time at which they are expressed into early and late effects. In brain, an additional set of toxicities may be expressed at an intermediate time, known as delayed early effects and principally manifesting as a somnolence syndrome [10]. True early toxicity includes hair loss and skin erythema, which occur within 3–4 weeks of commencing treatment. Late effects occurring many years after treatment completion include pituitary hormone failure, damage to visual and auditory pathways, and cognitive deficit. Radionecrosis, a radiation-induced pathology in normal brain, can occur at a range of time points between 6 months and several years after treatment. In much longer time frames, radiation-induced tumors—most commonly meningioma—are known to occur and, rarely, radiation-induced malignant tumors including glioma and sarcoma [11, 12]. Late effects of radiation on vasculature lead to increased risk of cerebrovascular accident.

Mechanisms of late toxicity are thought to be related to damage to the endothelial as well as the glial compartment in the central nervous system (CNS) [13, 14]. The majority of cells within the CNS are postmitotic and therefore not expected to exhibit sensitivity to radiation. The most sensitive differentiated cell type in the CNS is the oligodendrocyte population; hence, demyelination occurs following relatively low radiation doses and may account for some of the observed toxicity [15]. In the glial compartment, it has become clear recently that, even in adults, a population of stem-like cells exists in the CNS with regenerative potential. These cells inhabit very specific anatomic and biological niches in the CNS and are thought to be particularly sensitive to radiation. Anatomically, they have been located in periventricular and hippocampal regions although it is not clear if these are the only sites where these populations can reside. Biologically, these cells inhabit a relatively hypoxic environment in close relation to cerebral vasculature. Current models of late toxicity following radiotherapy to brain include the effects of loss of these cells, which may account for changes in cognitive function including memory deficits. In rodent models, loss of these cells is associated with cognitive deficit following exposure to radiation, and studies are ongoing to address whether specifically excluding stem cell-rich areas from radiation may reduce these long-term side effects. It is a concern that in the few clinical data sets that have related dose to stem cell areas to the expression of cognitive deficit, suggest that these effects are sensitive to very low doses [16].

To date, late effects of radiotherapy have been resistant to treatment. Hyperbaric oxygen and steroids have been investigated as a treatment modality for radionecrosis, but evidence in favor of their utility for brain lesions is limited [17]. Interestingly, some recent data suggest that the antiangiogenic agent bevacizumab may ameliorate brain radionecrosis in some patients [18].

Equally relevant are data pointing to interactions between radiotherapy and new agents that are coming in to use as concomitant or adjuvant treatment. Recent clinical studies provide some information suggesting that concomitant vascular endothelial growth factor (VEGF) inhibition may increase risk of optic neuropathy in patients treated for glioblastoma [19]. These data underline the importance of careful translational studies to address potential interactions of these agents in normal tissue as well as tumor models.

Developments in Radiotherapy Technology

The radiotherapy field has developed rapidly in the last few years as the technology available to deliver more accurate, flexible, and conformal treatment has become reality. The overall aim of this technological development is to achieve an improved therapeutic ratio, by limiting dose to non-tumor tissue and/or increasing the dose delivered to the tumor. This relies on optimization of radiotherapy delivery in 3-dimensional space and of very accurate definition of tumor and normal tissue in relation to the treatment beam.

Radiosurgery

Radiosurgery describes the use of highly focused radiotherapy, usually administered in a single dose with the aim of producing local tumor or tissue ablation. As discussed above, the limited potential of normal tissue to recover from large single radiation doses limits the applicability of this approach to small, well-defined targets in situations in which local normal tissue can be effectively excluded from the high-dose region. Historically, radiosurgery has been most commonly delivered using cobalt-based technology (Gamma Knife) in which multiple sources are targeted at a single point, producing a very high dose at the intersection of the radiation beams (isocenter) with a steep dose fall-off beyond. This delivers high, nonhomogenous dose through the target volume, resulting in ablation of irradiated tissue. The biology of these effects is very different to fractionated treatment more commonly used for tumor treatment, and the very high dose probably predominantly affects the vasculature. It has been demonstrated in preclinical studies that a hierarchical dose response may pertain in these circumstances, such that above certain dose levels, most effect is driven through apoptosis in endothelium [20].

In many oncological indications, delivery of ablative doses to small, well-defined lesions is not relevant as, particularly in glioma, the tumor volume is poorly defined due to the infiltrating nature of the disease and the target volumes are usually relatively large. Radiosurgery is therefore not commonly used for this indication although may be relevant in specific situations including re-treatment (discussed below). Radiosurgery is a standard approach to treating brain metastases as these lesions are well-defined small targets. Clinical studies that have addressed radiosurgery as a method to deliver boost doses to high-grade gliomas by targeting subregions to higher total doses have not demonstrated an advantage to this approach [21, 22].

Image-Guided Radiotherapy

Accurate radiotherapy delivery depends critically on imaging information. Magnetic resonance imaging (MRI) remains the standard imaging modality in brain since it provides best anatomic detail and gives most information regarding anatomical disruption due to tumor. For radiotherapy planning, MRI data are overlaid onto computed

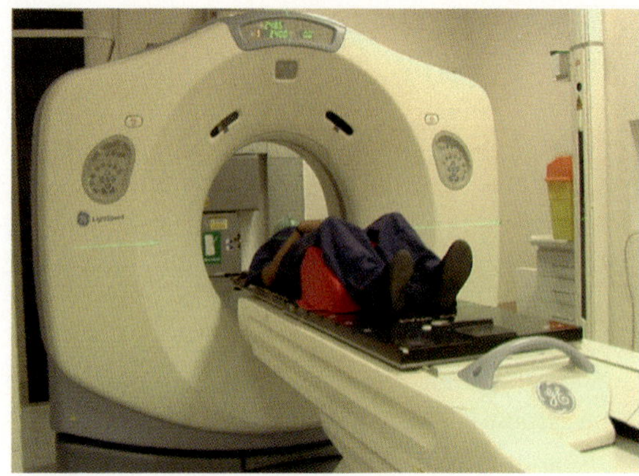

Fig. 10.2 A patient immobilized in a treatment shell undergoing a computed tomography scan for radiotherapy planning

tomography (CT) information, taken with the patient in the treatment position, as shown in Fig. 10.2. Conventionally, the gross tumor volume is defined for high-grade tumors by the gadolinium-enhancing region, and the additional margin for clinical target, which is also included in the high-radiation-dose region, is defined by T2-weighted MRI or fluid attenuated inversion recovery (FLAIR) signal, representing areas of edema and/or infiltration. In non-enhancing low-grade tumors, T2 or FLAIR signal is used to define gross target volume, and a further margin is added to encompass likely infiltration beyond this [23]. In all circumstances, an additional volume has to be treated to take account of variability in patient setup from day-to-day and inherent uncertainties in definitions of structures on imaging. This final margin (PTV) adds a significant volume to the region that is exposed to high doses and can only be reduced by reducing setup variation and/or improving day-to-day visualization of the target. A typical radiotherapy plan is shown in Fig. 10.3.

Image-guided radiotherapy aims to reduce the uncertainties in target definition and patient setup by using integrated imaging platforms that allow repeated imaging data sets to be captured and related to the radiotherapy treatment plan. For brain cancer, this relies on either X-ray imaging, while the patient is on the treatment bed or capturing CT data close to the treatment period. This provides good quality information on skull bony anatomy in three dimensions (3-D), which, in the

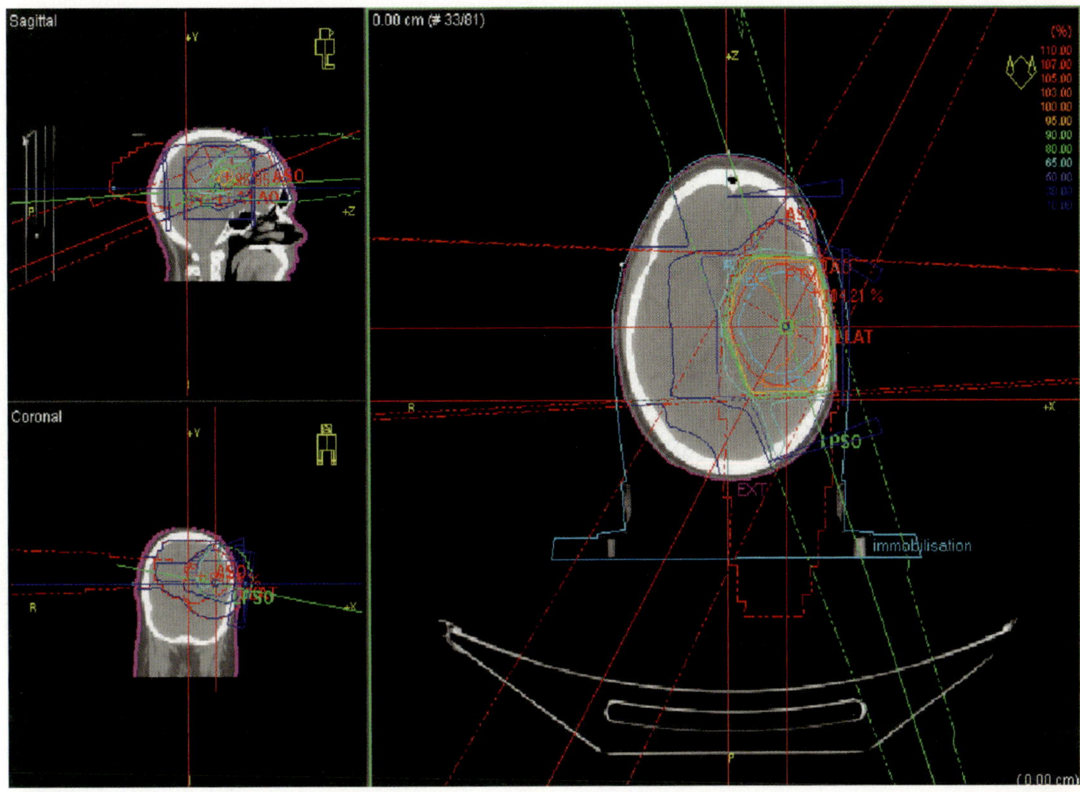

Fig. 10.3 A dose contour map for radiotherapy of a brain tumor target with the target outlined in *red* and dose contours represented by different colors as indicated

absence of significant change in soft tissue geometry during treatment, can be related to previously recorded MRI data. In the most advanced technology solutions, imaging data are captured in real time during treatment delivery, and variations in setup are corrected by coordinating patient positioning, for example, on a robotically controlled couch with radiation beam delivery. A linear accelerator with integrated imaging capability is shown in Fig. 10.4.

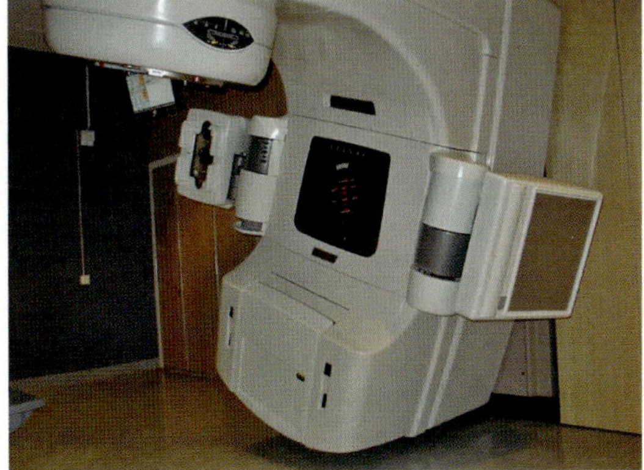

Fig. 10.4 A linear accelerator for delivery of external beam radiotherapy. The patient bed is positioned below the head of the machine, and integrated imaging panels for image-guided treatment are reflected back at either side

Stereotactic Radiotherapy

An alternative and complementary approach utilizes very accurate patient positioning to overcome a major element of uncertainty in setup. Historically, frame-based systems associated with fiducial markers to provide 3-D reference points (originally developed for neurosurgery) have been used in this context. The advantages of this system are that the rigid frame reduces setup variation to a few millimeters; however, invasive systems where the frame is located by titanium screws in the skull are not practical for more than a single treatment, and mouth-bite-based systems are not always well tolerated. Frameless stereotactic approaches are becoming more common, in which real-time imaging replaces the necessity for rigid positioning.

Intensity-Modulated Radiotherapy

Three-dimensional imaging and evaluation of dose distribution overlaid on CT and MRI data has been standard practice in brain cancer radiotherapy for many years and is capable of accurate, conformal treatment for many tumor types. The limitations of this approach are apparent when complex 3-D targets are defined or when radiosensitive organs at risk of radiation damage (OAR) are located in concavities close to tumor. In these circumstances, 3-D conformal treatment cannot deliver high doses to tumor and spare normal tissue. In brain, this is particularly relevant when high doses need to be delivered to regions close to the posterior orbit and skull base. Intensity-modulated radiotherapy (IMRT) delivers treatment using dynamic modulation of the radiation beam in 3-D; as the beam exits the treatment machine, it can be modulated using multileaf collimators that are moved within the beam path. This permits much more complex shapes to be treated in a more highly conformal way and specifically enables dose to be delivered around concave structures.

The applicability of IMRT for treatment of brain cancers has been evaluated in clinical and dosimetric studies, which show that it can be used to target these tumors effectively [24, 25]. Disadvantages include longer treatment time and higher whole-body dose, although these can be reduced by further developments in technology including approaches in which IMRT is delivered more rapidly in an arc mode.

Radiotherapy for High-Grade Glioma

Radiotherapy remains a central treatment modality in high-grade glioma following maximal debulking surgery. The data published some time ago established that the addition of radiotherapy adds significantly to outcome in this disease, and this has been confirmed by more recent meta-analysis; however, the exact contribution in disease subtypes is yet to be completely resolved [26].

In glioblastoma multiforme (GBM), the commonest adult high-grade glioma, the addition of radiotherapy contributes significantly to survival. Data from recent large studies confirm that, in patients under 70 with performance status (PFS) > WHO grade 70, median survival following surgery and radiotherapy is around 1 year. These data do not take account of subgroups of patients that may have different prognoses based either on clinical or molecular pathology classification. The relevance of radiotherapy to older patients with GBM has been addressed in separate studies, and these confirm the advantage of radiotherapy in this age group [27]. Whether radiotherapy or chemotherapy with, for example, temozolomide is the most effective treatment for these patients remains a matter of debate since the trial data are somewhat contradictory and a high incidence of grade 3 and grade 4 toxicities following temozolomide in some studies is a concern [28].

It is interesting that the most widely used molecular diagnostic, O(6)-methylguanine-DNA methyltransferase (MGMT) promoter methylation, initially investigated as a biomarker for sensitivity to temozolomide, also seems to predict for a better outcome in patients treated with radiotherapy alone. This suggests that this molecular characteristic may actually represent a prognostic category in this patient group [29].

Many attempts have been made to improve the effects of radiotherapy in this patient group. These have included dose escalation studies using boost doses with external beam radiotherapy, radiosurgical techniques, or brachytherapy with implantable radiation sources. None of these have shown definitive improvement in survival, but several studies have reported a high incidence of radionecrosis when doses >60 Gy are delivered to significant volumes of normal brain [30–33].

Biological approaches to dose escalation have also been investigated including hypoxic cell sensitization, S-phase sensitizing drugs, and hyperfractionation using more than one dose of radiation per day. Unfortunately, these approaches failed and may have been partly limited by the imaging and radiotherapy technology available at the time.

As discussed above, radiotherapy technology is changing rapidly, and the full impact of new delivery techniques now available have not yet been widely investigated in this patient group. It should be stated that, in view of the rather short prognosis for many patients with this diagnosis, more resource-intensive radiotherapy modalities may be harder to justify; nevertheless, there is evidence that we are beginning to be able to identify better prognostic subgroups in whom best use of technology would be appropriate including, for example, patients with high-grade tumors that harbor the IDH1 mutation [34].

In other high-grade astrocytic tumors, radiotherapy is of equally significant importance in treatment. In anaplastic astrocytoma (WHO Grade III astrocytoma), maximal surgery followed by external beam radiotherapy remains standard treatment. Whether combined modality treatment will improve outcome further as it has in GBM remains to be seen, and relevant studies are ongoing. In these patients, who are often younger than the GBM cohort, with longer median survivals, use of optimized radiotherapy delivery may be justifiable [35].

In oligodendroglial tumors, radiotherapy has been a mainstay of treatment for many years, and it is now clear that these patients, specifically those with the common genetic alteration of LOH 1p/19q, represent a good prognostic subgroup who do well with treatment using either radiotherapy or chemotherapy. The scheduling of different treatments remains open to debate, as there is no clear evidence that which modality used as primary treatment affects the overall outcome and most patients will have both during the course of their disease. One issue to bear in mind is that younger patients may be better advised to delay radiotherapy if there is concern regarding cognitive decline and that patients with very large volume tumors may do better with first-line chemotherapy in order to avoid treating very large target volumes with radiotherapy which will inevitably put more normal tissue at risk of toxicity.

Re-treatment with Radiotherapy at Relapse

There has been a resurgence of interest recently in using new radiotherapy technology to deliver second course of radiotherapy to patients with high-grade gliomas at relapse. This may be particularly appropriate in patients who have small-volume disease, a time interval since initial treatment of more than 1 year, and chemoresistant disease. In these circumstances, re-treatment with focal radiotherapy to doses equivalent to 50 Gy may be tolerated and in many single center, non-randomized studies has been associated with progression-free survival in the region of 6–9 months [36]. The radiobiology predicting risk of radionecrosis in this setting is not fully defined however, so these patients need to be carefully selected. This approach has not been tested in a randomized study and is critically dependent on how the remaining tumor target volume is defined. In this context, the use of amino acid positron emission tomography (PET) is the subject of ongoing investigation [37].

Radiotherapy for Low-Grade Glioma

Management of low-grade glioma remains controversial although it is clear that not all of these patients require immediate treatment. Surgery may be appropriate in a carefully selected proportion, but radiotherapy remains a standard approach in those deemed to need treatment and who are not suitable for radical surgery. A recent European study has addressed whether chemotherapy with temozolomide is as effective as radiotherapy as first-line treatment in these patients, but the results are not yet available. The limitations of radiotherapy in this patient group are often the extensive nature of these tumors, meaning that large volumes of brain need to be treated and the relatively young age at which these patients often need treatment [38]. It is not established that these tumors require such high doses as those used in high-grade tumors, and large studies have not demonstrated a dose response between 45 and 60 Gy

[39]. In principle, the lowest effective dose is recommended, and many centers treat to doses between 50 and 55 Gy.

Radiotherapy for Brain Metastases

Brain metastases are a far commoner malignant diagnosis in adult patients than primary brain tumors. The true incidence can be difficult to ascertain since the primary tumor is more often recorded as cause of death; however, there are recent data suggesting a very significant increase in incidence of this diagnosis. This is likely to be due in part to better diagnostic imaging as well as a higher proportion of patients with common tumors who are living longer with good control of systemic disease, leading to commoner diagnosis of brain relapse [40]. These patients are therefore becoming a commoner indication for radiotherapy and represent a group in which new approaches to radiotherapy may be relevant.

Historically, radiotherapy for brain metastases has been given in a palliative context, usually as a short course of whole brain radiotherapy. Typical doses of 20 Gy in 5 fractions or 30 Gy in 10 fractions have been used with little evidence to suggest that any specific regime is significantly more effective [41]. Partly as a result of the changing epidemiology of brain metastases, as a better performance status cohort presents for treatment, the approach to this disease is changing, with more use of radiosurgery as a management option, as described above. It is also apparent that the prognosis for this patient group as a whole is changing and that many patients may survive a year or more, hence raising concerns about long-term toxicity of radiotherapy.

Few studies have investigated the late side effects of radiotherapy in a prospective manner in these patients, and even fewer have collected these data alongside quality of life. It is a concern that in some recently published data sets there seems to be a correlation between use of whole brain radiotherapy and significant cognitive decline in good prognosis patients [42]. It should be stated though that these data are based on a rather specific element of neuropsychological testing, focusing on verbal memory, tested at limited time points, and there are no associated quality of life data, so it is not possible to evaluate the full impact on these patients. It is also noteworthy that in another prospective study that evaluated neurocognitive end points, although they also documented a decline in verbal memory, this was not associated with a reduction in quality of life [43]. Therefore, the impact of whole brain radiotherapy on important toxicities in this group is still a matter of debate.

The impact of whole brain radiotherapy (WBRT) when given in addition to surgery or radiosurgery has recently been addressed in a large European study. These patients, with 1–3 metastases and controlled systemic disease, were randomized to whole brain radiotherapy or surveillance after initial treatment with either surgery or radiosurgery [44]. The survival data clearly show that WBRT does not lengthen survival in these patients who have had radiosurgery or surgery. WBRT was associated with a reduced incidence of local recurrence in brain, but this did not impact on independent living. These data confirm that whole brain radiotherapy cannot be justified in the patient group in terms of survival improvement, and detailed quality of life data from this study are awaited. It should be noted however that previous data have suggested that surgery in addition to WBRT may improve survival in the cohort of patients with a solitary metastasis [45]. This information has led to an increased use of local treatment with surgery or radiosurgery in good prognosis patients with oligometastases.

As discussed above, the stem cell model is also influencing the approach to radiotherapy in this patient group since techniques such as arc-delivered IMRT can be used to treat whole brain and exclude potential stem cell-rich areas, which may be involved in the pathogenesis of cognitive decline after treatment [46]. Whether this approach is associated with reduced cognitive impairment and equivalent local control is the subject of ongoing investigations.

Future View

It is clear that radiotherapy is likely to remain a central treatment modality for many patients with primary and secondary brain tumors. It is likely that the technology will continue to develop and allow increasingly accurate delivery in 3-D space and for dose modification in real time. In combination with new imaging techniques, it is also foreseeable that more individualized treatment will be possible, taking into account biomarkers of tumor biology to design dose distribution. Optimized combination treatment will need to take into account what can be achieved with radiotherapy to further improve outcomes for these patients.

References

1. Rothkamm K, Lobrich M. Evidence for a lack of DNA double-strand break repair in human cells exposed to very low x-ray doses. Proc Natl Acad Sci U S A. 2003;100(9):5057–62.
2. Olive PL, Banath JP. Phosphorylation of histone H2AX as a measure of radiosensitivity. Int J Radiat Oncol Biol Phys. 2004;58(2):331–5.
3. Short SC, et al. DNA repair after irradiation in glioma cells and normal human astrocytes. Neuro Oncol. 2007;9(4):404–11.
4. Short SC. Rad51 inhibition is an effective means of targeting DNA repair in glioma models and CD133+ tumor-derived cells. Neuro Oncol. 2011;13(5):487–99.
5. Salomoni P, Calabretta B. Targeted therapies and autophagy: new insights from chronic myeloid leukemia. Autophagy. 2009;5(7):1050–1.
6. Nandi S, et al. Low-dose radiation enhances survivin-mediated virotherapy against malignant glioma stem cells. Cancer Res. 2008;68(14):5778–84.
7. Stupp R, et al. Radiotherapy plus concomitant and adjuvant temozolomide for glioblastoma. N Engl J Med. 2005;352(10):987–96.
8. Stupp R, et al. Effects of radiotherapy with concomitant and adjuvant temozolomide versus radiotherapy alone on survival in glioblastoma in a randomised phase III study: 5-year analysis of the EORTC-NCIC trial. Lancet Oncol. 2009;10(5):459–66.
9. Chalmers AJ, et al. Cytotoxic effects of temozolomide and radiation are additive- and schedule-dependent. Int J Radiat Oncol Biol Phys. 2009;75(5):1511–9.
10. Powell C, et al. Somnolence syndrome in patients receiving radical radiotherapy for primary brain tumours: a prospective study. Radiother Oncol. 2011;100(1):131–6.
11. Minniti G, et al. Risk of second brain tumor after conservative surgery and radiotherapy for pituitary adenoma: update after an additional 10 years. J Clin Endocrinol Metab. 2005;90(2):800–4.
12. Brada M, et al. Risk of second brain tumour after conservative surgery and radiotherapy for pituitary adenoma. BMJ. 1992;304(6838):1343–6.
13. Otsuka S, et al. Depletion of neural precursor cells after local brain irradiation is due to radiation dose to the parenchyma, not the vasculature. Radiat Res. 2006;165(5):582–91.
14. Benczik J, et al. Late radiation effects in the dog brain: correlation of MRI and histological changes. Radiother Oncol. 2002;63(1):107–20.
15. Panagiotakos G, et al. Long-term impact of radiation on the stem cell and oligodendrocyte precursors in the brain. PLoS One. 2007;2(7):e588.
16. Marsh JC, et al. Sparing of the neural stem cell compartment during whole-brain radiation therapy: a dosimetric study using helical tomotherapy. Int J Radiat Oncol Biol Phys. 2010;78(3):946–54.
17. Chuba PJ, et al. Hyperbaric oxygen therapy for radiation-induced brain injury in children. Cancer. 1997;80(10):2005–12.
18. Gonzalez J, et al. Effect of bevacizumab on radiation necrosis of the brain. Int J Radiat Oncol Biol Phys. 2007;67(2):323–6.
19. Sherman JH, et al. Optic neuropathy in patients with glioblastoma receiving bevacizumab. Neurology. 2009;73(22):1924–6.
20. Truman JP, et al. Endothelial membrane remodeling is obligate for anti-angiogenic radiosensitization during tumor radiosurgery. PLoS One. 2010;5(8):e12310.
21. Baumert BG, et al. EORTC 22972–26991/MRC BR10 trial: fractionated stereotactic boost following conventional radiotherapy of high grade gliomas. Clinical and quality-assurance results of the stereotactic boost arm. Radiother Oncol. 2008;88(2):163–72.
22. Baumert BG, et al. Fractionated stereotactic radiotherapy boost after post-operative radiotherapy in patients with high-grade gliomas. Radiother Oncol. 2003;67(2):183–90.
23. Stall B, et al. Comparison of T2 and FLAIR imaging for target delineation in high grade gliomas. Radiat Oncol. 2010;5:5.
24. Narayana A, et al. Intensity-modulated radiotherapy in high-grade gliomas: clinical and dosimetric results. Int J Radiat Oncol Biol Phys. 2006;64(3):892–7.
25. Khoo VS, et al. Comparison of intensity-modulated tomotherapy with stereotactically guided conformal radiotherapy for brain tumors. Int J Radiat Oncol Biol Phys. 1999;45(2):415–25.
26. Laperriere N, Zuraw L, Cairncross G. Radiotherapy for newly diagnosed malignant glioma in adults: a systematic review. Radiother Oncol. 2002; 64(3):259–73.
27. Donato V, et al. Elderly and poor prognosis patients with high grade glioma: hypofractionated radiotherapy. Clin Ter. 2007;158(3):227–30.

28. Chinot OL. Should radiotherapy be standard therapy for brain tumors in the elderly? Cons. Semin Oncol. 2003;30(6 Suppl 19):68–71.
29. Rivera AL, et al. MGMT promoter methylation is predictive of response to radiotherapy and prognostic in the absence of adjuvant alkylating chemotherapy for glioblastoma. Neuro Oncol. 2010;12(2):116–21.
30. Chamberlain MC, et al. Concurrent cisplatin therapy and iodine 125 brachytherapy for recurrent malignant brain tumors. Arch Neurol. 1995;52(2):162–7.
31. Prados MD, et al. Interstitial brachytherapy for newly diagnosed patients with malignant gliomas: the UCSF experience. Int J Radiat Oncol Biol Phys. 1992; 24(4):593–7.
32. Fulton DS, et al. Increasing radiation dose intensity using hyperfractionation in patients with malignant glioma. Final report of a prospective phase I-II dose response study. J Neurooncol. 1992;14(1):63–72.
33. Fulton DS, et al. Misonidazole combined with hyperfractionation in the management of malignant glioma. Int J Radiat Oncol Biol Phys. 1984;10(9):1709–12.
34. Combs SE, et al. Progostic significance of IDH-1 and MGMT in patients with glioblastoma:one step forward and one step back? Radiat Oncol. 2011; 13(6):115.
35. van den Bent MJ. Anaplastic oligodendroglioma and oligoastrocytoma. Neurol Clin. 2007;25(4):1089–109. ix-x.
36. Combs SE, et al. Recurrent low-grade gliomas: the role of fractionated stereotactic re-irradiation. J Neurooncol. 2005;71(3):319–23.
37. Grosu AL, et al. Reirradiation of recurrent high-grade gliomas using amino acid PET (SPECT)/CT/MRI image fusion to determine gross tumor volume for stereotactic fractionated radiotherapy. Int J Radiat Oncol Biol Phys. 2005;63(2):511–9.
38. van den Bent MJ. Can chemotherapy replace radiotherapy in low-grade gliomas? Time for randomized studies. Semin Oncol. 2003;30(6 Suppl 19):39–44.
39. Karim AB, et al. A randomized trial on dose response in radiation therapy of low grade glioma: EORTC study 22844. Int J Radiat Oncol Biol Phys. 1996; 36(3):549–56.
40. Smedby KE, et al. Brain metastases admissions in Sweden between 1987 and 2006. Br J Cancer. 2009;101(11):1919–24.
41. Rades D, et al. Comparison of short-course versus long-course whole-brain radiotherapy in the treatment of brain metastases. Strahlenther Onkol. 2008; 184(1):30–5.
42. Chang EL, et al. Neurocognition in patients with brain metastases treated with radiosurgery or radiosurgery plus whole brain radiotherapy: a randomised controlled trial. Lancet Oncol. 2009;10(11):1037–44.
43. Mehta MP, et al. Motexafin gadolinium combined with prompt whole brain radiotherapy prolongs time to neurologic progression in non-small-cell lung cancer patients with brain metastases: results of a phase III trial. Int J Radiat Oncol Biol Phys. 2009;73(4):1069–76.
44. Kocher M. Adjuvant whole-brain radiotherapy versus observation after radiosurgery or surgical resection of one to three cerebral metastases: results of the EORTC 22952–26001 study. J Clin Oncol. 2011;29(2): 134–41.
45. Patchell RA, et al. A randomized trial of surgery in the treatment of single metastases to the brain. N Engl J Med. 1990;322(8):494–500.
46. Kirby N, et al. Physics strategies for sparing neural stem cells during whole brain radiation treatments. Med Phys. 2011;38(10):5338–44.

Managing the Elderly Patient

11

Kathryn Graham and Anthony J. Chalmers

Abstract

Glioblastoma multiforme (*GBM*) is a devastating disease at any age. However, GBM has traditionally been associated with particularly poor outcomes in *elderly* patients, with reported median survival time of just a few months typical in many case series. The reasons for this are not entirely clear, but it has been suggested that the elderly are generally frailer and less able to cope with the *toxicity* associated with standard treatment approaches to GBM, notably *surgery* and *radiotherapy*. This may explain why single-modality therapy or best supportive care only becomes increasingly common with advancing age. More recently, however, there has been speculation that GBM may be a biologically more aggressive disease in the elderly and perhaps inherently more resistant to radiation. This has yet to be confirmed but has provoked interest in understanding precisely why age has such a negative effect on survival at a time when a significant increase in numbers of elderly GBM patients is predicted due to an aging population. Interestingly, since the introduction of the Stupp protocol, which promotes the use of *chemoradiotherapy* post surgical resection, several reports have emerged indicating that elderly patients can tolerate aggressive multimodality therapy with impressive median survival times of over a year in some cases. However, it is important to point out that *patient selection* is likely to be critical and the results in these series cannot be extrapolated to the general elderly population. Unsurprisingly, the gold standard treatment of GBM in the elderly has yet to be determined.

K. Graham, B.Sc. (Hons), MB ChB, MRCP, Ph.D., FRCR
Department of Clinical Oncology,
Beatson West of Scotland Cancer Centre,
Glasgow 9120 YN, United Kingdom

A.J. Chalmers, B.A. (Hons), BM BCh, MRCP, FRCR, Ph.D. (✉)
Institute of Cancer Sciences, University of Glasgow,
University Square, Glasgow G12 8QQ, UK
e-mail: anthony.chalmers@glasgow.ac.uk

C. Watts (ed.), *Emerging Concepts in Neuro-Oncology*,
DOI 10.1007/978-0-85729-458-6_11, © Springer-Verlag London 2013

As such, a number of *clinical trials* have been developed to specifically answer this question in patients over the age of 70, a group that has previously been excluded from pivotal trials in GBM. It is hoped that these studies will pinpoint clinical and/or molecular *prognostic factors* that will guide treatment of the individual elderly patient with the optimal combination of therapy.

Keywords
Glioblastoma multiforme (GBM) • Elderly • Toxicity • Surgery • Radiotherapy • Chemoradiotherapy • Patient selection • Clinical trials • Prognostic factors

Introduction

Glioblastoma (GBM) is an incurable and devastating malignancy, the incidence of which increases with age. Despite a modest improvement in overall survival from this disease in the past 20 years, stemming from advances in diagnostic and therapeutic techniques, the median survival of patients with GBM remains poor and is at best 14–18 months [1, 2]. Elderly patients, however, fare even worse and it has long been recognized that age is the single most important prognostic factor in GBM. Why this should be the case has been the subject of much debate for many years and, indeed, remains a controversial issue today. It has been suggested that the inferior outcomes seen in older patients are simply a result of less aggressive treatment or withholding treatment, most likely due to concerns over ability to cope with therapy and/or fear of inducing significant toxicity. A number of large population-based studies certainly concur with this viewpoint and demonstrate different patterns of care in elderly GBM patients compared with younger adults. As researchers achieve a greater understanding of the molecular basis of primary brain tumors, however, evidence is emerging that GBM may be a biologically different disease in the elderly. This may explain, or at least partly explain, why the prognosis in elderly patients is poor even in the context of multimodality therapy. Unsurprisingly, the gold standard treatment of GBM in the elderly has yet to be determined. Important factors have been the tendency for key studies in GBM to be limited to patients below a certain age and the culture of nihilism that has surrounded the management of these patients in many centers. There is now a call for a consensus approach to the management of GBM in the elderly, especially as the population is aging and clinicians are facing the prospect of treating a progressively older cohort of patients.

Definition of Elderly

The treatment of cancer in the elderly is of increasing importance in oncology. However, the definition of "elderly" can vary from study to study and from clinician to clinician. Historically, the term "elderly" was linked to the age of eligibility for retirement benefits, typically 65 years. Accordingly, a cutoff of 65 years has traditionally been the norm for geriatric medicine. As the life expectancy of the population continues to increase, however, a new definition of elderly may need to be sought. Some authors have advocated a distinction between the "young old" (65–74 years), the "older old" (75–84 years), and the "oldest old" (>85 years) [3]. In keeping with the changes in population dynamics, some recent GBM studies have included patients up to the age of 70 years rather than 65 years, most notably the pivotal Stupp trial [1, 2]. Until a revised definition has been agreed, there will continue to be discrepancies in establishing the upper age limits for clinical trials in GBM. For the purposes of this review, the term "elderly" refers to patients aged 65 years and above.

Epidemiology of GBM in the Elderly

Approximately 50 % of cases of GBM occur in patients aged >65 years [4]. While this proportion is likely to rise because of the aging population, there has been speculation that the actual incidence of GBM in the elderly is also increasing. In 1990 the National Cancer Institute published a report detailing a marked increase in the incidence of primary brain tumors in the elderly, including GBM [5]. This apparent increase was later attributed to an improvement in cancer detection, owing to more widespread use of imaging in older patients [6]. This explanation has not been universally accepted, especially as another series has also illustrated an increase in the age-adjusted incidence of GBM [7]. It is important to remember that discrepancies in local practice may mean that a rise in the number of cases may not necessarily be reflected in referral patterns to all tertiary treatment centers. Conversely, as public expectations of healthcare provision continue to rise, more patients may be referred who would previously have been managed conservatively.

Pathology of GBM in the Elderly

The last 10–15 years has seen many exciting molecular developments in brain tumor pathology, most notably the discovery of the prognostic and predictive power of loss of heterozygosity of 1p19q in oligodendroglioma [8–13] and the prognostic value of methylation of the O6-methylguanine-DNA methyltransferase (MGMT) promoter in GBM [14]. It has been suggested that these and other markers of glioma biology may be affected by age. Indeed, GBMs arising in the context of a previously diagnosed lower-grade glioma (secondary GBM) have a better prognosis than primary GBM, and the incidence of secondary GBM decreases with age. This correlates with the observation that the IDH-1 mutation that is commonly found in low-grade gliomas is not detected in the elderly [15] and has been proposed as a possible explanation for the poorer outcomes seen in older patients. However, secondary GBMs account for no more than 10 % of all cases [16], so this hypothesis is unlikely to account for the overall disparity in survival between younger adults and the elderly. Furthermore, a relatively recent analysis showed no difference in the prognosis of primary versus secondary GBMs once age-adjusted analysis was performed [16].

Following on from these findings, a potential correlation between age and biological aggressiveness in primary GBMs has been investigated in a number of clinicopathological series. Initial studies focused primarily on markers of proliferation and/or histological features and generated either negative or conflicting results [17–20]. Recently there has been interest in more sophisticated chromosomal/molecular analysis, particularly the significance of epidermal growth factor receptor (EGFR) amplification and MGMT methylation. The data regarding patterns of EGFR expression and outcome in GBM according to age is confusing (reviewed in [21]), but there is some evidence to suggest that MGMT may influence prognosis in older as well as younger adults. This will be discussed in more detail in the next section.

The Effect of Age on Prognosis in GBM

The single most important prognostic factor in GBM is age. Survival in GBM begins to decline at the age of 45 and decreases dramatically thereafter [16, 22]. Patients over the age of 65 have a 2-year survival of less than 5 % in historical series compared to over 20 % in patients below 50 years. Data from numerous retrospective, prospective, and epidemiological studies corroborate these findings [16, 23, 24]. In both the original and the recently updated Radiation Therapy Oncology Group (RTOG) recursive partitioning analyses (RPA) for patients with high-grade glioma, age over 50 years was the clinical factor with the greatest predictive significance for survival [25, 26]. However, determining an exact cutoff age above which prognosis is so poor as to justify the withholding of treatment is difficult. Recommendations vary according to the statistical model applied and do not take into account variability between patients. Indeed, a recent case series reported good outcomes in a small number of GBM patients treated aggressively who were all over the age of 80 years [27].

If selected elderly patients can respond favorably to treatment for GBM, why is the general prognosis reported for larger elderly cohorts so poor? It is unlikely that there is a single answer to this question and the evidence to date suggests that the reasons for adverse outcomes in this group are multifactorial. One of the major discriminators of prognosis in the RPA analyses was performance status [25, 26], and it is not entirely surprising to find that elderly patients with GBM tend to have a poorer level of functioning, both physically and cognitively. This can be at least partly explained by medical comorbidities, the incidence and severity of which increase with advancing age. However, it has also been suggested that older patients present with larger tumors [28], possibly as a result of age-related cerebral atrophy providing increased scope for tumor growth prior to the development of raised intracranial pressure. Tumor size has previously been correlated with reduced survival in both low-grade and high-grade gliomas [29, 30]. An example of GBM arising in the brain of an elderly patient is shown in Fig. 11.1. Regardless of whether the patient's poor performance status is attributable to tumor burden or comorbid medical conditions, frail patients have limited physiological reserve and are less likely to tolerate surgical and oncological interventions. Yet, not all elderly patients are frail and infirm. It is unclear why fit older patients still fare worse than their younger counterparts. Three main reasons have been put forward: reluctance to treat the elderly patient, increased resistance to chemotherapy and/or radiotherapy, and heightened treatment-related toxicity.

Age and Patterns of Care in GBM

It is widely recognized that older age may be an obstacle to receiving optimal medical care, particularly in oncology. Studies in women with breast cancer, for example, have demonstrated that elderly patients have reduced access to informational support at first diagnosis [31] and that this discrepancy follows through to lower referral rates to hospice/palliative medicine services at the end of life [32]. It is therefore important to question whether patterns of care differ between younger and older patients with GBM and, if so, whether the disparities are large enough to influence survival outcome. To this end, a number of large epidemiological studies comprising several 1,000 patients with GBM have been published. To date there is no evidence to support the existence of either a delay in diagnosis in the elderly [33–35] or a prolongation in the time between diagnosis and treatment [34]. This indicates that the elderly are not disadvantaged at the points of diagnosis or initiation of treatment.

The current standard of care for GBM, as established in the National Cancer Institute of Canada (NCIC) and the European Organization for the Research and Treatment of Cancer (EORTC) collaborative trial, is maximal surgical debulking followed by concurrent chemoradiotherapy and maintenance of temozolomide for 6 months [1, 2]. However, several large population-based analyses indicate that the probability of receiving multimodality therapy is reduced with increasing age and, in fact, patients over the age of 65 are significantly more likely to receive no treatment at all [16, 33, 34, 36–44]. For example, in a review of 715 adult GBM cases in Zurich, Switzerland, Kita et al. noted that best supportive care was often the only treatment offered to older patients and this increased with advancing age. Here, 27 % of patients aged 55–64 received supportive care only, compared with 44 % of those aged 65–74 and 75 % of those aged over 75 years [40]. In the case of elderly patients who do receive treatment, surgery rates are generally much lower, and they are more likely to have a biopsy as opposed to a definitive surgical procedure [36, 38]. Radiotherapy rates are also notably lower in elderly cohorts: approximately 65 % in patients over 70 [39] compared with over 90 % in younger adults [44]. A population-based study of over 3,000 GBM patients in Ontario, Canada, demonstrated that increasing age was also associated with lower mean radiation dose [38]. The most recently published United States-based

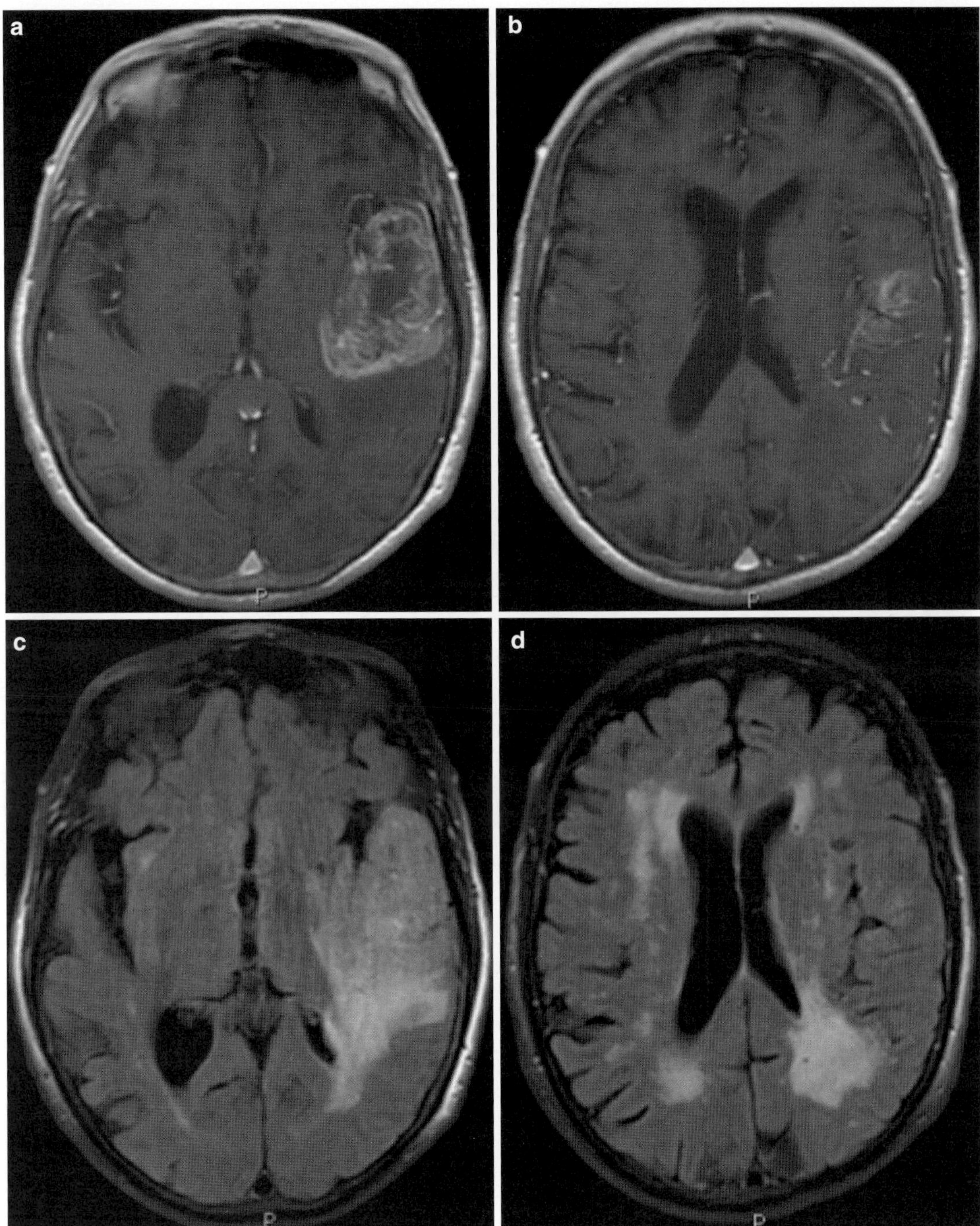

Fig. 11.1 Appearance of glioblastoma in an elderly brain. Contrast-enhanced T1-weighted (**a** and **b**) and FLAIR (**c** and **d**) MR images of the brain of an 81-year-old lady with presumed glioblastoma of the left parietal lobe. Note the presence of cerebral atrophy and abnormalities on FLAIR sequence suggestive of ischemic changes in the normal brain

Surveillance, Epidemiology and End Results (SEER) Program analysis, which reported on almost 3,000 patients over the age of 70 years treated between 1993 and 2005, demonstrates that this pattern of less aggressive treatment is not changing over time. While a higher proportion of patients in this study received some type of treatment, this was mainly single modality; less than half were treated with both surgery and radiotherapy [39]. Given that the addition of temozolomide chemotherapy to radiation has only become standard practice within the last 5 years, accurate data on patterns of care with respect to the use of chemotherapy are not yet available.

As these large studies were predominantly conducted in North America, it is important to consider the possibility that a financial barrier to medical treatment may exist for some patients. However, the SEER database is linked to Medicare, the health insurance provider for well over 90 % of elderly patients in the United States [45], so it is unlikely that discrepancies in treatment according to age are due to disparities in access to healthcare. This is reinforced by the findings of the Swiss cohort where 82 % of patients below the age of 65 years received active treatment (surgery followed by radiotherapy, surgery alone, or radiotherapy alone) as opposed to 47 % of patients above the age of 65 years despite the fact that Switzerland has a sophisticated healthcare system with unrestricted access [40]. Similarly, the German study by Lutterbach et al. remarked that access to healthcare was not determined by age [34].

These studies undoubtedly provide a valuable insight into the lower uptake (or offering) of treatment with advancing age and emphasize that this is a worldwide phenomenon. However, it is difficult to conclude that "inadequate treatment" is entirely responsible for poorer survival. This is especially relevant as not all groups collected data on survival. Interpretation of the data is also limited by lack of information on performance status and/or medical comorbidity. Another major drawback of these studies is that variations in referral patterns to tertiary treatment centers mean that they might not have included all patients with GBM. Thus, the proportion of patients receiving no treatment may actually be underestimated. In summary, there are clear age-related differences in the management of GBM patients, and this is probably reflected in the poorer outcome seen in elderly GBM patients, but it is highly likely that additional factors are also involved.

Age and Treatment Resistance in GBM

A factor that has been mooted as a potential reason for lower rates of radiotherapy uptake in the elderly is the apparent shortened survival advantage when compared with adult GBM patients [46–48]. While this difference could be partly explained by death due to other causes, it has also been suggested that age may influence the radiosensitivity of primary brain tumors. Some groups have attempted to address this by quantifying the radiological response of GBM to radiotherapy in younger versus older adults [49–51]. Using a simple assessment scale in patients who had measurable disease, age was found to be a predictor of poorer radiological response to radiation, although most of the imaging techniques would now be considered outdated. In addition, both performance status and extent of surgical resection were independent prognostic factors, which suggests inherent selection bias. It seems unlikely that age itself is a pivotal factor in determining responsiveness to ionizing radiation, but an association between intrinsic biological factors and age is plausible. Tumor radiosensitivity is complex and depends on myriad molecular characteristics and DNA repair mechanisms, many of which are altered in GBM. Until there is proof that GBM in the elderly represents a different biological spectrum of disease from that of younger patients, however, the view of age as a surrogate for radioresistance must remain speculative.

Since the introduction of the Stupp protocol, there has been further debate about treatment resistance in the elderly. This stems from historical reports that glioma cell lines from older patients are less chemosensitive [52, 53], although the agents tested were nitrosoureas rather than temozolomide. Following the emergence of MGMT methylation status as a predictor of response to temozolomide, a number of groups have tried to establish whether epigenetic silencing of this gene varies with age, since lower levels of MGMT methylation could perhaps explain the poorer outcome in elderly patients treated with this regimen. Intriguingly, recent data suggests that there are no significant differences in the proportion of MGMT-methylated tumors in older versus younger patients [27, 54–58]. In addition, a recent case series of 83 patients over the age of 70, all of whom received treatment with concurrent chemoradiotherapy for GBM, supports the importance of MGMT as a clinical marker in elderly patients as well as younger adults [55]. Here, MGMT-methylated patients had a median survival of 15 months and a 2-year survival of 28 %. Unmethylated patients had a much poorer outcome with a median survival of 10 months and 2-year survival of only 10 %. If these findings are confirmed in a larger series, it seems less likely that GBM in the elderly population is a biologically different disease from that seen in younger patients. It may be the case, however, that additional cytogenetic or molecular aberrations have yet to be identified, especially in unmethylated GBMs, which probably represent a heterogeneous group of tumors.

Age and Treatment Toxicity in GBM

Cancer treatment in the elderly is fraught with risks. Patients are typically frailer and less capable of tolerating radical procedures such as surgery. For instance, there is a higher risk of surgical complications and a tendency to require a longer hospital stay following surgery, which increases the risk of hospital-acquired infections [59]. There may be alterations in drug metabolism due to changes in body weight, liver mass, and the oxidative system. This in turn can affect the distribution and absorption of anesthetic agents, antibiotics, and anticonvulsants, not to mention chemotherapy. In addition, elderly patients are more likely to be subject to polypharmacy, which increases the risk of drug-drug interactions. Chemotherapy-related toxicity is often more pronounced. Hematological toxicity in particular is increased, possibly due to compromised stem cell reserve [60], and there is an elevated risk of neutropenia along with the associated infectious complications, hospitalizations, and mortality rates [61].

It is also well recognized that elderly patients generally cope less well with radiotherapy. Many anecdotal reports indicate that the elderly are more likely to suffer from radiation-related fatigue and somnolence in the short term. Unfortunately, most studies performed in this age group do not include late toxicity as an endpoint, mainly because most of these patients do not live long enough to develop neurological sequelae. It is therefore difficult to gauge the precise effects of radiotherapy on the elderly brain. There are certainly biological reasons why radiation might be more toxic in this population, most notably higher rates of cerebrovascular disease and diabetes. Small vessel damage is thought to be an important contributor to late radiation toxicity, particularly the most critical sequelae of brain irradiation: cerebral necrosis. While it is reasonable to predict that preexisting vasculopathy and/or hypertension would exacerbate and/or accelerate this process, there is little concrete scientific data to support this. Certainly, the lack of relevant animal models has hindered efforts to elucidate the mechanisms and risk factors that combine to produce late radiation toxicity in the brain. Although age and radiation necrosis cannot be definitively linked, it has been documented that age is a significant risk factor for the development of both cerebral atrophy and encephalopathy [62–64]. This probably explains the global neurocognitive decline that can follow brain irradiation in the elderly. Again, it is unclear whether vascular risk factors predispose for this phenomenon. Of note, it typically takes at least 6–9 months and often up to 1–2 years for these clinical effects to become apparent [62, 65, 66] so it could be argued that trying to gain a greater understanding of late radiation effects in the elderly is not necessary given the predicted short survival. Clearly, the risks of acute and subacute side effects still remain important issues when considering management of the elderly patient. However, if the

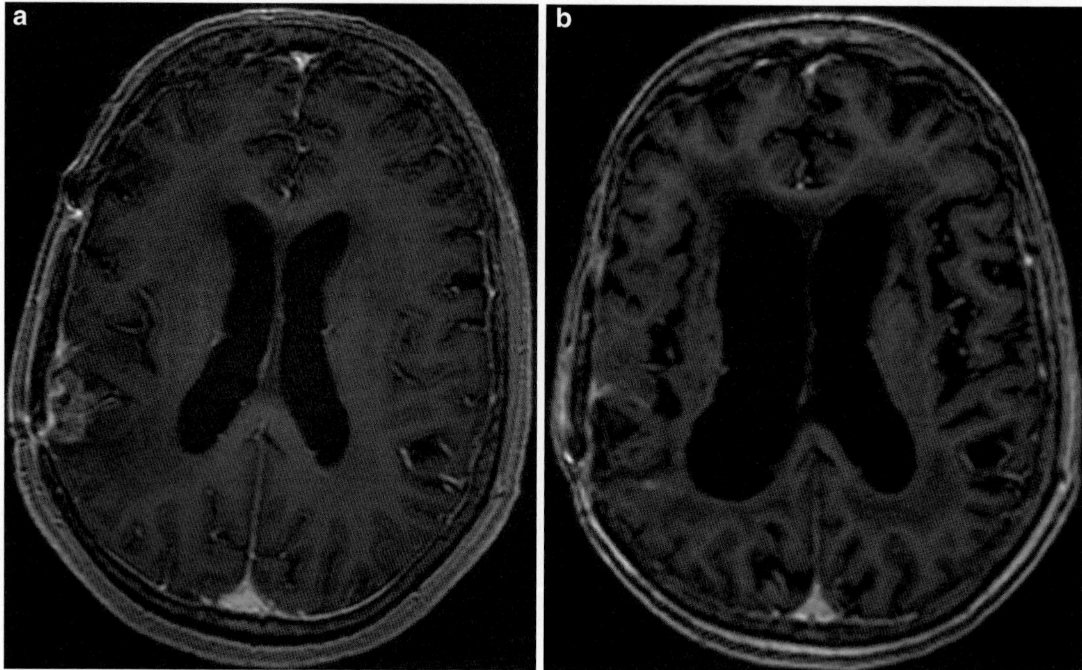

Fig. 11.2 Treatment-related effects in an elderly brain. Contrast-enhanced T1-weighted MR images of a 64-year-old lady with histologically confirmed glioblastoma of the right parietal lobe, prior to radical chemoradiotherapy (**a**) and 5 years post treatment (**b**). Of note, there is no evidence of tumor recurrence, but gross cerebral atrophy is present corresponding to a clinical picture of dementia and incontinence, presumed secondary to treatment

median overall survival for certain subgroups of elderly patients is pushed out beyond 6–9 months, minimizing late radiation effects will become a more pressing issue. An example of significant late radiation toxicity in a long-term survivor of GBM is demonstrated in Fig. 11.2.

Management of GBM in the Elderly Patient

GBM is a symptomatic disease associated with headaches, progressive loss of neurological function, and deterioration in cognitive abilities. Multimodality treatment of GBM is lengthy and potentially toxic, but equally treatment can improve survival and relieve some of the aforementioned symptoms of the disease. The key to successfully managing the elderly patient with GBM is to balance tumor-related symptomatology with the risks of treatment-related toxicity. This is likely to vary from patient to patient depending on their performance status, comorbid medical conditions, and expressed wishes. Accordingly, treatment of the elderly should encompass a broad spectrum, from best supportive care to maximal surgical debulking plus chemoradiotherapy. It must always be remembered, however, that quality of life is of paramount importance especially as the anticipated survival for most of these patients is likely to be a matter of months.

Surgery

Surgery is a critical aspect of the management of patients with GBM, as it delivers diagnostic information while simultaneously providing rapid relief of mass effect. In addition, the act of cytoreduction is thought to improve tolerance to adjuvant therapy. Firm evidence in support of this statement may be lacking, with the exception of a study that demonstrated a higher response rate to chemotherapy (and improved survival) following surgical debulking in the recurrent disease setting [67], but it is generally accepted that this is the case. It has already been outlined that surgical resection rates are generally lower in the elderly, and it has been suggested that less aggressive treatment may contribute to the poorer outcome seen in this patient population. The two are not necessarily linked, not least because the precise role of surgery in terms of survival has been a contentious issue for many years. Systematic reviews have repeatedly found no convincing evidence of a survival advantage of surgical resection over a biopsy (reviewed in [68]). However, a significant number of prospective and retrospective studies have indicated that maximal resection is associated with a longer survival (reviewed in [68]), and it has been argued that earlier studies used less effective surgical techniques. There is increasing consensus that more extensive surgery, in combination with increasingly sophisticated imaging techniques, can offer a survival advantage. What is more, it has very recently been suggested that maximal debulking can also increase the efficacy of adjuvant therapies [69]. So, if aggressive surgical resection can alleviate disease-related symptoms, increase tolerance to radiotherapy, and potentially improve survival and/or effectiveness of adjunctive therapies, why is this not offered routinely to elderly patients?

The most plausible explanation is the generally frailer and comorbid condition of older patients with newly diagnosed GBM [36, 38, 40]. The elderly are more likely to have concomitant cerebrovascular and systemic disease and poor physiological reserves. These factors can have a marked impact on surgical morbidity and mortality. Interestingly, elderly GBM patients are more likely to present with symptoms of cognitive dysfunction than their younger counterparts [33], which in itself is associated with a higher rate of perioperative complications [70]. In a German series of 44 patients with a primary brain tumor (all aged >80 years), 43 % of patients improved after surgery and 34 % remained stable. However, over 20 % of patients deteriorated and the overall perioperative mortality was 11 % [71]. By contrast, Kelly et al. found that postoperative mortality was only slightly higher than biopsy-related mortality, at 2.5 % vs. 2.2 % [72], although the latter study analyzed a younger age group (>65 years as opposed to >80 years). There is no denying that surgery can have profound negative effects in the elderly. Equally, significant improvements in functional status and quality of life following surgery have been documented by a number of sources indicating that its use can be justified in the elderly [72–74]. The same cannot be said for radiotherapy: poor performance status pre-irradiation predicts for poor performance status postirradiation [75].

Another possible explanation for less aggressive surgery (or no surgery) in older patients may be the dearth of definitive randomized phase III evidence of a survival benefit, particularly as the accumulating data in this field is largely based on studies of younger adults. To this end, researchers have attempted to address the question of whether surgical resection improves survival specifically in the elderly. This has been performed mainly through case series and by subgroup analysis on the data from the large population-based cohorts, although a small single-institution randomized trial was undertaken in Finland in the 1990s [76]. However, performance status is an important potential confounding factor in studies of therapeutic management of elderly GBM patients, and the results must be interpreted with caution. Patients undergoing surgery are likely to be healthier, and they may have more localized and/or superficial lesions that may be biologically more favorable. Moreover, the rate of adjuvant therapy (and in particular data regarding completion of adjuvant therapy) is not always clear. The fact that virtually all of these studies, with the exception of the Finnish trial, are retrospective adds to the complexity of the available evidence and hints at inherent selection bias.

The Finnish group examined a total of 23 patients with malignant glioma aged over 65 years and randomly assigned patients to biopsy only or surgical resection followed by radiotherapy +/− chemotherapy (a further seven were excluded due to low-grade malignancy or benign pathology). The median survival time was significantly longer in patients who underwent surgical resection

Table 11.1 Surgical resection versus biopsy in elderly GBM patients

Group	Age	*N*	Median OS (months)	OS benefit?
Randomized data				
Vuorinen 2003 [76]	≥65	23	Resection 5.6 Biopsy 2.8	✓
Retrospective series				
Kelly 1994 [72]	≥65	128	Resection 6.8 Biopsy 3.8	✓
Mohan 1998 [78]	≥70	102	Maximal resection 17.3 Subtotal resection 7.2 Biopsy 3.4	✓
Chaichana 2011 [79]	≥65	80	Maximal resection 4.9 Subtotal resection 5.7 Biopsy 4.0	✓
Ewelt 2011 [80]	≥65	103	Maximal resection 13.9 Subtotal resection 7.0 Biopsy 2.2	✓
Zachenhofer 2011 [77]	≥65	20	Maximal resection 8.2 Subtotal resection 7.8 Biopsy 7.8	✗

N number of patients, *OS* overall survival

compared with patients who underwent biopsy alone (5.6 months vs. 2.8 months, respectively) [76]. Of note, the median age was similar in both groups, but preoperative Karnofsky performance status (KPS) was higher in the craniotomy group [77] compared with the biopsy-only group [70]. While this unique study is a valuable contribution to the literature, it is too small to provide any definitive conclusions.

The small number of retrospective case series published to date has yielded inconsistent results, as shown in Table 11.1 [72, 76–80]. The first report, by Kelly et al., compared outcomes of surgical resection in 40 patients aged over 65 years with outcomes of biopsy only in a further 88 patients, a proportion of whom went on to have adjuvant treatment. Both groups had comparable median age of approximately 70 years and KPS approaching 85 %. Intriguingly, while the authors reported their findings as only a "modest improvement," survival was almost doubled in the group who had undergone surgical resection (6.3 months vs. 3.6 months) [72]. Mohan et al. also reported a significant impact on survival of complete versus partial resection versus biopsy in a study of GBM patients over the age of 65 (17.2 months vs. 7.2 months vs. 3.4 months, respectively) [78]. Interestingly, in 2011, there were three separate reports on the effect of extent of surgery in elderly GBM patients aged over 65 years. All three were retrospective, single-institution studies featuring between 20 and 103 patients. Both Chaichana et al. and Ewelt et al. reported a positive effect of surgical resection on overall survival as opposed to biopsy only, although the benefit was only several weeks in the former [79, 80]. The resected patient group in the Chaichana study was compared with a historical series of patients who had undergone biopsy only so matching for KPS index was permitted [79]. Conversely, the decision for resection was strongly based on KPS in the Ewelt cohort [80]. The third study did not show any advantage of surgery (either maximal resection or subtotal resection) over biopsy [77].

It is important to note that there were only 20 elderly patients in the Zachenhofer series, so it is perhaps not entirely surprising that the findings were negative [77]. Interpreting the results of such small studies can prove troublesome. However, it can also be difficult to dissect out meaningful results from the data produced by the larger population-based cohorts due to previously mentioned discrepancies in KPS level and surgical bias. For instance, Scott et al. remarked that surgery was associated with increased cancer-specific survival compared with no treatment in their review of almost 3,000 elderly GBM patients, but the authors acknowledged that information on performance status was lacking [39]. Analysis of some smaller cohorts containing up to several 100 patients has demonstrated conflicting findings. For example, both Pierga et al. and Chang et al. reported a 5–6 month survival advantage for tumor resection versus no resection [81, 82]. Of note, both of these groups suspected that their findings were influenced by a strong selection bias [81, 82]. Conversely, surgery was not found to have any bearing on survival in other published series [75, 83].

Taken together, the limited data that is available suggests that surgical resection as opposed to biopsy alone is tolerated, at least in fit elderly patients with a KPS of ≥70 and an accessible lesion, and may be associated with a small survival advantage. Less-fit patients may also benefit symptomatically and functionally, and surgery may render a proportion of these patients suitable for adjuvant treatment, but the risks of surgery must always be carefully considered in the context of poor physiological condition and medical comorbidities. Technological developments such as functional magnetic resonance imaging (fMRI), intraoperative neurofunctional monitoring, and neuronavigation have rendered neurosurgical procedures safer and more effective. To what extent this will influence surgical management of the elderly patient, fit or unfit, has yet to be determined. It is possible that some centers that are currently reluctant to operate on the elderly may continue to refrain from radical procedures. In order to promote a more standardized approach to the elderly population, it will be necessary to pinpoint preoperative factors that influence survival. Chaichana et al. have provided some insight into identifying which patients are more likely to benefit from aggressive surgery [29]. Their retrospective review of over 100 patients with an average age of 73 years indicated that the presence of more than one risk factor had a significantly negative impact on survival. The risk factors comprised KPS <80, chronic obstructive pulmonary disease, motor deficit, language deficit, cognitive deficit, and tumor size larger than 4 cm. While the authors accept that this study did not allow for the effect of adjuvant therapy and requires prospective evaluation, it is an interesting exploration of the potential value of prognostic factors and may be the first step in developing a surgical algorithm for the management of GBM in the elderly.

Radiotherapy

Robust evidence of a survival benefit following aggressive surgery in the elderly GBM patient has yet to be shown, but the survival advantage of postoperative treatment, in the form of radiotherapy, has been demonstrated. While a number of retrospective case series hinted at the value of radiotherapy in this context, there is now randomized phase III data available to substantiate this [48]. Historical series using a variety of dose/fractionation schedules illustrated a median survival of 4–12 months, as shown in Table 11.2 [72, 78, 83–90], although it should be noted that survival of over 9 months was only elicited in studies with fewer than 30 patients [87, 90]. In addition, several of these case series included anaplastic astrocytoma as well as GBM [83–85, 88]. This may have resulted in an overestimation of the actual survival time. In order to accurately assess the effect of radiation on survival, Keime-Guibert et al. randomized 81 elderly patients over the age of 70 with newly diagnosed anaplastic astrocytoma or GBM to radiotherapy plus best supportive care or best supportive care alone in the postoperative setting; surgical resection and biopsy were both permitted. The trial was discontinued at the first interim analysis because the radiotherapy arm was found to be significantly more effective. The median survival was 29.1 weeks for patients undergoing radiotherapy as opposed to 16.9 weeks for those patients in receipt of best supportive care only. This was in spite of the lower radiation dose applied in this study (50.4 Gy in 28 fractions) in contrast to the standard dose/fractionation regimen for GBM (60 Gy in 30 fractions). The authors did not report any difference in health-related quality of life or

Table 11.2 Summary of radiotherapy trials in elderly patients with GBM

Group	Age	*N*	Fractionation	Median OS (months)
Randomized data comparing surgery and radiotherapy with surgery alone				
Keime-Guibert 2007 [48]	≥70	81	50.4/28	Surgery + RT 7.3 Surgery 4.3
Historical radiotherapy case series in the elderly				
Ampil 1992 [84]	≥65	21	60/33	4.0
Kelly 1994 [72]	≥65	96	NR	4.2
Hoegler 1997 [83]	≥70	23	37.5/15	8.0
Mohan 1998 [78]	≥70	58	Various	7.3
Villa 1998 [85]	≥70	85	60/30	4.2
Jeremic 1999 [86]	≥60	44	45/15	9.0
Brandes 2003 [87]	≥65	24	59.4/33	11.2
Glantz 2003 [88]	≥70	54	60/33	4.1
Muacevic 2003 [89]	≥65	123	60/30	5.6
Idbaih 2008 [90]	≥70	28	40/15	11.7
Scott 2011 [99]	≥70	206	Various	4.5

N number of patients, *OS* overall survival, *NR* not reported, *RT* radiotherapy

cognitive status between the treatment groups, indicating that radiotherapy was well tolerated in this patient population. Indeed, no severe adverse events were recorded, and only 6 patients (15 %) did not complete the course of radiotherapy.

Even in the face of a 3-month survival advantage from radiation, it is clear from the various SEER analyses and other population-based studies that radiotherapy treatment is not always delivered to elderly patients. This is most likely due to concerns over patient frailty and ability to cope with a protracted course of treatment. However, it is important to consider the possibility that clinicians may have an age cutoff above which they feel radiation is not applicable due to poor tolerability and/or minimal perceived benefit. The patients themselves may decline treatment over fears of excessive toxicity and negative impact on their quality of life. For many elderly patients, the prospect of 2–3 weeks of radiotherapy planning followed by 6 weeks of radiotherapy treatment with daily hospital visits is daunting. In reality the length of treatment-free survival may amount to no more than a number of weeks, and this must be taken into account when selecting and counseling potential treatment candidates.

It is for these reasons that hypofractionated radiotherapy has been advocated in the elderly and/or frail patient with GBM. Hypofractionation has the advantage of reducing the time frame (and potentially reducing the morbidity) of treatment while maintaining comparable survival outcomes to more lengthy conventional radiotherapy. The most commonly studied regimen in the management of GBM is 40 Gy in 15 fractions. Radiobiologically, this dose should provide similar tumor control to 60 Gy in 30 fractions. While increasing the dose of radiation per fraction does pose an increased risk of neurotoxicity, it has already been pointed out that the most critical toxicities typically occur at least 6–9 months post treatment, if not longer [62, 65, 66]. Thus, patients with an expected prognosis of well under 1 year are unlikely to be at high risk of experiencing problems relating to radiation necrosis. However, as both old age and large fraction size are known risk factors for radiation-induced encephalopathy [63, 64], a hypofractionated regimen may exacerbate this particular outcome. Although the literature in this field suggests a significant time to onset in excess of 1 year, there are many anecdotal reports of generalized neurocognitive decline in elderly patients similar to that seen with encephalopathy at earlier time points.

In terms of effectiveness, single-arm historical case series of hypofractionated regimens, including 40 Gy in 15 fractions, did not demonstrate

Table 11.3 Hypofractionated radiotherapy compared with standard fractionation in GBM

Group	Age	N	Fractionation	Median OS (months)	Outcome of hypofractionation
Randomized data					
Roa 2004 [96]	≥60	100	60/30	5.1	Equivalent
			40/15	5.6	
Nonrandomized data					
Bauman 1994 [94]	All ages	92	>50	10	Inferior
			30/10	6	
			No RT	1	
Ford 1997 [91]	All ages	59	60/30	4	Equivalent
			36/12		
Mohan 1998 [78]	≥70	102	≥55	7.3	Inferior
			<45	4.5	
			No RT	1.2	
Hulshof 2000 [92]	All ages	155	66/33	7	Equivalent
			40/8	5.6	
			28/4	6.6	
McAleese 2003 [95]	All ages	136	60/30	7.5–9.5	Inferior
			30/6	5	
Lutterbach 2005 [93]	≥60	96	60/30	5.6	Equivalent
			42/12	7.3	

N number of patients, *OS* overall survival

inferior survival in elderly populations [83, 86, 90], as already shown in Table 11.2. At the same time, various hypofractionated schedules have been analyzed in comparison with standard fractionation approaches up to a total dose of 66 Gy in a combination of retrospective and prospective studies, as outlined in Table 11.3 [77, 91–96]. Three studies illustrated equivalent survival with hypofractionated and conventional regimens [34, 91–93]. However, only Lutterbach et al. specifically looked at "elderly" patients, although this is debatable as the age cutoff was 60 years [93]. Both Ford et al. and Hulshof et al. included younger patients in their analyses, albeit in the case of Hulshof et al., almost one half of the 155 patients were aged over 60 years [91, 92]. Three further series demonstrated a worse outcome in the hypofractionated arms [78, 94, 95]. It should be noted that while Mohan et al. included only patients aged over 70 [78], the other groups had wider entry criteria and accepted younger patients provided that their KPS level was sufficiently low (either aged 50–70 years with KPS 50–90 or any age with KPS <50) [94, 95]. Hence, a proportion of patients selected for short-course radiotherapy in the two latter-mentioned studies were generally frail and deemed not fit for long-course treatment. It is likely that a significant number of these patients had a poorer prognosis at the outset and this may have skewed the results. Both groups used matched controls as part of their analyses, but attempting to retrospectively match patients is not always accurate.

To answer this clinical question in a more controlled way, a randomized phase III trial was established. Two regimens were tested in 100 patients over the age of 60: 40 Gy in 15 fractions versus the standard 60 Gy in 30 fractions [96]. The median survival for both groups was comparable (5.6 months vs. 5.1 months, respectively) suggesting that hypofractionated radiotherapy in elderly patients with GBM is equivalent to conventional radiotherapy. However, this trial has been subject to a number of criticisms. Firstly, whether the age of 60 is a valid threshold for the term "elderly" is controversial. Secondly, the patients in this study were of relatively poor performance status and had not been optimally debulked. Thirdly, late neurological toxicity was not assessed, although this was probably irrelevant as

Table 11.4 Chemoradiotherapy versus radiotherapy in elderly GBM patients

Group	Age	Chemotherapy	Sequencing	N	Median OS (months)
Mohan 1998 [78]	≥70	BCNU, PCV	Adjuvant	16	RT + chemo 8.0
				86	RT 4.9
Pierga 1999 [81]	≥70	BCNU, PCV	Adjuvant	12	RT + chemo 13.5
				18	RT 6.3
Brandes 2003 [87]	≥60	PCV, TMZ	Adjuvant	54	RT + chemo 14.9
				24	RT 11.2
Patwardhan 2004 [98]	≥59	BCNU, TMZ, Gliadel	Adjuvant	9	RT + chemo 13.6
				6	RT 5.5
Kimple 2010 [97]	≥70	Etoposide, TMZ, irinotecan	Concurrent + adjuvant	14	RT + chemo 11.6
				4	RT 6.5
Scott 2011 [99]	≥70	CCNU, TMZ, carboplatin	Concurrent + adjuvant	29	RT + chemo 13.3
				45	RT 7.2

N number of patients, *OS* overall survival, *RT* radiotherapy, *TMZ* temozolomide

survival rates were just under 6 months. Nonetheless, hypofractionated regimens, particularly 40 Gy in 15 fractions and 30 Gy in 6 fractions, have become standard practice in many centers for the treatment of elderly and/or frail patients who are unlikely to tolerate a conventional course of treatment.

Chemoradiotherapy

In 2005, a new standard of care for GBM was defined in a phase III trial, which demonstrated that the addition of concurrent and adjuvant temozolomide chemotherapy to radical radiotherapy was associated with significantly superior survival [1]. The caveat is that this trial had an upper age limit of 70 years. However, combining chemotherapy with radiotherapy in the elderly has been widely practiced using a number of cytotoxic drugs given concomitantly and/or in the adjuvant phase. Nitrosoureas and temozolomide are the predominant cytotoxic agents, although platinum, topoisomerase inhibitors, and even targeted therapies have also been employed. Results of several case series have tended to show a superior outcome with chemoradiotherapy compared with radiotherapy alone, as illustrated in Table 11.4 [78, 81, 87, 97–99]. Median overall survival reached over a year in some cases, although it is very likely that only the fittest patients who had also undergone optimal debulking were selected for triple-modality treatment, which may have significantly influenced the outcome.

Many centers have a policy of restricting the Stupp protocol to patients aged below 70 years, so data on the safety and effectiveness of this regimen in older patients is limited. However, a handful of single-institution series published recently have documented outcomes and toxicity in elderly cohorts [54, 55, 57, 100–103]. This data is shown in Table 11.5 [2, 54, 55, 57, 100–103]. As with the earlier chemoradiotherapy series featuring an array of drugs and/or scheduling, median survival times of over 1 year have been reported. It is important to point out that the definition of "elderly" in these published series varies between 60 years and 70 years, and again, it is extremely likely that the patients in these case series were selected on the basis of general fitness. In fact, several groups remarked that combination treatment appeared to be most advantageous in patients with higher KPS [100, 101]. Interestingly, a trend benefit analysis of the original Stupp data by age showed a decreasing benefit with increasing age, with hazard ratios of 0.63 for patients aged 50–60 years ($P < 0.05$), 0.72 for patients aged 60–65 years ($P = 0.096$), and 0.80 for patients aged 65–70 years ($P = 0.34$) [104].

An important question if chemoradiotherapy is to become more common practice in elderly patients is whether this regimen will be tolerated. This is relevant both in the short term, due to the acute toxicity of chemotherapy and potential

Table 11.5 Stupp protocol-based chemoradiotherapy studies in elderly GBM patients

Group	Age	*N*	Median OS (months)	1	2	3	G3/G4 toxicity
Combs [a]2008 [100]	≥65	43	11	88 %	12 %	NR	Combined 9 %
Minniti 2008 [101]	≥70	32	10.6	94 %	NR	NR	Concomitant 6 % Adjuvant 27 %
Sjiben 2008 [57]	≥65	19	8.5	NR	NR	NR	Combined 42 %
Brandes 2009 [54]	≥65	58	13.7	100 %	NR	NR	Concomitant 19 % Adjuvant 46 %
Stupp 2009 [2]	60–70	83	10.9	NR	NR	NR	NR
Fiorica 2010 [102]	≥65	42	10.2	69 %	52 %	14 %	Concomitant 5 % Adjuvant 7 %
Gerstein 2010 [103]	≥65	51	11.5	59 %	20 %	NR	Combined 41 %
Minniti 2011 [55]	≥70	83	12.8	NR	NR	NR	Combined 27 %

N number of patients, *OS* overall survival, *1* patients completing CRT, *2* patients commencing adjuvant temozolomide, *3* patients completing 6 cycles of adjuvant temozolomide, *NR* not reported
[a]50 mg/m^2 temozolomide during concomitant phase in this study

exacerbation of acute radiation toxicity, and also in the long term especially if prolonged survival reveals excess late radiation effects. Age is a known risk factor for reduced chemotherapy tolerance. The pharmacokinetics of individual agents should always be borne in mind when administering chemotherapy to elderly patients. Fortunately, temozolomide is metabolized by nonenzymatic processes that are less subject to variability between individuals [105], and it has a relatively favorable toxicity profile [106]. It does, however, cause noncumulative myelosuppression, particularly thrombocytopenia. Although this section focuses on the adverse effects of chemotherapy, it is also important to point out that elderly patients often have a degree of mucosal atrophy and reduced gastrointestinal motility. Hence, absorption of oral chemotherapy agents such as temozolomide may be impaired [107], and the effectiveness of this regimen may be compromised.

Examination of toxicity in the recently published elderly Stupp protocol data sets indicates that the incidence of grade 3 or 4 events is extremely variable. While Combs et al. and Fiorica et al. report levels of severe toxicity at less than 10 % [100, 102], a significantly higher level of grade 3 or 4 events of approximately 40 % has been reported by three other groups [54, 57, 103]. Notably, just over 80 % of patients in the cohort reported by Combs et al. received 50 mg/m^2 of daily temozolomide at the outset during the concurrent phase as opposed to the standard dose of 75 mg/m^2 which may have contributed to the lower levels of toxicity in this series [100]. Most of the described toxicity was hematological, and this probably explains why adjuvant chemotherapy was not given in the majority of patients, although it is often difficult to elicit this information from the published material. In fact, most of the studies provided relatively clear information on the percentage of patients who completed concomitant chemoradiotherapy without a dose reduction and/or stopping chemotherapy, but not the percentage of patients who (i) commenced adjuvant chemotherapy or (ii) completed 6 cycles of adjuvant chemotherapy, as illustrated in Table 11.5. Two pertinent questions stem from this missing data. Firstly, is this regimen as well tolerated in elderly patients as some authors would lead us to believe if very few patients can actually complete? Secondly, is concomitant chemotherapy the key active component, and is there any additional benefit from adjuvant temozolomide in the elderly?

In relation to the first question, significant toxicity has certainly been documented [54, 57, 103], and this has already been alluded to. After hematological toxicity, neurological sequelae were the next most common problem. It is concerning that a prospective phase II study of concurrent temozolomide in patients over the age of 65 reported grade 2 deterioration in mental status in 31 % and grade 3 deterioration in a further 25 %, leading to significant disability [108]. Moreover, grade 3 leukoencephalopathy occurred in 6 %. Of note, the median interval between start of treatment and development of neurological toxicity was 6 months in this study, whereas time to progression was 9.5 months, indicating a correlation with treatment rather than disease progression. In another series, 25 % of patients experienced grade 3 or grade 4 deterioration in mental state during or just after radiotherapy, and the rate of grade 3 encephalopathy was 10 % [54]. Hence, the neurological and neurocognitive sequelae of combined treatment may be profound. Any responses to the second question would be speculative and likely to remain so for some time, as there are no plans to compare concomitant versus concomitant plus adjuvant chemotherapy in the near future.

On balance, the reported series to date suggest that there is a potential benefit of aggressive multimodality therapy in the elderly, but caution should be exercised because (i) this benefit is likely to be smaller compared with younger patients, (ii) the treatment may be considerably more toxic, and (iii) patient selection is crucial. Should this type of approach become more commonplace, then the issue of pseudoprogression is likely to be raised. At present it is unknown whether this phenomenon is any more or less common in the elderly. The only confirmed risk factor is MGMT status [109]. As yet, there is no definitive evidence that methylation of MGMT in GBM varies significantly with age [27, 54–58]. It may be the case that the aging cerebral vasculature may be more subject to radiation-induced disruption and dysfunction, in which case a higher incidence of pseudoprogression is a distinct possibility in the elderly. If so, will the degree of pseudoprogression be more profound? This is potentially concerning as it has been proposed that severe cases of pseudoprogression may predispose to necrosis [110]. The elderly may therefore be at higher risk of toxicity from combined chemoradiotherapy, both in the short term and in the long term. Conversely, pseudoprogression is thought to perhaps indicate improved clinical outcome [109, 111–113], but clinicians might be more likely to pull out of treatment earlier in an elderly patient with a scan suggestive of progression. Hence, some older patients with a response to treatment may be denied ongoing effective therapy. There is no doubt that this is an interesting topic for future study, and as the significance of this phenomenon becomes clearer, it is likely that imaging and/or markers of pseudoprogression will be incorporated into clinical trials of multimodality therapy.

Chemotherapy Versus Radiotherapy

While there has been a recent flurry of publications advocating aggressive multimodality therapy in the elderly [55, 99–102], it is interesting that the most up-to-date clinical trials featuring elderly GBM patients have focused on de-intensification protocols, mainly comparing radiotherapy with chemotherapy. This is not an entirely new concept as a number of phase II studies and retrospective series using temozolomide as an alternative to radiotherapy have previously been reported. These demonstrated median survival durations of just over 6 months, with acceptable toxicity [88, 114–117]. In some cases, imaging was used to evaluate measurable disease, and partial responses or stable disease was elicited in up to 70 % of patients [115–117]. Certainly, there are advantages of opting for chemotherapy over radiotherapy as this allows patients to be treated at home for the most part, only attending hospital for a clinic visit every 4 weeks (if following the standard 28-day cycle of temozolomide). On the other hand, careful blood monitoring is required and compliance may be an issue, especially if there is evidence of cognitive deficit.

The Nordic Brain Tumor Study Group randomized 342 patients over the age of 60 years to conventional radiotherapy (60 Gy in 30 fractions), hypofractionated radiotherapy (34 Gy in 10 fractions), or temozolomide (200 mg/m^2 daily for 5/28 days for 6 cycles). Preliminary results suggest that the three arms are equivalent, although evaluation is confounded by crossover from radiotherapy to temozolomide and vice versa. However, the median overall survival was relatively short across the various arms (6–9 months), despite the fact that 60 years was the minimum age and 75 % had a good performance status of 0–1 [118]. Meanwhile, the Neuro-Oncology Working Group of the German Cancer Society (NOA) conducted a two-arm study to investigate the efficacy of chemotherapy versus conventional radiotherapy alone. NOA-08 randomized 412 patients, all over the age of 65 years, to 60 Gy in 30 fractions or temozolomide (100 mg/m^2 daily, 1 week on/1 week off, until progression). Early results indicate that radiation may have on advantage for radiation over chemotherapy although, once again, median overall survival rates were disappointing at less than 9 months; survival was measured at 293 days in the radiotherapy arm versus 245 days in the chemotherapy arm [119]. Another drawback of the chemotherapy arm in this trial was the prospect of remaining on treatment until disease progression or death. This would probably be unappealing to the majority of elderly patients.

At present, there is no substantive data to support the use of temozolomide over radiation in the elderly GBM patient although chemotherapy remains a viable alternative in patients who refuse radiotherapy.

Intracavitary Chemotherapy

GBM is a unique disease in that chemotherapy can be safely applied into the surgical cavity at the time of debulking. Gliadel wafers are biodegradable polymers containing 3.85 % carmustine. Compared with surgery alone, implantation of these wafers at the time of repeat surgery may prolong survival [120], but this practice remains controversial and does not form part of routine treatment at many centers. There is also data to suggest that this approach may be beneficial at the time of first surgery [121, 122], although this is based on patients of all ages. Chaichana et al. have recently published the findings of a sizeable case–control study of elderly patients aged over 65 [123]. Altogether, 88 patients had intracavitary carmustine wafers inserted at initial surgery, and half of these patients were matched with controls who had not undergone implantation. Reassuringly, there was no increase in perioperative morbidity and mortality in the carmustine wafer cohort. In terms of efficacy, a survival advantage of 2–3 months was demonstrated for the carmustine group. However, this study was not a randomized controlled trial, and as such intracavitary treatment cannot be considered standard practice. Nonetheless, this approach merits further investigation as it may be considerably less toxic and better tolerated than the Stupp protocol in elderly patients.

Best Supportive Care

GBM is undoubtedly a devastating disease, characterized by progressive loss of neurological function and changes in cognitive ability and personality. Indeed, a proportion of sufferers will be unsuitable for any oncological treatment at the outset due to significant disability. For these patients, best supportive care is of paramount importance. Steroids can relieve some of the pressure symptoms associated with tumor growth. However, the use of glucocorticoids must be considered in the context of their potentially devastating side effects such as emotional lability, insomnia, proximal myopathy, weight gain, immunosuppression, venous thrombosis, and hyperglycemia. It is imperative that the patient is closely monitored in the community, especially if there is a history of heart failure and/or diabetes. The use of analgesia and anticonvulsants may also be required with subsequent risks of toxicity and drug-drug interactions. Patients often require extensive physical assistance and close supervision which can result in marked personal and economic stresses on the caregiver. Early contact with hospice and/or community palliative medicine team is advised as well as information on support groups for both the patient and the carer.

Recurrent Disease

As the median survival of elderly GBM patients is just a few months and a significant number do not receive any treatment at the outset, there is virtually no data on how best to manage the elderly patient with recurrent disease. The decision to treat should be centered around the individual patient, and various factors must be taken into account, including performance status, response to initial therapy, time since diagnosis, and whether the recurrence is local or diffuse. Therapeutic options are similar to those of the general adult population and include further surgery, systemic chemotherapy with temozolomide or nitrosoureas, targeted agents such as vascular endothelial growth factor receptor (VEGFR) inhibitors, and radiotherapy. However, re-irradiating the aging brain of an elderly patient would be a daunting prospect for most radiation oncologists. The former options are therefore more likely to be carried out in clinical practice.

Future Perspectives

The numbers of elderly GBM patients are ever increasing, but these patients have largely been excluded from the pivotal, practice-changing trials. It has now been recognized by the neuro-oncology community that the optimal management of GBM in older patients needs to be determined. Realistically, this could be achieved in one of two ways: either by including all age groups in future clinical trials or alternatively devising separate trial protocols for those aged over 65 years (or perhaps over 70 years). As elderly patients are generally frailer and less able to tolerate traditional oncological therapies, it seems reasonable to consider them as a separate group and devise protocols accordingly. Indeed, there has already been some progress in this direction, as shown by the temozolomide versus radiotherapy studies that had a minimum age criteria of 70. Some would say, however, that fit elderly patients are being undertreated by this approach and that triple-modality therapy should be an option. The NCIC and EORTC have recognized this and designed a randomized trial that compares radiotherapy alone with radiotherapy plus concurrent and adjuvant temozolomide for up to 1 year in the over 65 age group. The radiotherapy regimen in this study is 40 Gy in 15 fractions over 3 weeks, based on the Roa data that showed equivalence to 60 Gy in 30 fractions over 6 weeks [96]. The primary objective of this trial is to assess the impact of concomitant therapy on survival. Toxicity data will be particularly interesting, especially as chemotherapy is being combined with a higher dose of radiation per fraction than in the Stupp protocol. The only other prospective studies to date examining the effect of multimodal treatment in the elderly have been small single-institution trials that focused on hypofractionated radiotherapy followed by adjuvant chemotherapy as opposed to a concomitant regimen [124, 125]. Of note, in both of these studies, the median survival was around 9 months, yet chemotherapy was planned for up to 12 months. It is unclear why the NCIC-EORTC groups opted for 12 months of treatment, given that a substantial proportion of their patients are unlikely to be alive at this point. Whether this length of treatment is acceptable and/or appropriate for the majority of elderly patients will only be realized when the final data is available for survival and quality of life analysis.

The NCIC-EORTC trial has incorporated molecular analysis into the protocol, and it is hoped that this additional information will help select out those patients who are most likely to benefit from multimodality therapy. It has already been mentioned that MGMT status appears to be as common in elderly GBM patients as in their younger counterparts and may have prognostic value in the elderly despite their overall poorer outcome [55]. Some investigators have proposed that the negative effect of age can be counteracted by methylation of MGMT [126], although there is no substantive evidence to support this. Clearly, this area requires further clarification and large-scale prospective evaluation such as that provided by the NCIC-EORTC study is key. It is increasingly likely that more GBM studies in the future will include molecular testing, comprising not only MGMT analysis but also a more rigorous examination of the various genetic alterations and/or molecular signaling pathways that might contribute to clinical outcomes, with the ultimate aim of developing a more individualized approach to therapy. To this end, a number of targeted agents are currently under investigation in GBM, but the possibility of individualized treatment, particularly in the elderly population, is some way off.

The most widely studied targeted agent in GBM is bevacizumab, a humanized monoclonal antibody that inhibits VEGF activity. Although this agent is not yet widely available, it is licensed for use in the recurrent setting in some parts of the world. This is based on phase II data demonstrating an increase in 6-month progression-free survival when bevacizumab was administered in combination with irinotecan [127, 128]. Although elderly patients were not excluded from these trials, the median age in both studies was less than 55, suggesting a higher proportion of younger patients. Intriguingly, a retrospective analysis of a single-institution study showed that patients over the age of 55 gained the most benefit from single-agent bevacizumab in the context of recurrent disease [129]. Antiangiogenic treatment may be a useful therapeutic tool in the elderly, but this premise is based on very preliminary data. Further work is required to establish whether VEGF inhibition has a role in the management of primary and/or recurrent GBM in the elderly, either alone or in combination with radiotherapy and/or chemoradiotherapy. Although targeted agents are generally not as toxic as traditional oncological therapies, VEGF inhibitors are not without adverse effects. Indeed, there is some evidence from other tumor types to suggest that toxicity is more pronounced in the elderly when bevacizumab is combined with chemotherapy [130]. A number of other targeted agents are also under investigation; many of these are tyrosine kinase inhibitors directed against growth factor receptors including, but not necessarily exclusive to, VEGF. Even if some of these agents prove too toxic to be combined with standard concomitant therapy, especially in the elderly, they may have a role either in combination with radiotherapy or as single-agent treatment in frail patients, provided that there is sufficient evidence of efficacy.

DNA repair is an important mechanism of radiation resistance, and a number of novel agents are available that target components of the DNA damage response. Of the compounds under development, the most advanced are inhibitors of poly(ADP-ribose) polymerase (PARP), some of which have been used as single agents in the treatment of BRCA-mutated breast and ovarian cancer, with remarkable success and minimal toxicity [131, 132]. A large body of preclinical data has also established PARP inhibitors as effective radiosensitizers and early-phase clinical trials in combination with radiotherapy are now underway [133]. Of particular relevance to the treatment of GBM, the radiosensitizing effects of PARP inhibitors are observed only in actively replicating cells [134]. Since the cells of the normal brain are non-replicating, this raises the prospect of tumor-specific radiosensitization for GBM. As single-agent PARP inhibitors such as olaparib are extremely well tolerated, these compounds may be particularly well suited to the treatment of elderly GBM patients, in combination with either radical or short-course radiotherapy [135].

While there is currently much interest in trying to develop new therapeutic targets in GBM, it is important to remember that the most critical component of management is radiotherapy. Numerous studies have not demonstrated a benefit of dose escalation, so there is little to be gained by further exploration of this route, especially in elderly patients where this is concern about the tolerability of radiation. This does not mean that there is no room for improvement in terms of delivery of radiotherapy to older patients. The advent of intensity-modulated radiotherapy (IMRT) has provided radiation oncologists with a greater ability to sculpt the dose around a target volume. IMRT is often used to spare a specific organ at risk, such as the spinal cord or optic chiasm, and has the advantage of delivering a highly conformal, homogeneous dose to the target volume while simultaneously sparing normal tissue. Hence, techniques such as IMRT may be of particular benefit in elderly patients, by minimizing radiation dose to normal brain and improving tolerance to treatment. An example of a more favorable dose distribution using IMRT compared with conventional radiotherapy is illustrated in Fig. 11.3.

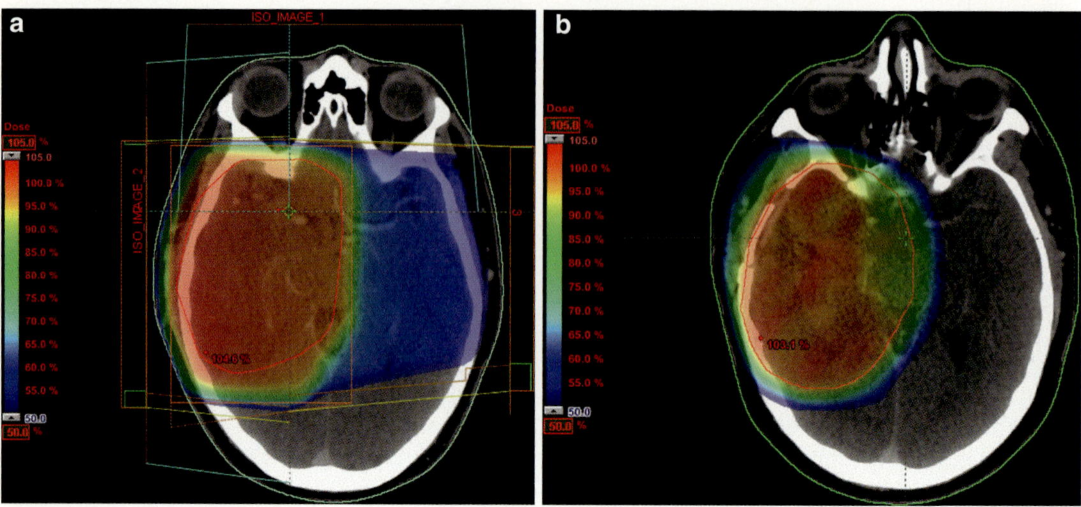

Fig. 11.3 Normal brain sparing with intensity-modulated radiotherapy (IMRT). Contrast-enhanced computed tomography (CT) planning images of two patients with right temporal lobe tumors treated with radical chemoradiotherapy using different techniques: conventional 3-field arrangement (**a**) and IMRT (**b**). The PTV in each case is indicated in *red*, and color dose wash demonstrates the 50 % isodose (*blue* shading). Note the improved sparing of surrounding normal brain in the IMRT-treated case

Conclusion

Managing elderly GBM patients effectively can be challenging, as they are often frailer and less able to tolerate standard multimodality therapy. However, a subgroup of elderly patients is less impaired in terms of neurological function and performance status and can cope with "aggressive" management. Underpinning the use of multimodality treatment is debulking surgery. Recent reports are challenging the widely held view that elderly patients do not tolerate neurosurgical intervention, and evidence is emerging that tumor resection can improve performance status in this patient group. A more interventional neurosurgical approach brings a number of potential benefits: (i) rapid and effective relief of raised intracranial pressure and possible improvement in performance status, (ii) high-quality tissue for diagnosis and molecular classification that might help to predict prognosis and guide nonsurgical treatment, (iii) the potential for use of local cytotoxic agents, and (iv) a possible improvement in tolerance of subsequent radiotherapy and/or chemotherapy. While not all of these statements are yet supported by high-level evidence, it is the opinion of the authors that selected elderly patients will derive significant, cumulative benefits from more aggressive neurosurgical management. Still, it is important to be mindful that even the fittest elderly GBM patients may not necessarily derive the same survival advantage as younger patients. Ultimately, the key to successful management of GBM in the elderly population is to differentiate between those patients who are most likely to benefit from multimodality therapy and those who would be better served by de-intensification protocols. Currently, elderly patients constitute approximately half of all patients with GBM, and this proportion is likely to increase significantly over the coming years. It is therefore imperative that we achieve a greater understanding of how to select patients for the various treatment approaches appropriately. Hopefully the implementation of carefully designed clinical trials in the elderly will identify prognostic factors, clinical and/or molecular, that will guide treatment with the optimal combination of conventional and novel therapies.

Acknowledgements We would like to thank Mairi Mackinnon and Aoife O'Regan for providing the images.

References

1. Stupp R, Mason WP, van den Bent MJ, Weller M, Fisher B, Taphoorn MJ. Radiotherapy plus concomitant and adjuvant temozolomide for glioblastoma. N Engl J Med. 2005;352(10):987–96. Epub 2005/03/11.
2. Stupp R, Hegi ME, Mason WP, van den Bent MJ, Taphoorn MJ, Janzer RC. Effects of radiotherapy with concomitant and adjuvant temozolomide versus radiotherapy alone on survival in glioblastoma in a randomised phase III study: 5-year analysis of the EORTC-NCIC trial. Lancet Oncol. 2009;10(5): 459–66. Epub 2009/03/10.
3. Yancik R. Epidemiology of cancer in the elderly. Current status and projections for the future. Rays. 1997;22(1 Suppl):3–9. Epub 1997/01/01.
4. Fisher JL, Schwartzbaum JA, Wrensch M, Wiemels JL. Epidemiology of brain tumors. Neurol Clin. 2007;25(4):867–90. vii. Epub 2007/10/30.
5. Greig NH, Ries LG, Yancik R, Rapoport SI. Increasing annual incidence of primary malignant brain tumors in the elderly. J Natl Cancer Inst. 1990;82(20): 1621–4. Epub 1990/10/17.
6. Jukich PJ, McCarthy BJ, Surawicz TS, Freels S, Davis FG. Trends in incidence of primary brain tumors in the United States, 1985–1994. Neuro Oncol. 2001;3(3):141–51. Epub 2001/07/24.
7. Hess KR, Broglio KR, Bondy ML. Adult glioma incidence trends in the United States, 1977–2000. Cancer. 2004;101(10):2293–9. Epub 2004/10/12.
8. Cairncross JG, Ueki K, Zlatescu MC, Lisle DK, Finkelstein DM, Hammond RR. Specific genetic predictors of chemotherapeutic response and survival in patients with anaplastic oligodendrogliomas. J Natl Cancer Inst. 1998;90(19):1473–9. Epub 1998/10/17.
9. Cairncross G, Berkey B, Shaw E, Jenkins R, Scheithauer B, Brachman D. Phase III trial of chemotherapy plus radiotherapy compared with radiotherapy alone for pure and mixed anaplastic oligodendroglioma: Intergroup Radiation Therapy Oncology Group Trial 9402. J Clin Oncol. 2006;24(18):2707–14. Epub 2006/06/20.
10. Hashimoto N, Murakami M, Takahashi Y, Fujimoto M, Inazawa J, Mineura K. Correlation between genetic alteration and long-term clinical outcome of patients with oligodendroglial tumors, with identification of a consistent region of deletion on chromosome arm 1p. Cancer. 2003;97(9):2254–61. Epub 2003/04/25.
11. Ino Y, Betensky RA, Zlatescu MC, Sasaki H, Macdonald DR, Stemmer-Rachamimov AO. Molecular subtypes of anaplastic oligodendroglioma: implications for patient management at diagnosis. Clin Cancer Res. 2001;7(4):839–45. Epub 2001/04/20.
12. Thiessen B, Maguire JA, McNeil K, Huntsman D, Martin MA, Horsman D. Loss of heterozygosity for loci on chromosome arms 1p and 10q in oligodendroglial tumors: relationship to outcome and chemosensitivity. J Neurooncol. 2003;64(3):271–8. Epub 2003/10/16.
13. van den Bent MJ, Carpentier AF, Brandes AA, Sanson M, Taphoorn MJ, Bernsen HJ. Adjuvant procarbazine, lomustine, and vincristine improves progression-free survival but not overall survival in newly diagnosed anaplastic oligodendrogliomas and oligoastrocytomas: a randomized European Organisation for Research and Treatment of Cancer phase III trial. J Clin Oncol. 2006;24(18):2715–22. Epub 2006/06/20.
14. Hegi ME, Diserens AC, Gorlia T, Hamou MF, de Tribolet N, Weller M. MGMT gene silencing and benefit from temozolomide in glioblastoma. N Engl J Med. 2005;352(10):997–1003. Epub 2005/03/11.
15. Nobusawa S, Watanabe T, Kleihues P, Ohgaki H. IDH1 mutations as molecular signature and predictive factor of secondary glioblastomas. Clin Cancer Res. 2009;15(19):6002–7. Epub 2009/09/17.
16. Ohgaki H, Dessen P, Jourde B, Horstmann S, Nishikawa T, Di Patre PL. Genetic pathways to glioblastoma: a population-based study. Cancer Res. 2004;64(19):6892–9. Epub 2004/10/07.
17. McKeever PE, Junck L, Strawderman MS, Blaivas M, Tkaczyk A, Cates MA. Proliferation index is related to patient age in glioblastoma. Neurology. 2001;56(9):1216–8. Epub 2001/05/09.
18. Kleihues P, Ohgaki H. Phenotype vs genotype in the evolution of astrocytic brain tumors. Toxicol Pathol. 2000;28(1):164–70. Epub 2000/02/11.
19. Burger PC, Green SB. Patient age, histologic features, and length of survival in patients with glioblastoma multiforme. Cancer. 1987;59(9):1617–25. Epub 1987/05/01.
20. Barker FG, Prados MD, Chang SM, Davis RL, Gutin PH, Lamborn KR. Bromodeoxyuridine labeling index in glioblastoma multiforme: relation to radiation response, age, and survival. Int J Radiat Oncol Biol Phys. 1996;34(4):803–8. Epub 1996/03/01.
21. Huang PH, Xu AM, White FM. Oncogenic EGFR signaling networks in glioma. Sci Signal. 2009;2(87):re6. Epub 2009/09/10.
22. Wrensch M, Minn Y, Chew T, Bondy M, Berger MS. Epidemiology of primary brain tumors: current concepts and review of the literature. Neuro Oncol. 2002;4(4):278–99. Epub 2002/10/03.
23. Lacroix M, Abi-Said D, Fourney DR, Gokaslan ZL, Shi W, DeMonte F. A multivariate analysis of 416 patients with glioblastoma multiforme: prognosis, extent of resection, and survival. J Neurosurg. 2001;95(2):190–8. Epub 2002/01/10.
24. Surawicz TS, Davis F, Freels S, Laws Jr ER, Menck HR. Brain tumor survival: results from the National Cancer Data Base. J Neurooncol. 1998;40(2):151–60. Epub 1999/01/19.
25. Curran Jr WJ, Scott CB, Horton J, Nelson JS, Weinstein AS, Fischbach AJ. Recursive partitioning analysis of prognostic factors in three Radiation Therapy Oncology Group malignant glioma trials. J Natl Cancer Inst. 1993;85(9):704–10. Epub 1993/05/05.
26. Li J, Wang M, Won M, Shaw EG, Coughlin C, Curran Jr WJ. Validation and simplification of the radiation

therapy oncology group recursive partitioning analysis classification for glioblastoma. Int J Radiat Oncol Biol Phys. 2011;81(3):623–30. Epub 2010/10/05.
27. Piccirilli M, Bistazzoni S, Gagliardi FM, Landi A, Santoro A, Giangaspero F. Treatment of glioblastoma multiforme in elderly patients. Clinico-therapeutic remarks in 22 patients older than 80 years. Tumori. 2006;92(2):98–103. Epub 2006/05/27.
28. Simpson JR, Horton J, Scott C, Curran WJ, Rubin P, Fischbach J. Influence of location and extent of surgical resection on survival of patients with glioblastoma multiforme: results of three consecutive Radiation Therapy Oncology Group (RTOG) clinical trials. Int J Radiat Oncol Biol Phys. 1993;26(2):239–44. Epub 1993/05/20.
29. Chaichana KL, Chaichana KK, Olivi A, Weingart JD, Bennett R, Brem H. Surgical outcomes for older patients with glioblastoma multiforme: preoperative factors associated with decreased survival. Clinical article. J Neurosurg. 2011;114(3):587–94. Epub 2010/10/05.
30. Kreth FW, Faist M, Rossner R, Volk B, Ostertag CB. Supratentorial World Health Organization Grade 2 astrocytomas and oligoastrocytomas. A new pattern of prognostic factors. Cancer. 1997;79(2):370–9. Epub 1997/01/15.
31. Maly RC, Leake B, Silliman RA. Health care disparities in older patients with breast carcinoma: informational support from physicians. Cancer. 2003;97(6): 1517–27. Epub 2003/03/11.
32. Gagnon B, Mayo NE, Hanley J, MacDonald N. Pattern of care at the end of life: does age make a difference in what happens to women with breast cancer? J Clin Oncol. 2004;22(17):3458–65. Epub 2004/07/28.
33. Lowry JK, Snyder JJ, Lowry PW. Brain tumors in the elderly: recent trends in a Minnesota cohort study. Arch Neurol. 1998;55(7):922–8. Epub 1998/07/25.
34. Lutterbach J, Bartelt S, Momm F, Becker G, Frommhold H, Ostertag C. Is older age associated with a worse prognosis due to different patterns of care? A long-term study of 1346 patients with glioblastomas or brain metastases. Cancer. 2005;103(6):1234–44. Epub 2005/01/25.
35. Steinfeld AD, Donahue B, Walker I. Delay in the diagnosis of glioblastoma multiforme: is age a factor? Cancer Invest. 1996;14(4):317–9. Epub 1996/01/01.
36. Barnholtz-Sloan JS, Williams VL, Maldonado JL, Shahani D, Stockwell HG, Chamberlain M. Patterns of care and outcomes among elderly individuals with primary malignant astrocytoma. J Neurosurg. 2008;108(4):642–8. Epub 2008/04/02.
37. Iwamoto FM, Reiner AS, Panageas KS, Elkin EB, Abrey LE. Patterns of care in elderly glioblastoma patients. Ann Neurol. 2008;64(6):628–34. Epub 2008/12/25.
38. Paszat L, Laperriere N, Groome P, Schulze K, Mackillop W, Holowaty E. A population-based study of glioblastoma multiforme. Int J Radiat Oncol Biol Phys. 2001;51(1):100–7. Epub 2001/08/23.
39. Scott J, Tsai YY, Chinnaiyan P, Yu HH. Effectiveness of radiotherapy for elderly patients with glioblastoma. Int J Radiat Oncol Biol Phys. 2011;81(1):206–10. Epub 2010/08/03.
40. Kita D, Ciernik IF, Vaccarella S, Franceschi S, Kleihues P, Lutolf UM. Age as a predictive factor in glioblastomas: population-based study. Neuroepidemiology. 2009;33(1):17–22. Epub 2009/03/28.
41. Rosenthal MA, Drummond KJ, Dally M, Murphy M, Cher L, Ashley D. Management of glioma in Victoria (1998–2000): retrospective cohort study. Med J Aust. 2006;184(6):270–3. Epub 2006/03/22.
42. Scoccianti S, Magrini SM, Ricardi U, Detti B, Buglione M, Sotti G. Patterns of care and survival in a retrospective analysis of 1059 patients with glioblastoma multiforme treated between 2002 and 2007: a multicenter study by the Central Nervous System Study Group of Airo (italian Association of Radiation Oncology). Neurosurgery. 2010;67(2):446–58. Epub 2010/07/21.
43. Magrini SM, Ricardi U, Santoni R, Krengli M, Lupattelli M, Cafaro I. Patterns of practice and survival in a retrospective analysis of 1722 adult astrocytoma patients treated between 1985 and 2001 in 12 Italian radiation oncology centers. Int J Radiat Oncol Biol Phys. 2006;65(3): 788–99. Epub 2006/05/10.
44. Chang SM, Parney IF, Huang W, Anderson Jr FA, Asher AL, Bernstein M. Patterns of care for adults with newly diagnosed malignant glioma. JAMA. 2005;293(5):557–64. Epub 2005/02/03.
45. Warren JL, Klabunde CN, Schrag D, Bach PB, Riley GF. Overview of the SEER-Medicare data: content, research applications, and generalizability to the United States elderly population. Med Care. 2002;40(8 Suppl):IV-3–18. Epub 2002/08/21.
46. Walker MD, Green SB, Byar DP, Alexander Jr E, Batzdorf U, Brooks WH. Randomized comparisons of radiotherapy and nitrosoureas for the treatment of malignant glioma after surgery. N Engl J Med. 1980;303(23):1323–9. Epub 1980/12/04.
47. Walker MD, Alexander Jr E, Hunt WE, MacCarty CS, Mahaley Jr MS, Mealey Jr J. Evaluation of BCNU and/or radiotherapy in the treatment of anaplastic gliomas. A cooperative clinical trial. J Neurosurg. 1978;49(3):333–43. Epub 1978/09/01.
48. Keime-Guibert F, Chinot O, Taillandier L, Cartalat-Carel S, Frenay M, Kantor G. Radiotherapy for glioblastoma in the elderly. N Engl J Med. 2007;356(15): 1527–35. Epub 2007/04/13.
49. Barker 2nd FG, Chang SM, Larson DA, Sneed PK, Wara WM, Wilson CB. Age and radiation response in glioblastoma multiforme. Neurosurgery. 2001;49(6): 1288–97.
50. Gaspar LE, Fisher BJ, MacDonald DR, LeBer DV, Halperin EC, Schold Jr SC. Malignant glioma—timing of response to radiation therapy. Int J Radiat Oncol Biol Phys. 1993;25(5):877–9. Epub 1993/04/02.
51. Murovic J, Turowski K, Wilson CB, Hoshino T, Levin V. Computerized tomography in the prognosis of

malignant cerebral gliomas. J Neurosurg. 1986; 65(6):799–806. Epub 1986/12/01.
52. Rosenblum ML, Gerosa M, Dougherty DV, Reese C, Barger GR, Davis RL. Age-related chemosensitivity of stem cells from human malignant brain tumours. Lancet. 1982;1(8277):885–7. Epub 1982/04/17.
53. Grant R, Liang BC, Page MA, Crane DL, Greenberg HS, Junck L. Age influences chemotherapy response in astrocytomas. Neurology. 1995;45(5):929–33. Epub 1995/05/01.
54. Brandes AA, Franceschi E, Tosoni A, Benevento F, Scopece L, Mazzocchi V. Temozolomide concomitant and adjuvant to radiotherapy in elderly patients with glioblastoma: correlation with MGMT promoter methylation status. Cancer. 2009;115(15):3512–8. Epub 2009/06/11.
55. Minniti G, Salvati M, Arcella A, Buttarelli F, D'Elia A, Lanzetta G. Correlation between O6-methylguanine-DNA methyltransferase and survival in elderly patients with glioblastoma treated with radiotherapy plus concomitant and adjuvant temozolomide. J Neurooncol. 2011;102(2):311–6. Epub 2010/08/06.
56. Gerstner ER, Yip S, Wang DL, Louis DN, Iafrate AJ, Batchelor TT. Mgmt methylation is a prognostic biomarker in elderly patients with newly diagnosed glioblastoma. Neurology. 2009;73(18):1509–10. Epub 2009/11/04.
57. Sijben AE, McIntyre JB, Roldan GB, Easaw JC, Yan E, Forsyth PA. Toxicity from chemoradiotherapy in older patients with glioblastoma multiforme. J Neurooncol. 2008;89(1):97–103. Epub 2008/04/10.
58. Weller M, Hentschel B, Felsberg J, et al. Impact of MGMT promoter methylation in glioblastoma of the elderly. J Clin Oncol. 2011;29(158): 2001.
59. Taylor ME, Oppenheim BA. Hospital-acquired infection in elderly patients. J Hosp Infect. 1998;38(4):245–60. Epub 1998/05/29.
60. Baraldi-Junkins CA, Beck AC, Rothstein G. Hematopoiesis and cytokines. Relevance to cancer and aging. Hematol Oncol Clin North Am. 2000;14(1):45–61. viii. Epub 2000/02/19.
61. Wasil T, Lichtman SM. Clinical pharmacology issues relevant to the dosing and toxicity of chemotherapy drugs in the elderly. Oncologist. 2005;10(8):602–12. Epub 2005/09/24.
62. Swennen MH, Bromberg JE, Witkamp TD, Terhaard CH, Postma TJ, Taphoorn MJ. Delayed radiation toxicity after focal or whole brain radiotherapy for low-grade glioma. J Neurooncol. 2004;66(3):333–9. Epub 2004/03/16.
63. Corn BW, Yousem DM, Scott CB, Rotman M, Asbell SO, Nelson DF. White matter changes are correlated significantly with radiation dose. Observations from a randomized dose-escalation trial for malignant glioma (Radiation Therapy Oncology Group 83–02). Cancer. 1994;74(10):2828–35. Epub 1994/11/15.
64. Tsuruda JS, Kortman KE, Bradley WG, Wheeler DC, Van Dalsem W, Bradley TP. Radiation effects on cerebral white matter: MR evaluation. AJR Am J Roentgenol. 1987;149(1):165–71. Epub 1987/07/01.
65. Schultheiss TE, Kun LE, Ang KK, Stephens LC. Radiation response of the central nervous system. Int J Radiat Oncol Biol Phys. 1995;31(5): 1093–112.
66. Kim JH, Brown SL, Jenrow KA, Ryu S. Mechanisms of radiation-induced brain toxicity and implications for future clinical trials. J Neurooncol. 2008; 87(3):279–86. Epub 2008/01/23.
67. Keles GE, Lamborn KR, Chang SM, Prados MD, Berger MS. Volume of residual disease as a predictor of outcome in adult patients with recurrent supratentorial glioblastomas multiforme who are undergoing chemotherapy. J Neurosurg. 2004;100(1):41–6. Epub 2004/01/28.
68. Lanzetta G, Minniti G. Treatment of glioblastoma in elderly patients: an overview of current treatments and future perspective. Tumori. 2010;96(5):650–8. Epub 2011/02/10.
69. Stummer W, van den Bent MJ, Westphal M. Cytoreductive surgery of glioblastoma as the key to successful adjuvant therapies: new arguments in an old discussion. Acta Neurochir (Wien). 2011;153(6): 1211–8. Epub 2011/04/12.
70. Oszvald A, Guresir E, Setzer M, Vatter H, Senft C, Seifert V. Glioblastoma therapy in the elderly and the importance of the extent of resection regardless of age. J Neurosurg. 2012;116(2):357–64.
71. Pietila TA, Stendel R, Hassler WE, Heimberger C, Ramsbacher J, Brock M. Brain tumor surgery in geriatric patients: a critical analysis in 44 patients over 80 years. Surg Neurol. 1999;52(3):259–63. discussion 63–4. Epub 1999/10/08.
72. Kelly PJ, Hunt C. The limited value of cytoreductive surgery in elderly patients with malignant gliomas. Neurosurgery. 1994;34(1):62–6. discussion 6–7. Epub 1994/01/01.
73. Imperato JP, Paleologos NA, Vick NA. Effects of treatment on long-term survivors with malignant astrocytomas. Ann Neurol. 1990;28(6):818–22. Epub 1990/12/01.
74. Kondziolka D, Flickinger JC, Bissonette DJ, Bozik M, Lunsford LD. Survival benefit of stereotactic radiosurgery for patients with malignant glial neoplasms. Neurosurgery. 1997;41(4):776–83. discussion 83–5. Epub 1997/10/08.
75. Meckling S, Dold O, Forsyth PA, Brasher P, Hagen NA. Malignant supratentorial glioma in the elderly: is radiotherapy useful? Neurology. 1996;47(4):901–5. Epub 1996/10/01.
76. Vuorinen V, Hinkka S, Farkkila M, Jaaskelainen J. Debulking or biopsy of malignant glioma in elderly people—a randomised study. Acta Neurochir. 2003;145(1):5–10. Epub 2003/01/25.
77. Zachenhofer I, Maier R, Eiter H, Muxel B, Cejna M, DeVries A. Overall survival and extent of surgery in adult versus elderly glioblastoma patients: a population based retrospective study. Wien Klin Wochenschr. 2011;123(11–12):364–8. Epub 2011/05/19.
78. Mohan DS, Suh JH, Phan JL, Kupelian PA, Cohen BH, Barnett GH. Outcome in elderly patients

undergoing definitive surgery and radiation therapy for supratentorial glioblastoma multiforme at a tertiary care institution. Int J Radiat Oncol Biol Phys. 1998;42(5):981–7. Epub 1998/12/30.

79. Chaichana KL, Garzon-Muvdi T, Parker S, Weingart JD, Olivi A, Bennett R. Supratentorial glioblastoma multiforme: the role of surgical resection versus biopsy among older patients. Ann Surg Oncol. 2011;18(1):239–45. Epub 2010/08/11.
80. Ewelt C, Goeppert M, Rapp M, Steiger HJ, Stummer W, Sabel M. Glioblastoma multiforme of the elderly: the prognostic effect of resection on survival. J Neurooncol. 2011;103(3):611–8. Epub 2010/10/19.
81. Pierga JY, Hoang-Xuan K, Feuvret L, Simon JM, Cornu P, Baillet F. Treatment of malignant gliomas in the elderly. J Neurooncol. 1999;43(2):187–93. Epub 1999/10/26.
82. Chang EL, Yi W, Allen PK, Levin VA, Sawaya RE, Maor MH. Hypofractionated radiotherapy for elderly or younger low-performance status glioblastoma patients: outcome and prognostic factors. Int J Radiat Oncol Biol Phys. 2003;56(2):519–28. Epub 2003/05/10.
83. Hoegler DB, Davey P. A prospective study of short course radiotherapy in elderly patients with malignant glioma. J Neurooncol. 1997;33(3):201–4. Epub 1997/07/01.
84. Ampil F, Fowler M, Kim K. Intracranial astrocytoma in elderly patients. J Neurooncol. 1992;12(2):125–30. Epub 1992/02/01.
85. Villa S, Vinolas N, Verger E, Yaya R, Martinez A, Gil M. Efficacy of radiotherapy for malignant gliomas in elderly patients. Int J Radiat Oncol Biol Phys. 1998;42(5):977–80. Epub 1998/12/30.
86. Jeremic B, Shibamoto Y, Grujicic D, Milicic B, Stojanovic M, Nikolic N. Short-course radiotherapy in elderly and frail patients with glioblastoma multiforme. A phase II study. J Neurooncol. 1999;44(1):85–90. Epub 1999/12/03.
87. Brandes AA, Vastola F, Basso U, Berti F, Pinna G, Rotilio A. A prospective study on glioblastoma in the elderly. Cancer. 2003;97(3):657–62. Epub 2003 /01/28.
88. Glantz M, Chamberlain M, Liu Q, Litofsky NS, Recht LD. Temozolomide as an alternative to irradiation for elderly patients with newly diagnosed malignant gliomas. Cancer. 2003;97(9):2262–6. Epub 2003 /04/25.
89. Muacevic A, Kreth FW. Quality-adjusted survival after tumor resection and/or radiation therapy for elderly patients with glioblastoma multiforme. J Neurol. 2003;250(5):561–8. Epub 2003/05/09.
90. Idbaih A, Taillibert S, Simon JM, Psimaras D, Schneble HM, Lopez S. Short course of radiation therapy in elderly patients with glioblastoma multiforme. Cancer Radiother. 2008;12(8):788–92. Epub 2008/12/03.
91. Ford JM, Stenning SP, Boote DJ, Counsell R, Falk SJ, Flavin A. A short fractionation radiotherapy treatment for poor prognosis patients with high grade glioma. Clin Oncol (R Coll Radiol). 1997;9(1):20–4. Epub 1997/01/01.
92. Hulshof MC, Schimmel EC, Andries Bosch D, Gonzalez Gonzalez D. Hypofractionation in glioblastoma multiforme. Radiother Oncol. 2000;54(2): 143–8. Epub 2000/03/04.
93. Lutterbach J, Ostertag C. What is the appropriate radiotherapy protocol for older patients with newly diagnosed glioblastoma? J Clin Oncol. 2005; 23(12):2869–70. Epub 2005/04/20.
94. Bauman GS, Gaspar LE, Fisher BJ, Halperin EC, Macdonald DR, Cairncross JG. A prospective study of short-course radiotherapy in poor prognosis glioblastoma multiforme. Int J Radiat Oncol Biol Phys. 1994;29(4):835–9. Epub 1994/07/01.
95. McAleese JJ, Stenning SP, Ashley S, Traish D, Hines F, Sardell S. Hypofractionated radiotherapy for poor prognosis malignant glioma: matched pair survival analysis with MRC controls. Radiother Oncol. 2003;67(2):177–82. Epub 2003/06/19.
96. Roa W, Brasher PM, Bauman G, Anthes M, Bruera E, Chan A. Abbreviated course of radiation therapy in older patients with glioblastoma multiforme: a prospective randomized clinical trial. J Clin Oncol. 2004;22(9):1583–8. Epub 2004/03/31.
97. Kimple RJ, Grabowski S, Papez M, Collichio F, Ewend MG, Morris DE. Concurrent temozolomide and radiation, a reasonable option for elderly patients with glioblastoma multiforme? Am J Clin Oncol. 2010;33(3):265–70. Epub 2009/10/14.
98. Patwardhan RV, Shorter C, Willis BK, Reddy P, Smith D, Caldito GC. Survival trends in elderly patients with glioblastoma multiforme: resective surgery, radiation, and chemotherapy. Surg Neurol. 2004;62(3):207–13. discussion 14–5. Epub 2004/09/01.
99. Scott JG, Suh JH, Elson P, Barnett GH, Vogelbaum MA, Peereboom DM. Aggressive treatment is appropriate for glioblastoma multiforme patients 70 years old or older: a retrospective review of 206 cases. Neuro Oncol. 36;13(4):428–36. Epub 2011/03/03.
100. Combs SE, Wagner J, Bischof M, Welzel T, Wagner F, Debus J. Postoperative treatment of primary glioblastoma multiforme with radiation and concomitant temozolomide in elderly patients. Int J Radiat Oncol Biol Phys. 2008;70(4):987–92. Epub 2007/10/31.
101. Minniti G, De Sanctis V, Muni R, Filippone F, Bozzao A, Valeriani M. Radiotherapy plus concomitant and adjuvant temozolomide for glioblastoma in elderly patients. J Neurooncol. 2008;88(1):97–103.
102. Fiorica F, Berretta M, Colosimo C, Stefanelli A, Ursino S, Zanet E. Glioblastoma in elderly patients: safety and efficacy of adjuvant radiotherapy with concomitant temozolomide. Arch Gerontol Geriatr. 2010;51(1):31–5. Epub 2009/07/25.
103. Gerstein J, Franz K, Steinbach JP, Seifert V, Fraunholz I, Weiss C. Postoperative radiotherapy and concomitant temozolomide for elderly patients with glioblastoma. Radiother Oncol. 2010;97(3): 382–6. Epub 2010/09/21.
104. Brandes AA, Franceschi E. Primary brain tumors in the elderly population. Curr Treat Options Neurol. 2011;13(4):427–35. Epub 2011/04/13.

105. Lichtman SM. Pharmacokinetics and pharmacodynamics in the elderly. Clin Adv Hematol Oncol. 2007;5(3):181–2. Epub 2007/05/24.
106. Brada M, Stenning S, Gabe R, Thompson LC, Levy D, Rampling R. Temozolomide versus procarbazine, lomustine, and vincristine in recurrent high-grade glioma. J Clin Oncol. 2010;28(30):4601–8. Epub 2010/09/22.
107. Baker SD, Grochow LB. Pharmacology of cancer chemotherapy in the older person. Clin Geriatr Med. 1997;13(1):169–83. Epub 1997/02/01.
108. Stummer W, Nestler U, Stockhammer F, Krex D, Kern BC, Mehdorn HM. Favorable outcome in the elderly cohort treated by concomitant temozolomide radiochemotherapy in a multicentric phase II safety study of 5-ALA. J Neurooncol. 2011;103(2):361–70. Epub 2010/10/06.
109. Brandes AA, Franceschi E, Tosoni A, Blatt V, Pession A, Tallini G. MGMT promoter methylation status can predict the incidence and outcome of pseudoprogression after concomitant radiochemotherapy in newly diagnosed glioblastoma patients. J Clin Oncol. 2008;26(13):2192–7. Epub 2008/05/01.
110. Brandsma D, Stalpers L, Taal W, Sminia P, van den Bent MJ. Clinical features, mechanisms, and management of pseudoprogression in malignant gliomas. Lancet Oncol. 2008;9(5):453–61. Epub 2008/05/03.
111. Roldan GB, Scott JN, McIntyre JB, Dharmawardene M, de Robles PA, Magliocco AM. Population-based study of pseudoprogression after chemoradiotherapy in GBM. Can J Neurol Sci. 2009;36(5):617–22. Epub 2009/10/17.
112. Taal W, Brandsma D, de Bruin HG, Bromberg JE, Swaak-Kragten AT, Smitt PA. Incidence of early pseudo-progression in a cohort of malignant glioma patients treated with chemoirradiation with temozolomide. Cancer. 2008;113(2):405–10. Epub 2008/05/20.
113. Sanghera P, Perry J, Sahgal A, Symons S, Aviv R, Morrison M. Pseudoprogression following chemoradiotherapy for glioblastoma multiforme. Can J Neurol Sci. 2010;37(1):36–42. Epub 2010/02/23.
114. Chamberlain MC, Chalmers L. A pilot study of primary temozolomide chemotherapy and deferred radiotherapy in elderly patients with glioblastoma. J Neurooncol. 2007;82(2):207–9. Epub 2006/10/05.
115. Chinot OL, Barrie M, Frauger E, Dufour H, Figarella-Branger D, Palmari J. Phase II study of temozolomide without radiotherapy in newly diagnosed glioblastoma multiforme in an elderly populations. Cancer. 2004;100(10):2208–14. Epub 2004/05/13.
116. Gallego Perez-Larraya J, Ducray F, Chinot O, Catry-Thomas I, Taillandier L, Guillamo JS. Temozolomide in elderly patients with newly diagnosed glioblastoma and poor performance status: an ANOCEF phase II trial. J Clin Oncol. 2011;29(22):3050–5. Epub 2011/06/29.
117. Laigle-Donadey F, Figarella-Branger D, Chinot O, Taillandier L, Cartalat-Carel S, Honnorat J. Up-front temozolomide in elderly patients with glioblastoma. J Neurooncol. 2010;99(1):89–94. Epub 2010/01/09.
118. Malmstrom M, Gronberg BH, Stupp R, et al. Glioblastoma in elderly patients: a randomized controlled trial comparing survival in patients treated with 6-week radiotherapy versus hypofractionated radiotherapy over 2 weeks versus temozolomide single-agent chemotherapy. J Clin Oncol. 2010;28(18S): 2002.
119. Wick W, Engel C, Combs SE, et al. NOA-08 randomized phase III trial of 1-week-on/1week-off temozolomide versus involved-field radiotherapy in elderly (older than age 65) patients with newly diagnosed anaplastic astrocytoma or glioblastoma. J Clin Oncol. 2010;28(18S): 2001.
120. Brem H, Piantadosi S, Burger PC, Walker M, Selker R, Vick NA. Placebo-controlled trial of safety and efficacy of intraoperative controlled delivery by biodegradable polymers of chemotherapy for recurrent gliomas. The Polymer-brain Tumor Treatment Group. Lancet. 1995;345(8956):1008–12. Epub 1995/04/22.
121. Valtonen S, Timonen U, Toivanen P, Kalimo H, Kivipelto L, Heiskanen O. Interstitial chemotherapy with carmustine-loaded polymers for high-grade gliomas: a randomized double-blind study. Neurosurgery. 1997;41(1):44–8. discussion 8–9. Epub 1997/07/01.
122. Westphal M, Hilt DC, Bortey E, Delavault P, Olivares R, Warnke PC. A phase 3 trial of local chemotherapy with biodegradable carmustine (BCNU) wafers (Gliadel wafers) in patients with primary malignant glioma. Neuro Oncol. 2003;5(2):79–88. Epub 2003/04/04.
123. Chaichana KL, Zaidi H, Pendleton C, McGirt MJ, Grossman R, Weingart JD. The efficacy of carmustine wafers for older patients with glioblastoma multiforme: prolonging survival. Neurol Res. 2011;33(7):759–64. Epub 2011/07/16.
124. Muni R, Minniti G, Lanzetta G, Caporello P, Frati A, Enrici MM. Short-term radiotherapy followed by adjuvant chemotherapy in poor-prognosis patients with glioblastoma. Tumori. 2010;96(1):60–4. Epub 2010/05/05.
125. Minniti G, De Sanctis V, Muni R, Rasio D, Lanzetta G, Bozzao A. Hypofractionated radiotherapy followed by adjuvant chemotherapy with temozolomide in elderly patients with glioblastoma. J Neurooncol. 2009;91(1):95–100. Epub 2008/09/02.
126. Gorlia T, van den Bent MJ, Hegi ME, Mirimanoff RO, Weller M, Cairncross JG. Nomograms for predicting survival of patients with newly diagnosed glioblastoma: prognostic factor analysis of EORTC and NCIC trial 26981–22981/CE.3. Lancet Oncol. 2008;9(1):29–38. Epub 2007/12/18.
127. Friedman HS, Prados MD, Wen PY, Mikkelsen T, Schiff D, Abrey LE. Bevacizumab alone and in combination with irinotecan in recurrent glioblastoma. J Clin Oncol. 2009;27(28):4733–40. Epub 2009/09/02.
128. Vredenburgh JJ, Desjardins A, Herndon 2nd JE, Dowell JM, Reardon DA, Quinn JA. Phase II trial of bevacizumab and irinotecan in recurrent malignant glioma. Clin Cancer Res. 2007;13(4):1253–9. Epub 2007/02/24.

129. Nghiemphu PL, Liu W, Lee Y, Than T, Graham C, Lai A. Bevacizumab and chemotherapy for recurrent glioblastoma: a single-institution experience. Neurology. 2009;72(14):1217–22. Epub 2009/04/08.
130. Ramalingam SS, Dahlberg SE, Langer CJ, Gray R, Belani CP, Brahmer JR. Outcomes for elderly, advanced-stage non small-cell lung cancer patients treated with bevacizumab in combination with carboplatin and paclitaxel: analysis of Eastern Cooperative Oncology Group Trial 4599. J Clin Oncol. 2008;26(1):60–5. Epub 2008/01/01.
131. Tutt A, Robson M, Garber JE, Domchek SM, Audeh MW, Weitzel JN. Oral poly(ADP-ribose) polymerase inhibitor olaparib in patients with BRCA1 or BRCA2 mutations and advanced breast cancer: a proof-of-concept trial. Lancet. 2010;376(9737):235–44. Epub 2010/07/09.
132. Audeh MW, Carmichael J, Penson RT, Friedlander M, Powell B, Bell-McGuinn KM. Oral poly (ADP-ribose) polymerase inhibitor olaparib in patients with BRCA1 or BRCA2 mutations and recurrent ovarian cancer: a proof-of-concept trial. Lancet. 2010;376(9737):245–51. Epub 2010/07/09.
133. Chalmers AJ, Lakshman M, Chan N, Bristow RG. Poly(ADP-ribose) polymerase inhibition as a model for synthetic lethality in developing radiation oncology targets. Semin Radiat Oncol. 2010;20(4):274–81. Epub 2010/09/14.
134. Dungey FA, Loser DA, Chalmers AJ. Replication-dependent radiosensitization of human glioma cells by inhibition of poly(ADP-Ribose) polymerase: mechanisms and therapeutic potential. Int J Radiat Oncol Biol Phys. 2008;72(4):1188–97. Epub 2008/10/29.
135. Chalmers AJ. Overcoming resistance of glioblastoma to conventional cytotoxic therapies by the addition of PARP inhibitors. Anticancer Agents Med Chem. 2010;10(7):520–33. Epub 2010/10/01.

12 Brain Tumor Presentation in Children and Young People

David A. Walker and Sophie H. Wilne

Abstract

Brain tumors in children (<18 years) frequently present after protracted delays, complicating initial management and risking life and disability. These experiences cause great anxiety for the patient and their family. The current priority on early cancer diagnosis is the focus of this chapter.

The HeadSmart campaign, launched in 2011, seeks to shorten symptom interval (median) for children from 3 to 1 month (3 into 1). The campaign is ongoing which we hope will join the ranks of successful media campaigns linked to child health, i.e., meningitis and sudden infant death (Back to Sleep).

Keywords

Brain tumor • Diagnostic delay • Symptom interval • Children and young people

Introduction

This chapter sets out to identify whether the specialist neuro-oncology community which is primarily focused upon the delivery of complex care for the brain tumor patient in childhood can influence health systems to promote enhanced public and professional awareness and faster diagnosis of brain tumors in the broader health community. Faster diagnosis of brain tumors intuitively would offer the opportunity to:

- Intervene earlier in the disease's progress and thereby reduce the risk of acquired neurological disability due to tumor-related brain injury prior to, or at the time of, surgery or radiation therapy.
- Reduce the number of initial operations conducted as urgent or emergency procedures due to severe raised intracranial pressure with the associated enhanced mortality and morbidity risks.
- Reduce patients' and families' anxieties about the consequences of avoidable delays in

D.A. Walker, BMedSci, BM, BS, FRCP, FRCPHCH (✉)
Children's Brain Tumour Research Centre,
School Medical Sciences,
University of Nottingham,
Nottingham, NG7 2RD, UK
e-mail: david.walker@nottingham.ac.uk

S.H. Wilne, BA Hons (Cantab), MBBS, M.D.
Department of Paediatric Oncology,
Nottingham Children's Hospital,
Nottingham University Hospitals NHS Trust,
University of Nottingham,
Nottingham, NG7 2UH, UK

C. Watts (ed.), *Emerging Concepts in Neuro-Oncology*,
DOI 10.1007/978-0-85729-458-6_12, © Springer-Verlag London 2013

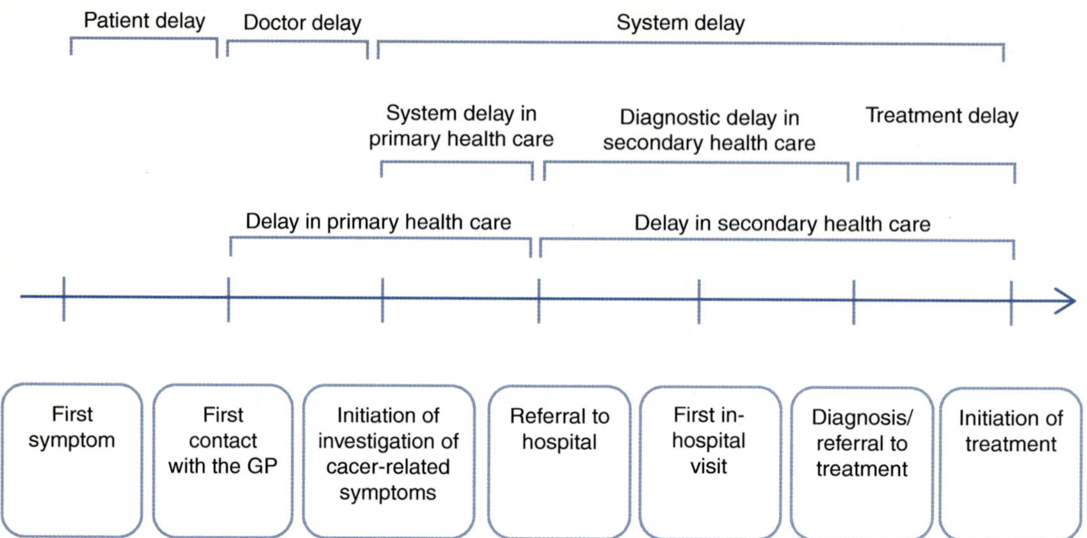

Fig. 12.1 National awareness and early diagnosis initiative pathway

diagnosis, whether they are due to patient and family delays or physician and health system delays.

- Enhance the public's confidence in the health services.

Symptom Interval

Symptom interval (SI) is defined as time from symptom onset identified by the patient and family until the commencement of treatment after diagnosis. Four separate components of delay are recognized (see Fig. 12.1):

1. Patient delay, while the patient and family become aware of symptoms and decide whether to seek advice from their general practitioner (GP)
2. Primary health system delay, linked to GP recognition of symptoms justifying referral and the processes associated with the initiation of investigation or referral
3. Secondary health system delays linked to time taken to make arrangements for assessment, physician/surgical recognition of symptoms, and time taken to initiate and perform diagnostic investigations
4. The pretreatment interval which is the time taken to initiate the first treatment after diagnosis

Strategies, aimed at reducing the overall SI, require each interval component to be addressed specifically. Neal proposed that to reduce the patient delay, attempts should be made to increase awareness of symptoms and clarify how and when to act on these; to reduce primary care delay, attempts should be made to enhance awareness of potential cancer symptoms among primary care clinicians and by changing culture toward one where potential cancer symptoms are considered suspicious, until proven otherwise, while thresholds for referral or requesting GP-initiated investigations should be lowered [1]. To reduce secondary health system delays, revising and implementing new urgent cancer referral guidance based upon primary care-based research into the meaning of symptoms and symptom complexes and promoting and monitoring fast-track referral pathways for diagnostic investigations when the symptoms do not fulfill the urgent referral criteria are proposed as an appropriate strategy. Finally, the most effective diagnostic process, e.g., brain MR or CT scan, should be prioritized routinely

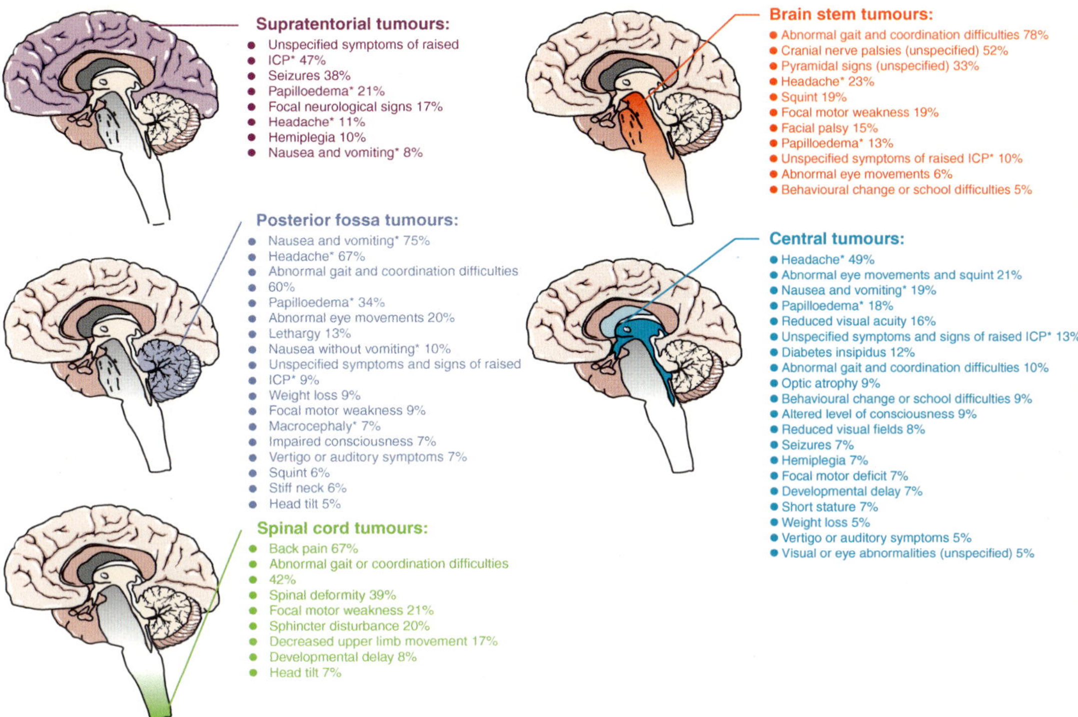

Fig. 12.2 Central nervous system (CNS) tumor presentation (Wilne et al. [3]; used with permission)

rather than advising other less sensitive or less specific tests, i.e., EEG.

Raising Concerns

Having worked in this area, focusing upon the need for more rapid diagnosis in children under 18 years, it has been possible to compare the reporting of SI in the UK in different eras and in different health systems [2]. In the UK, public concern was expressed repeatedly in the media and in parliament, that delays in diagnosis were occurring to the disadvantage of children. Furthermore, there was concern that the health system did not seem responsive to the needs of children with brain tumors and their families. This was summarized in a national manifesto presented in the Houses of Parliament in October 2010 as part of a parliamentary lobbying process (*A manifesto for everyone affected by a brain tumour*, Oct 2010). There are consistent reports from all health systems of delays in diagnosis for individuals where early death or significant disability may have been avoidable. On the other hand, there are other occasions where the rapid onset of symptoms is driven by the aggressive biology or critical location of the tumor where enhanced speed of intervention could not conceivably contribute to better outcomes.

Creating the Evidence-Based Clinical Guidance

With these provisos, we engaged in a process of reviewing existing evidence for variations in prediagnostic symptom interval within the UK and from different international health systems. We identified the symptom clusters and classified them by age group within childhood because of the developmental differences in clinical presentation and examination findings and their interpretation when compared to adult practice. This resulted in a strong evidence base for symptomatology. We went on to use this symptom evidence base in Fig. 12.2 conjunction with data from a multicenter cohort study [2] in a Delphi consensus process with over 150 clinicians experienced in seeing children as well as experienced parents, both groups who had been involved previously with the diagnosis of a child with brain tumor, and, as a result, we developed a series of consensus statements, which were put together as a clinical guideline, meeting national standards for clinical guidelines (http://www.sign.ac.uk/guidelines/fulltext/50/index.html). The end point of the clinical guideline was focused upon obtaining the result of brain scan (MRI or CT with contrast) to diagnose or exclude a brain tumor.

This clinical guideline was reviewed and endorsed by the Royal College of Paediatrics and Child Health and published on their website in 2007 (http://www.rcpch.ac.uk/oncology#pathways) [3]. We were aware that there were existing referral guidelines for cancer published by National Institute for Health and Clinical Excellence (NICE), which provided referral guidance for general practitioners in the UK in 2005 (http://www.nice.org.uk/cg027). These guidelines were produced without a systematic literature review and by a less formal consensus process; furthermore, they used, as their end point, referral to a secondary pediatric service rather than performing a brain scan as the critical step in making a diagnosis. They were published but have not, to date, been the subject of a program of dissemination or audit.

Disseminating the Guidance

Having developed the clinical guidance and obtained professional support from the RCPCH, we went on to seek support from other professional bodies associated with the wide range of specialties seeing children with symptoms in this area. These included emergency medicine, ophthalmologists, optometrists and opticians, general practitioners, community pediatricians, neurologists, and neurosurgeons. We were advised that complex clinical guidance of this type required a specific strategy aimed at its dissemination to the profession. We were also advised that the parental role in selecting children for assessment means that there is a need to raise awareness of these issues in the public.

The challenge was to design and manage a process of guideline dissemination, which was nationwide, directed at the public and the profession, providing graded advice which both raised awareness of the risk of brain tumor but did not create undue anxiety yet provided practical advice on when to perform a scan, when to observe and for how long to do so, and when to reassure.

Policy Strategy

The intention to disseminate new guidance and influence public and professional awareness would be optimized if the messages and their timing were compatible with government priorities within national policy. The National Cancer Plan [4] and its follow-up documents, the Cancer Reform Strategy [5] and Improving Outcomes: A Strategy for Cancer [6], all highlight the need to streamline access to clinical services and enhance awareness so as to speed up diagnosis as a strategy to improve survival rates. The National Patient Safety Agency recently published a thematic review in their National Reporting and Learning Service titled Delayed diagnosis of cancer [7]. This latter document scopes the safety issues, identifies possible solutions, and makes patient safety recommendations as well as recommendations for practitioners and health policy makers. It concludes that to minimize delays in diagnosis, it is necessary to:

- Have accessible a diagnostic tool for use in primary care, adapted for clinical guidelines.
- Identify, review, and disseminate current good practice in the processes of ordering, managing, and tracking tests and test results.
- Review and develop methods for empowering patients on a cancer diagnostic pathway.
- Develop a model for stronger leadership and improved safety reporting and learning,

including significant event audit at a local and national level.

- Improve routine monitoring of delayed diagnosis.

It also highlights the need to maximize survival rates by tackling service deficiencies across the whole cancer patient journey. One of the key commitments of the Cancer Reform Strategy in England was to establish a National Awareness and Early Diagnosis Initiative (NAEDI). The National Awareness and Early Diagnosis Initiative (NAEDI) was launched in November 2008 by CRUK in collaboration with the Department of Health and seeks to generate research into this aspect of care (http://info.cancerresearchuk.org/spotcancerearly/naedi/AboutNAEDI/) with a view to test the hypothesis that "delays lead to patients being diagnosed with more advanced disease and thus experiencing poor 1-year and 5-year survival rates, resulting in deaths that could potentially have been avoided" [8]. The Cancer Reform Strategy recognized that excellent progress had been made on early detection of cancer through screening but also that more needs to be done to promote early diagnosis in the large majority of patients who present with symptoms.

Members of the NAEDI steering group have identified multiple strands of evidence linking the poor cancer survival rates observed in the United Kingdom in the EUROCARE studies to advanced stage at diagnosis and to delays occurring between the onset of symptoms and the start of treatment. However, the evidence base is complex and is still incomplete [9, 10].

The NAEDI Pathway

To assist thinking about the issues related to late diagnosis of cancer, members of the NAEDI steering group have adopted a provisional "NAEDI pathway" (Fig. 12.1). This should provide a framework for testing various hypotheses regarding late diagnosis and its impact. The first step in the pathway proposes that low awareness of the signs and symptoms of cancer among the public in general or in specific subgroups, combined with negative beliefs about cancer, will lead to late presentation to primary care services and to low uptake of cancer-screening services. In addition to this, there may be perceived or actual barriers to accessing primary care services. Ultimately, delayed presentation by patients to primary care services may result in emergency presentations to hospital.

The second step in the pathway involves delays occurring within primary care. These may occur for a variety of reasons, including failure to consider cancer as a possible diagnosis and having inadequate access to diagnostic tests to confirm or exclude cancer as the underlying cause of a patient's symptoms. The difficulties that general practitioners face in this regard should not be underestimated. In England, an average GP will see seven or eight new cases of cancer (excluding non-melanoma skin cancer) each year but will see hundreds, or possibly thousands, of patients with symptoms that could possibly be due to cancer. When should the GP reassure, observe, request investigations, or refer to specialist services?

Delays following referral to specialist services have been well documented in the United Kingdom, with major efforts being made to streamline services to achieve defined waiting time targets. However, relatively little work has been undertaken to measure the relative contributions of patient delay, doctor delay, and system delay to overall delay for different cancer sites in this country. Studies of this type have, however, been undertaken in Denmark, another country with survival rates below the European average. A paper in this supplement summarizes the findings from Denmark [11].

The key hypothesis underpinning NAEDI is that delays lead to patients being diagnosed with more advanced disease and thus experiencing poor 1-year and 5-year survival rates, resulting in deaths that could potentially have been avoided. This could potentially account for at least some of the differences in outcomes observed within the United Kingdom between rich and poor [8].

Finally, the recent publication of Getting it Right for Children and Young People by Sir Ian Kennedy highlights the need to overcome cultural barriers linked to children's requirements in the NHS [12]. He identifies a disconnect between the biological rate of development of brain, which is critical to future health, development and functioning of these children later in their lives, and health spending in the early years. The report identifies the gap between UK health service performance in children's key outcomes and those observed in mainland Europe, with the UK underperforming. It also identifies the need to enhance the priority and joined-up planning for children's health services and policy.

HeadSmart

This is the name of the awareness campaign associated with the dissemination program for the RCPCH brain tumor referral guidelines.

The need to disseminate the messages about symptomatology and selection of children for scanning, observation, or reassurance was considered central to the plan to implement the new referral guidance. We envisaged a model consultation process where the parent and child or young person is empowered by the messages contained in a symptom card to seek medical advice sooner and more positively within the consultation. Similarly, the program aimed at the general practitioner or pediatrician is directed at enhancing their awareness of the symptoms and signs as well as access to the decision support website (www.headsmart.org.uk) focusing upon the need to select patients for referral for scanning, clinical review, or reassurance.

The messages and guidance were deliberately designed to be both specific in the selection of children for scanning while also as clear in selecting children for subsequent review or reassurance. The health messages are summarized on the credit card-sized information card (Fig. 12.3) and further advertised with posters and a short version guideline document and poster, meeting the AGREE criteria. The associated website contains all the information to support the campaign presented for professionals and for the public. It is structured to offer quick advice and strongly linked to relevant additional health message websites for symptoms and signs described, e.g., headache, vomiting, abnormal eye movements, or squint. The website was also designed to be a source for distribution of campaign messages and materials so that downloads of symptom checklists, posters, and campaign information could be shared as part of the dissemination campaign.

Finally, an interactive end's CAL package has been incorporated into the HeadSmart website, and an education module has been developed to offer relevant professional training using scenarios, interactive training and training materials for general practitioners, and general and emergency pediatricians.

Marketing

The public and professional focus of the campaign, together with the intention of creating impact across the UK, justified the need for a marketing campaign. Our target was to reduce the median symptom interval from 13 to 5 weeks. This target was selected as the critical performance measure using Quality Improvement methodology (http://www.health.org.uk/areas-of-work/topics/quality-improvement/quality-improvement/). We engaged marketing help to assist with strategy, design, and launch. A critical element of this was the design phase for the materials delivering the message. The symptom checklists were reduced to credit card size (Fig. 12.3), which have drawn widespread appreciation for their handiness and succinct messaging. The interface with the media was organized in advance with press releases planned and a launch day organized to maximize exposure in TV, radio, and printed media. An estimated 14 million people were contacted on this first launch day. There was a strong emphasis on partnership between professionals and the public in the content. Subsequently, plans have been made to sustain and evaluate the impact. Early feedback has identified symptomatology of which the public and profession were relatively unaware; this has identified messages about symptoms that need further emphasis. The professional survey identified

Fig. 12.3 HeadSmart symptom cards (Used with permission). www.headsmart.org.uk

that the actual current median symptom interval coincided exactly with the professional perception of the likely symptom interval (Fig. 12.4), highlighting the importance of perception in determining outcomes from complex health systems.

At the time of writing, we are considering the feasibility of trying to change the professionals' perception of the "accepted" symptom interval from 3 months to our target of 1 month in the belief that if this is changed, the system will be accelerated as a consequence. Our slogan will be "3 into 1."

Evaluation

The success of the campaign will be assessed by the change in the symptom interval, the level of awareness of the campaign, and its messages in public and professional surveys, as well as the effect on 1-year survival rates, disability rates and health service diagnostics, and other measures of clinical service usage.

The campaign launch was successful in establishing awareness of the campaign in 11 % of the UK population and 70 % of the professional respondents within 4 months of the launch as judged by public and professional surveys. Targets for public and professional awareness of the campaign have been set at 70 % and 100 %, respectively. At this early stage, there is evidence of enhanced professional awareness of the less well-recognized symptoms (e.g., head tilt, altered puberty). Symptom interval data was collected by a national network of clinical champions working in the children's cancer treatment centers across the UK. Preliminary data from this network, compared to previously published multicenter or regional studies, suggests an early trend to a shortening SI.

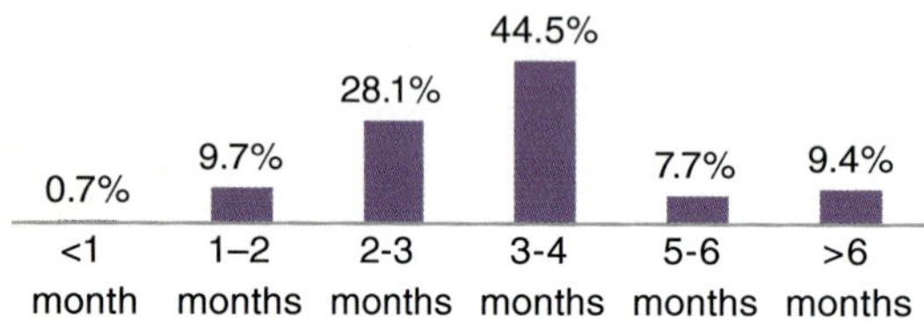

Fig. 12.4 Professional prediction of symptom interval for childhood CNS tumor

Route to Diagnosis and Multiple Referrals

Data from 294 children (median age 6.6 year, range 0.11–17.7) is available spanning launch of the campaign. The median SI is 8.0 weeks (0–398 weeks). The median symptom onset to consultation with a healthcare professional interval is 2.4 weeks (0–123 weeks), and the median consultation to diagnosis interval is 2.6 weeks (0–398 weeks). Imaging that identified the tumor took place as an outpatient in 34.1 %, an inpatient in 40.1 %, and from the emergency department in 23.2 %. 2.9 % of children were referred via a "2-week wait" cancer referral, and 2.2 % were found as incidental findings following scans for sinuses, after head injury and medical screening. Tumor diagnoses were representative of population registries. The median symptom interval prior to and after the launch of the HeadSmart campaign is shown in Fig. 12.5. These changes reach statistical significance and are suggestive that the HeadSmart Campaign is associated with an early improvement.

Conclusions

It is too soon to say whether a tertiary specialist's initiative to change a combined primary and secondary care system of referral in a rare childhood disease can be successful. Any such judgment needs careful assessment using unbiased

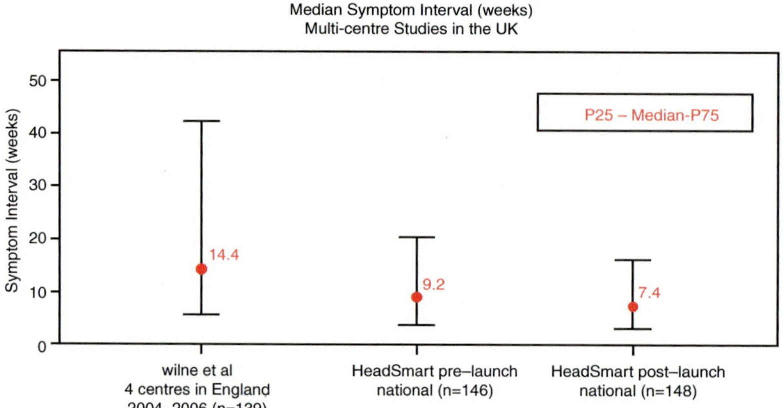

Fig. 12.5 Symptom interval data. Median symptom interval (weeks); Multicenter studies in the United Kingdom. Significant difference across three cohorts (Kruskal-Wallis test, $p<0.01$); no significant difference between pre- and postlaunch (Mann–Whitney U test, $p=0.136$)

methods of measurement of health service usage and outcomes. At this stage, we have prioritized public and professional awareness of the disease symptomatology, created a system of contemporaneous monitoring of symptom interval as a proxy for health system performance and reporting, and identified the need to change public and professional perception of what is an acceptable and expected symptom interval by setting a target (3 into 1).

By focusing upon these facets of the problem, we aim to influence the health system at large. Whether this will serve as a model for other serious childhood conditions where slow recognition can have adverse outcomes remains to be seen, e.g., speed of presentation of newly diagnosed diabetes in childhood. The Meningitis Trust's "Glass Test" and "The Back to Sleep" campaigns were highly successful in changing clinical practice and improving outcomes in paediatric practice; we hope that "HeadSmart: be brain tumour aware" can join their ranks.

References

1. Neal R. Do diagnostic delays in cancer matter. Br J Cancer. 2009;101:S9–12.
2. Wilne S, Collier J, et al. Presentation of childhood CNS tumours: a systematic review and meta-analysis. Lancet Oncol. 2007;8:685–95.
3. Wilne S, Koller K, et al. The diagnosis of brain tumours in children: a guideline to assist healthcare professionals in the assessment of children who may have a brain tumour. Arch Dis Child. 2010;95: 534–9.
4. The National Cancer Plan. Department of Health. London: HMSO; 2000.
5. Cancer Reform Strategy. Department of Health. London: NHS; 2007.
6. Improving Outcomes: a Strategy for Cancer. Department of Health. London: Macmillan; 2011.
7. National Reporting & Learning Service. Delayed diagnosis of cancer: thematic review. London: National Patient Safety Agency; 2011.
8. Richards M. The National Awareness and Early Diagnosis Initiative in England: assembling the evidence. Br J Cancer. 2009;101:S1–4.
9. Sant M, Allemani C, et al. Stage at diagnosis is a key explanation of differences in breast cancer survival across Europe. Int J Cancer. 2003;106:416–22.
10. Berrino F, De A, et al. Survival for eight major cancers and all cancers combined for European adults diagnosed in 1995–1999: results of the EUROCARE 4 study. Lancet Oncol. 2007;8:773–83.
11. Olesen F, Hansen R, et al. Delay in diagnosis: the experience in Denmark. Br J Cancer. 2009;101:S5–8.
12. Kennedy I. Getting it right for children and young people. London: Department of Health; 2010.

13 How Can We Improve Clinical Trial Recruitment in Neuro-Oncology?

Sarah J. Jefferies

Abstract

Current surgical and oncological treatments for patients presenting with tumors of the central nervous system (CNS) only achieve long-term cure in a minority of patients. Clinical trials are a well-established mechanism to assess new approaches in comparison to our current clinical standard treatment. Optimizing entry into clinical trials can be influenced by a number of factors. This chapter explores the challenges that occur in optimizing patient entry into neuro-oncology trials. Clinical trial participation may be affected by patient factors, clinician factors, study protocol issues, access difficulties, and the research setting. Patients with CNS tumors also may present with neurological difficulties, physical and cognitive, which make them a vulnerable population and thus crucial that the highest ethical standards are maintained in clinical trial design. Understanding the obstacles to clinical recruitment is crucial in optimizing future clinical trial designs for neuro-oncology patients.

Keywords

Central nervous system • High-grade glioma • Randomized controlled trials • Clinical equipoise • Randomization process • Recruitment • Ethical • National Cancer Research Network

Clinical trials are a methodology by which new treatments in medicine can be robustly assessed. The overall aim of most clinical trials, in the management of cancer, is to refine and improve cancer treatments. For the purpose of this chapter, the discussion will consider primary tumors affecting the central nervous system (CNS). Although they are relatively rare, primary CNS tumors contribute significantly to disability and disease-attributable death. This is demonstrated in the high number of years of life lost for patients with CNS tumors [1]. Despite clinical research efforts in neuro-oncology, survival for the majority of patients with malignant glioma has not significantly changed over the last 20 years. Patients who present with glioblastoma multiforme (GBM) have a median survival

S.J. Jefferies, MBBS, B.Sc., MRCP, FRCR, Ph.D.
Oncology Department, Addenbrooke's Foundation Trust, Hills Rd, Cambridge CB2 0QQ, UK
e-mail: sarah.jefferies@addenbrookes.nhs.uk

C. Watts (ed.), *Emerging Concepts in Neuro-Oncology*,
DOI 10.1007/978-0-85729-458-6_13, © Springer-Verlag London 2013

of 10–12 months, whereas for patients with anaplastic astrocytoma, the median survival is 3.5 years [2, 3]. The introduction of concomitant and adjuvant temozolomide with radiotherapy has improved survival for a small proportion of patients with a diagnosis of GBM [4].

Many new approaches to treatment for tumors of the CNS are currently being explored in clinical trials. As well as having clinical trial availability, it is also essential that the right studies are undertaken and efforts are made to optimize both access and recruitment. This chapter will explore the potential barriers to clinical trial participation and ways in which the situation may be improved in the future for patients with CNS tumors.

Research Setting

There is a wide variation in the ability to recruit patients into clinical trials dependent on the underlying diagnosis. Very high rates of participation are often demonstrated in pediatric oncology [5]. This contrasts sharply with the situation in the adult population. In 2000–2001, only two local research networks in the United Kingdom (UK) recruited a total of 7.5 % of patients into clinical trials [6]. The development of the National Cancer Research Network (NCRN) infrastructure in the UK has demonstrated an increase in recruitment to 12 % of incident cancer patients into clinical trials each year (32,000 patients) as reported in 2007–2008. As well as increasing recruitment overall, study delivery was also improved with more clinical trials meeting the recruitment target (74 % compared with 39 % before NCRN was established).

Timely and complete accrual to clinical trials of cancer treatments is crucial if treatments are to be improved, but it must be ensured that recruitment is conducted in a way that is appropriate for patients and clinicians and is ethically sound. High recruitment is not, therefore, the ultimate goal, but rather it is "optimal recruitment." This must take into account that patients are able to make informed decisions and decide themselves whether to enter the trial or not [7]. It is important to factor accrual rates into the design of clinical trials as poor accrual results in clinical trials failing to complete on time and/or recruit sufficient patients for a meaningful result.

In order to effectively address a research question for primary CNS tumors, due to the relatively small incidence, it is often necessary to undertake trial design which will require a multicenter approach, and in some instances, this may need an international collaboration. This can be a time-consuming process requiring a significant amount of administration within each hospital trust and can involve multiple cancer research networks and international research agreements. Despite this recruitment strategy, the numbers ultimately recruited per center in the UK may still be very small. As consequence of the small number of patients with CNS tumors considered for clinical trials, many clinicians do not have access to a dedicated research nurse support. This compounds the difficulty in recruiting patients into clinical trials.

Clinical Equipoise

The randomized controlled trial (RCT) has been a key technical advance in the development of evidence-based medicine. RCTs do raise some complex ethical issues. RCTs incorporate methods to maximize the validity of studies; concurrent control or comparison groups, randomization and stratification reducing potential differences between the study arms, and placebo and blinding may be utilized to reduce the likelihood of bias [8].

The ethical acceptability of RCTs which utilize randomization to assign treatment hinges on the concept of clinical equipoise. This is the premise that there is uncertainty about the relative therapeutic merits of treatments [9, 10]. Clinical equipoise assumes reasonable professional disagreement about the relative merits of each of the treatments proposed [11, 12]. RCTs highlight the tensions between the clinician's therapeutic obligations and the investigator's obligation to the individuals who will benefit from the information derived from the study [13]. In fact this dilemma also applies to nonrandomized studies, as it is present in all studies that pose risks justified by knowledge gains rather than benefits to individuals. However, clinical equipoise has been broadly accepted as the ethical criteria for clinical trials and has been endorsed by international research ethics guidelines [14].

Clinician Factors

Certain elements of the way that recruitment is conducted in cancer trials have been proposed to be associated with high accrual [15]. An important issue is that the research question should be considered important by both the clinician and patient. It has been shown that recruitment is increased if communication between the clinician and patient is of high quality and performed by an experienced clinician [15]. It is important to ensure that the information about the trial is delivered in a personal, tailored, and timely approach.

It has been proposed that the focus of the consultation in which a clinical trial is discussed should be made clear in advance to the patient [7]. If a patient believes that a consultation's purpose is to discuss specific results or treatment, they may perceive the discussion of a trial as an unwelcome deviation from their treatment plan. Throughout recruitment, patients should feel that the clinician gives priority to their care over the scientific imperative of the trial and that if trial continuation brought significant physical or emotional cost, the clinician would withdraw the patient from the trial.

Trial recruitment processes vary greatly in terms of timing, cancer types, and intervention with each trial having its own specific procedural and ethical challenges. Prior to obtaining the patients consent to participate in a clinical trial, they should be aware of the following: (1) that the trial is different from routine clinical practice, (2) how the treatment is allocated, and (3) of the risks, benefits, and the right to withdraw from the trial [15].

When explaining clinical trials, doctors often stress their dual role and the importance of integration of good clinical care with a clinical trial [15]. Some researchers have called for greater integration of clinical care and research in order to improve accrual [16], but others have voiced concerns about the ethics of doctors recruiting patients for whom they have clinical responsibility [16–20]. Because the motivation of the investigator (to answer an empirical question) and the clinician (to manage the clinical care of the patient) is different, concerns have revolved around the potential for clinicians to unduly influence patient consent when recruiting their own patients [16–20]. It has therefore been suggested that doctors should declare their dual roles [16–20]. The advent of the neuro-oncology multidisciplinary team meeting can also help identify patients who may be eligible for clinical trials which represents a collective decision-making process.

To date there is little convincing evidence that an individual patient when treated in the context of a clinical trial within a hospital trust will have better health outcomes than a similar patient receiving the same treatment as standard care within a different hospital trust. However, there is stronger support in the literature for the hypothesis that hospital trusts which are research active rather than those which are not have better health outcomes [21–24]. This may be due to research-active environments introducing new treatments earlier than their counterparts.

Randomization Process

Most clinical trials assessing cancer treatments involve a delayed randomization process, which allows time for the tailored and timely dissemination of trial information including the rationale and underlying processes. This assists patients in distinguishing between standard clinical care and the scientific rationale for the clinical trial. Previous research has shown that people report the advancement of science as being an important driver in agreeing to take part in trials [25–27]. Clinical trials that address areas where there is a greater uncertainty regarding the outcomes of the different treatment arms often prove to be more challenging in terms of recruitment. This is particularly the case if there is a non intervention arm in the clinical trial [28–30].

Clinical Trial Protocols

For many patients an important concept to understand is that there is often little personal benefit from clinical trial participation [27, 28]. The protocols may discourage clinical trial participation because of an insufficiently interesting question for the patient. Also the toxicity associated with the study treatment may be anticipated to be more significant than standard treatment. The complexity and difficulty of the protocol can make the study significantly more arduous than standard treatment [31–35] which may reduce participation. Some clinical trials require specialist interventions that can only be administered at specific institutions and may therefore require traveling long distances [36]. Other general barriers include the financial aspect of conducting clinical trials and the impact on the National Health System [37–42]. Negative publicity about conduct of trials also fuels mistrust in the clinical research enterprise and can have an adverse impact on clinical trial participation.

An illustration of the complex nature of problems with recruitment into a clinical trial was demonstrated in the National Surgical Adjuvant Project for Breast and Bowel Cancers clinical trial which aimed to compare segmental mastectomy and postoperative radiotherapy or segmental mastectomy alone, with total mastectomy. Due to low rates of accrual, a questionnaire was sent to all the principal investigators to assess why eligible patients were not entered into the trial. The findings were that the greatest concern and barrier to recruitment was from clinicians who had concerns for the doctor–patient relationship created by the randomization process. There were also difficulties with informed consent and discussions of uncertainties, perceived conflict between the roles of scientist and clinician, and practical difficulties in following procedures required on the study [43].

Neurological Factors

Particular challenges arise for patients who have a neuro-oncology diagnosis [44–47]. Patients may have neurological deficits that may present a physical barrier to entry into the study. Neurocognitive difficulties, as a direct consequence of the primary tumor location or following surgery, may also affect the ability of a patient to be considered for a clinical trial. This is compounded by the diagnosis of a glioma rendering patients unable to drive for a significant period of time which can make the attendance for visits for the clinical study difficult. The time and distance may be even greater if the trials are run from specialist neuro-oncology centers. Traveling costs can become very significant over the course of a clinical trial, and even though they are factored into trial design, often the remuneration does not cover the entire traveling costs.

Central Nervous System Tumor Pathology

There is a limited evidence base for the factors that predict clinical trial participation within neuro-oncology. It has been shown that patients with GBM are more likely to participate in trials than patients with less aggressive tumors. Patients with GBM currently still have a relatively poor prognosis with a median survival of 10–12 months which may influence both clinicians and patients to favor enrollment in a clinical trial [47]. This is consistent with findings from Verheggen et al., who reported that the perception of a threat was a factor that correlated with trial participation [31]. Increased entry into clinical trials has also been correlated with increasing severity of illness [48].

Radiology of CNS Tumors

Many research studies require an increase in frequency of standard magnetic resonance imaging (MRI) and may also require certain specialist MRI sequences. The former can lead to problems in funding studies as the imaging costs can be very expensive and may not be formally funded within the study. The latter may limit study sites to those centers that can provide specialist neuroradiology imaging. As well as performing specific sequences, it has become clear that interpretation of imaging requires specific neuroradiology input particularly in the assessment of response. Antiangiogenic agents and chemoradiation with temozolomide in the treatment of GBM have demonstrated MRI changes that include pseudoresponse and pseudoprogression. Recent image response criteria have been updated to include both of these phenomena [49]. As well as MRI, there is a need to prospectively incorporate physiological imaging such as diffusion-weighted imaging, MR spectroscopy, and cerebro vascular perfusion imaging into the clinical trial design.

Patient Age

Malignant glioma is more prevalent in the older population, and the incidence in the elderly appears to be increasing [50]. Age is a significant factor for clinical trial participation, with younger patients more likely to enroll. This is consistent with what has been shown repeatedly in the literature for other disease sites [51, 52]. Despite the increasing number of older people in the population and the greater frequency of malignant disease in this age group, there is substantial underrepresentation of patients aged over 65 in cancer treatment trials [53, 54]. In a survey of American oncologists, 50 % felt that some patients were not suitable for clinical trials based on age alone [55].

Age is particularly relevant among patients with glioblastoma as increasing age is an adverse prognostic factor for survival [53, 54]. Trials have been designed with regard to age-specific therapies for older patients with glioma [56, 57]. It is a common misunderstanding that older patients may not be able to tolerate or benefit from treatment in clinical trials. Giovanazzi-Bannon et al. showed that there was no significant difference between elderly patients and younger patients for several clinical trial end points, such as treatment delays and toxicities [58]. Studies have been designed to specifically address the molecular, cytogenetic, and biologic factors that may be correlated to age and that may influence outcome and may help target specific agents to the older population. Evaluating how well older patients tolerate treatments also may be valuable in understanding the likely impact of clinical trial results to the general population. Thus, there is a definite need to improve accrual rates for older patients with GBM as well as to develop more protocols designed specifically for older patients.

Access to Clinical Trials

Further areas that may affect patient participation in clinical trials include patient education and access to health care. Ideally, a prospective evaluation of the potential barriers to clinical trial participation is needed. Information may become available via clinical information systems entering patient's eligibility at neuro-oncology multidisciplinary meetings and subsequent audit of entry into clinical trials.

Strategies to improve recruitment to clinical trials have been described in the literature. Positive publicity through support groups or in the lay press can help educate patients and families about the potential benefits of clinical trial participation. Expanding neuro-oncology clinical trials so that access is available throughout a cancer network is one strategy. Recognition of the time, cost, and training necessary to recruit patients into clinical trials is also important in addressing some barriers to clinical trial participation.

Targeted Therapies

There have been many studies in CNS tumors evaluating the role of single-agent targeted therapy which have demonstrated minimal efficacy [59]. The rationale underlying these studies was the knowledge of potential targets existing in glioma and the ease and tolerability of administration of the agents. This ignored the key aspects of drug distribution, mechanism of action, and biological activity before proceeding with phase II studies.

The optimum rationale would be an integrated phase 0/I/II correlative study protocol for new agents. Phase 0 trials generally evaluate pharmacokinetic and pharmacodynamic profiles of new drugs in a small group of patients prior to initiating classic phase I tolerability studies [60]. These early studies can also examine the biological effects of targeted agents [61]. For phase II studies, tissue analysis for the presence of the target and other relevant biological markers would allow subsequent identification of the patients most likely to benefit. This approach is not unique to neuro-oncology; other researchers have reached similar conclusions with regard to more stringent studies of therapeutic agents in smaller populations of patients prior to initiating standard phase I studies [61–65].

The most challenging issue for CNS tumors is the safe acquisition of tumor samples at diagnosis. In addition, because of the increased risk of morbidity associated with serial biopsies, the development of surrogate markers of activity is more critical in the design of trials. Validated imaging markers would be particularly beneficial. MR spectroscopy imaging, diffusion- and perfusion-weighted imaging, and positron emission tomography (PET) with novel imaging probes are techniques that may be incorporated into future trials

Many patients that have a CNS tumor will require treatment with anticonvulsant medication. Hepatic enzyme-inducing antiepileptic drugs (EIAEDs) can alter the pharmacokinetic parameters and toxicity profiles of therapeutic agents. This has been shown for both cytotoxic and molecularly targeted agents [66, 67]. Trial designs in neuro-oncology have since evolved to test the agent in patients not taking EIAEDs first to determine efficacy. Only if there is evidence of efficacy in this patient population is there a reason to perform phase I testing in patients on EIAEDs to establish the appropriate dose for phase II evaluation. This strategy optimizes both patient and financial resources and expedites the assessment of an agent's clinical utility. EIAEDs can prove to be a barrier to entry to clinical studies as when patients are established on these preparations, many weeks may be needed for conversion to non-EIAEDs and allow a washout period.

Current Developments

Overall, there is much work to be done in the field of neuro-oncology to improve patient participation in clinical trials. It is the responsibility of all principal investigators to examine their programs and try to address some of these barriers to recruitment. An effort to improve accrual rates will help determine the value of new therapies as expeditiously as possible. Recently, there has been a significant increase in potential treatments for CNS tumors. This has led to a rapid rise in the number of clinical trials that are available. In the context of the challenges of recruitment and enrollment of patients, this has led to a new problem of having competing trials for a paucity of patients who may be eligible for them all. This creates an ethical and practical dilemma.

The options for management of this problem include offering patients information on all ongoing studies allowing them to make an informed decision about which trial to undertake. An alternative strategy is for consideration at the time of the neuro-oncology multidisciplinary team meeting as to which trial may be most suitable for a particular patient. Both of these options introduce the risk selection bias of patients for particular studies. Increasingly, eligibility may be dictated by the access to imaging, surgical procedure undertaken, and/or molecular analysis of the tumor.

CNS tumor patients comprise a vulnerable population, and it remains incumbent on neuro-oncology teams to maintain the highest ethical standards when addressing the issue of clinical trials. Changes in clinical trial design may be required to mitigate the conflicts created by competition for patients.

Significant effort has been expended on the part of patients and clinical trial personnel to design and conduct early clinical trials in neuro-oncology. There will need to be close and ongoing interaction of basic scientists and clinicians to translate information learned both in the laboratory and at the bedside to effectively identify the patient population most likely to benefit and to assess efficacy. A major priority is the acquisition of tissue for all patients at the time of initial diagnosis and at the time of recurrence when applicable. Such tissue allows genomic characterization of individual tumors and provides a tissue reference for future clinical trials.

Further development of noninvasive biomarkers will be a key component of this effort. Identifying enriched populations will ensure that patients have quick access to the most effective strategy for their individual tumors, without being exposed to unnecessary treatments. With the budgetary constraints present in the current era, it is imperative that the prioritization of effort and allocation of resources allow for rapid evaluation of new agents.

Conclusions

The factors considered within this chapter are the influences on clinical trial participation among patients with CNS tumors. A balance of ethical and optimal recruitment is necessary for successful clinical trials. Embedding of trials into the best clinical practice is most likely to yield optimum results. Accurate presentation of studies at the outset is paramount. The clinician–patient interface remains the most important aspect in trial recruitment.

For many years, trial recruitment has been slow due to the relatively small numbers of patients, and the total number of studies has also been limited. The more recent situation is that there has been a marked increase in the number of studies that are evaluating treatments for glioma. This requires careful management to provide equitable access for patients. Despite the increase in the trial activity, certain groups remain underrepresented, and future strategies should be implemented to improve recruitment of patients, especially among the elderly.

The formation of a multidisciplinary approach is essential with close interaction of basic scientists, neurosurgeons, and neuro-oncologists in the development and completion of translational studies. Education is also essential and support through local, national, and international neuro-oncology platforms should be encouraged. These collaborations support colleagues who share the same passion for taking care of patients, help them to remain engaged, and can keep them focused on this challenging field.

References

1. Burnet NG, Jefferies SJ, Benson RJ, Hunt DP, Treasure FP. Years of life lost (YLL) from cancer is an important measure of population burden — and should be considered when allocating research funds. Br J Cancer. 2005;92(2):241–5.
2. Fine HA, Dear KB, Loeffler JS, Black PM, Canellos GP. Meta-analysis of radiation therapy with and without adjuvant chemotherapy for malignant gliomas in adults. Cancer. 1993;71(8):2585–97.
3. Curran Jr WJ, Scott CB, Horton J, et al. Recursive partitioning analysis of prognostic factors in three Radiation Therapy Oncology Group malignant glioma trials. J Natl Cancer Inst. 1993;85(9):704–10.
4. Stupp R, Mason WP, van den Bent MJ, et al. Radiotherapy plus concomitant and adjuvant temozolomide for glioblastoma. N Engl J Med. 2005;352(10): 987–96.
5. Pritchard-Jones K, Dixon-Woods M, Naafs-Wilstra M, Valsecchi MG. Improving recruitment to clinical trials for cancer in childhood. Lancet Oncol. 2008;9(4):392–9.
6. Stead M, Cameron D, Lester N, Parmar M, Haward R, Kaplan R, Maughan T, Wilson R, Campbell H, Hamilton R, Stewart D, O'Toole L, Kerr D, Potts V, Moser R, Darbyshire J, Selby P. Strengthening clinical cancer research in the United Kingdom. Br J Cancer. 2011;104:1529–34. http://ncrndev.org.uk.
7. Goffman E. Encounters: two studies in the sociology of interaction. Indianapolis: Bobbs-Merill; 1961.
8. Joffe S, Truog RD. Equipoise and randomization. In: Emanuel EJ, Grady C, Crouch RA, Lie R, Miller F, editors. The oxford textbook of clinical research ethics. Oxford University Press: Oxford; 2008. p. 245–60.
9. Fried C. Medical experimentation: personal integrity and social policy. Amsterdam: North-Holland Publishing; 1974.
10. Freedman B. Equipoise and the ethics of clinical research. N Engl J Med. 1987;317:141–5.
11. Miller PB, Weijer C. Rehabilitating equipoise. Kennedy Inst Ethics J. 2003;13(2):93–118.
12. London AJ. Clinical equipoise: foundational requirement or fundamental error? In: Steinbock B, editor. The oxford handbook of bioethics. Oxford University Press: New York; 2007. p. 571–96.

13. Joffe S, Miller FG. Bench to bedside: mapping the moral terrain of clinical research. Hastings Cent Rep. 2008;38:30–42.
14. Council for International Organizations of Medical Sciences. International ethical guidelines for biomedical research involving human subjects. Geneva: Council for International Organizations of Medical Sciences. 2002. http://www.cioms.ch/publications/layout_guide2002.pdf.
15. Brody H, Miller FG. The clinician-investigator: unavoidable but manageable tension. Kennedy Inst Ethics J. 2003;13(4):329–46.
16. Vickers AJ, Scardino PT. The clinically-integrated randomized trial: proposed novel method for conducting large trials at low cost. Trials. 2009;10:14.
17. Wasan AD, Taubenberger SP, Robinson WM. Reasons for participation in pain research: can they indicate a lack of informed consent? Pain Med. 2009;10(1): 111–9.
18. Hui EC. The centrality of patient-physician relationship to medical professionalism: an ethical evaluation of some contemporary models. Hong Kong Med J. 2005;11(3):222–3.
19. Hui EC. Doctors as fiduciaries: a legal construct of the patient-physician relationship. Hong Kong Med J. 2005;11(6):527–9.
20. Bramstedt KA. A guide to informed consent for clinician-investigators. Cleve Clin J Med. 2004;71(11): 907–10.
21. du Bois A, Rochon Lamparters C, Pfisterer J. Pattern of care and impact of participation in clinical studies on the outcome in ovarian cancer. Gynecol Oncol. 2005;15:183–91.
22. Karjalainen S, Palva I. Do treatment protocols improve end results? A study of survival of patients with multiple myeloma in Finland. Br Med J. 1989;299: 1069–72.
23. Majundar SR, Roe MT, Peterson ED, Chen AY, Gibler WB, Armstrong PW. Better outcomes for patients treated at hospitals that participate in clinical trials. Arch Intern Med. 2008;168:657–62.
24. Guilfoyle MR, Weerakkody RA, Oswal A, Oberg I, Jeffery C, Haynes K, Kullar PJ, Greenberg D, Jefferies SJ, Harris F, Price SJ, Thomson S, Watts C. Implementation of neuro-oncology service reconfiguration in accordance with NICE guidance provides enhanced clinical care for patients with glioblastoma multiforme. Br J Cancer. 2011;104(12): 1810–5.
25. Madsen SM, Mirza MR, Holm S, Hilsted KL, Kampmann K, Riis P. Attitudes towards clinical research amongst participants and nonparticipants. J Intern Med. 2002;251(2):156–68.
26. Cassileth BR, Lusk EJ, Miller DS, Hurwitz S. Attitudes toward clinical trials among patients and the public. JAMA. 1982;248(8):968–70.
27. Wendler D, Krohmal B, Emanuel EJ, Grady C. Why patients continue to participate in clinical research. Arch Intern Med. 2008;168(12):1294–9.
28. Donovan J, Mills N, Smith M, Brindle L, Jacoby A, Peters T, Frankel S, Neal D, Hamdy F. Quality improvement report: improving design and conduct of randomised trials by embedding them in qualitative research: protecT (prostate testing for cancer and treatment) study. BMJ. 2002;325(7367):766–70.
29. Fallowfield LJ, Jenkins V, Brennan C, Sawtell M, Moynihan C, Souhami RL. Attitudes of patients to randomised clinical trials of cancer therapy. Eur J Cancer. 1998;34(10):1554–9.
30. Ellis PM, Butow PN, Tattersall MHN, Dunn SM, Houssami N. Randomized clinical trials in oncology: understanding and attitudes predict willingness to participate. J Clin Oncol. 2001;19(15):3554–61.
31. Verheggen FWSM, Nieman F, Jonkers R. Determinants of patient participation in clinical studies requiring informed consent: why patients enter a clinical trial. Patient Educ Couns. 1998;35:111–25.
32. Engelking C. Facilitating clinical trials. The expanding role of the nurse. Cancer. 1991;67 Suppl 6:1793–7.
33. Johansen MA, Mayer DK, Hoover Jr HC. Obstacles to implementing cancer clinical trials. Semin Oncol Nurs. 1991;7(4):260–7.
34. Ganz PA. Clinical trials. Concerns of the patient and the public. Cancer. 1990;65 Suppl 10:2394–9.
35. Cox K, Avis M. Psychosocial aspects of participation in early anticancer drug trials. Report of a pilot study. Cancer Nurs. 1996;19(3):177–86.
36. Winn RJ, Miransky J, Kerner JF, Kennelly L, Michaelson RA, Sturgeon SR. An evaluation of physician determinants in the referral of patients for cancer clinical trials in the community setting. Prog Clin Biol Res. 1984;156(4):63–73.
37. Fireman BH, Fehrenbacher L, Gruskin EP, Ray GT. Cost of care for patients in cancer clinical trials. J Natl Cancer Inst. 2000;92(2):136–42.
38. Fireman BH, Quesenberry CP, Somkin CP, Jacobson AS, Baer D, West D. Cost of care for cancer in a health maintenance organization. Health Care Financ Rev. 1997;18:51–76.
39. Wagner JL, Alberts SR, Soan JA, Cha S, Killian J, O'Connoll MJ. Incremental costs of enrolling cancer patients in clinical trials: a population-based study. J Natl Cancer Inst. 1999;91:847–53.
40. Brown ML. Cancer patient care in clinical trials sponsored by the National Cancer Institute: what does it cost? J Natl Cancer Inst. 1999;91:818–9.
41. Mechanic RE, Dobson A. The impact of managed care on clinical research: a preliminary investigation. Health Aff. 1996;15:72–89.
42. Fisher WB, Cohen SJ, Hammond MK, Turner S, Loehrer PJ. Clinical trials in cancer therapy: efforts to improve patient enrollment by community oncologists. Med Pediatr Oncol. 1991;19(3):165–8.
43. Taylor K, Margolese R, Soskolne C. Physicians' reasons for not entering eligible patients in a randomized clinical trial of surgery for breast cancer. N Engl J Med. 1984;310:1363–7.

44. Lara LN, Higdon R, Lim N. Prospective evaluation of cancer clinical trial accrual patterns: identifying potential barriers to enrollment. J Clin Oncol. 2001; 19(6):1728–33.
45. Ross S, Grant A, Counsell C, Gillespie W, Russell I, Prescott R. Barriers to participation in randomised controlled trials: a systematic review. J Clin Epidemiol. 1999;52(12):1143–56.
46. Lovato LC, Hill K, Hertert S, Hunninghake DB, Probstfield JL. Recruitment for controlled clinical trials: literature summary and annotated bibliography. Control Clin Trials. 1997;18(4):328–52.
47. Hunninghake DB, Darby CA, Probstfield JL. Recruitment experience in clinical trials: literature summary and annotated bibliography. Control Clin Trials. 1987;8 Suppl 4:6–30.
48. Britton A, McKee M, Black N, McPherson K, Sanderson C, Bain C. Threats to applicability of randomised trials: exclusions and selective participation. J Health Serv Res Policy. 1999;4(2):112–21.
49. Wen PY, Macdonald DR, Reardon DA, et al. Updated response assessment criteria for high-grade gliomas: response assessment in neuro-oncology working group. J Clin Oncol. 2010;28(11):1963–72.
50. Mao Y, Desmeules M, Semenciw RM, Hill G, Gaudette L, Wigle DT. Increasing brain cancer rates in Canada. CMAJ. 1991;145(12):1583–91.
51. Cottin V, Arpin D, Lasset C, et al. Small-cell lung cancer: patients included in clinical trials are not representative of the patient population as a whole. Ann Oncol. 1999;10(7):809–15.
52. McCusker J, Wax A, Bennett JM. Cancer patient accessions into clinical trials: a pilot investigation into some patient and physician determinants of entry. Am J Clin Oncol. 1982;5(2):227–36.
53. Halperin EC. Malignant gliomas in older adults with poor prognostic signs. Getting nowhere, and taking a long time to do it. Oncology (Williston Park). 1995;9(3):229–34. 237–238.
54. Whittle IR, Denholm SW, Gregor A. Management of patients aged over 60 years with supratentorial glioma: lessons from an audit. Surg Neurol. 1991;36(2): 106–11.
55. Benson AB, Pregler JP, Bean JA, Rademaker AW, Eshler B, Anderson K. Oncologists' reluctance to accrue patients onto clinical trials: an Illinois Cancer Center study. J Clin Oncol. 1991;9(11):2067–75.
56. Mohan DS, Suh JH, Phan JL, Kupelian PA, Cohen BH, Barnett GH. Outcome in elderly patients undergoing definitive surgery and radiation therapy for supratentorial glioblastoma multiforme at a tertiary care institution. Int J Radiat Oncol Biol Phys. 1998;42(5):981–7.
57. Bauman GS, Gaspar LE, Fisher BJ, Halperin EC, Macdonald DR, Cairncross JG. A prospective study of short-course radiotherapy in poor prognosis glioblastoma multiforme. Int J Radiat Oncol Biol Phys. 1994;29(4):835–9.
58. Giovanazzi-Bannon S, Rademaker A, Lai G, Benson 3rd AB. Treatment tolerance of elderly cancer patients entered onto phase II clinical trials: an Illinois Cancer Center study. J Clin Oncol. 1994;12(11):2447–52.
59. Chang SM, Wen P, Cloughesy T, et al. Phase II study of CCI-779 in patients with recurrent glioblastoma multiforme. Invest New Drugs. 2005;23:357–61.
60. Marchetti S, Schellens JH. The impact of FDA and EMEA guidelines on drug development in relation to phase 0 trials. Br J Cancer. 2007;97:577–81.
61. Kummar S, Kinders R, Rubinstein L, et al. Compressing drug development timelines in oncology using phase "0" trials. Nat Rev Cancer. 2007;7: 131–9.
62. Lee JW, Figeys D, Vasilescu J. Biomarker assay translation from discovery to clinical studies in cancer drug development: quantification of emerging protein biomarkers. Adv Cancer Res. 2007;96:269–98.
63. Workman P. The opportunities and challenges of personalized genome-based molecular therapies for cancer: targets, technologies, and molecular chaperones. Cancer Chemother Pharmacol. 2003;52:S45–56.
64. Booth CM, Calvert AH, Giaccone G. Design and conduct of phase II studies of targeted anticancer therapy: recommendations from the task force on methodology for the development of innovative cancer therapies (MDICT). Eur J Cancer. 2008;44:25–9.
65. Booth CM, Calvert AH, Giaccone G. Endpoints and other considerations in phase I studies of targeted anticancer therapy: recommendations from the task force on Methodology for the Development of Innovative Cancer Therapies (MDICT). Eur J Cancer. 2008;44:19–24.
66. Chang SM, Kuhn JG, Rizzo J, et al. Phase I study of paclitaxel in patients with recurrent malignant glioma: a North American Brain Tumor Consortium report. J Clin Oncol. 1998;16:2188–94.
67. Cloughesy TF, Kuhn J, Robins HI, et al. Phase I trial of tipifarnib in patients with recurrent malignant glioma taking enzyme- inducing antiepileptic drugs: a North American Brain Tumor Consortium Study. J Clin Oncol. 2005;23:6647–56.

Index

C. Watts (ed.), *Emerging Concepts in Neuro-Oncology*,
DOI 10.1007/978-0-85729-458-6, © Springer-Verlag London 2013

Printed by Printforce, the Netherlands